Refining and Casting of Steel

Refining and Casting of Steel

Editors
Karel Gryc
Jan Falkus

MDPI • Basel • Beijing • Wuhan • Barcelona • Belgrade • Manchester • Tokyo • Cluj • Tianjin

Editors
Karel Gryc
Institute of Technology and Business in České Budějovice
Czech Republic

Jan Falkus
AGH University of Science and Technology
Poland

Editorial Office
MDPI
St. Alban-Anlage 66
4052 Basel, Switzerland

This is a reprint of articles from the Special Issue published online in the open access journal *Metals* (ISSN 2075-4701) (available at: https://www.mdpi.com/journal/metals/special_issues/Refine_Cast_Steel).

For citation purposes, cite each article independently as indicated on the article page online and as indicated below:

LastName, A.A.; LastName, B.B.; LastName, C.C. Article Title. *Journal Name* **Year**, *Article Number*, Page Range.

ISBN 978-3-03943-434-3 (Hbk)
ISBN 978-3-03943-435-0 (PDF)

© 2020 by the authors. Articles in this book are Open Access and distributed under the Creative Commons Attribution (CC BY) license, which allows users to download, copy and build upon published articles, as long as the author and publisher are properly credited, which ensures maximum dissemination and a wider impact of our publications.

The book as a whole is distributed by MDPI under the terms and conditions of the Creative Commons license CC BY-NC-ND.

Contents

About the Editors vii

Preface to "Refining and Casting of Steel" ix

Karel Gryc and Jan Falkus
Refining and Casting of Steel
Reprinted from: *Metals* **2020**, *10*, 295, doi:10.3390/met10020295 1

Dana Baricová, Alena Pribulová, Peter Futáš, Branislav Buľko and Peter Demeter
Change of the Chemical and Mineralogical Composition of the Slag during Oxygen Blowing in the Oxygen Converter Process
Reprinted from: *Metals* **2018**, *8*, 844, doi:10.3390/met8100844 7

Andre N. Assis, Mohammed A. Tayeb, Seetharaman Sridhar and Richard J. Fruehan
Phosphorus Equilibrium Between Liquid Iron and $CaO\text{-}SiO_2\text{-}MgO\text{-}Al_2O_3\text{-}FeO\text{-}P_2O_5$ Slags: EAF Slags, the Effect of Alumina and New Correlation
Reprinted from: *Metals* **2019**, *9*, 116, doi:10.3390/met9020116 21

Fabian Andrés Calderón-Hurtado, Rodolfo Morales Dávila, Kinnor Chattopadhyay and Sa úl García-Hernández
Fluid Flow Turbulence in the Proximities of the Metal-Slag Interface in Ladle Stirring Operations
Reprinted from: *Metals* **2019**, *9*, 192, doi:10.3390/met9020192 33

Yandong Li, Tongsheng Zhang and Huamei Duan
Influence of Al on Evolution of the Inclusions in Ti-Bearing Steel with Ca Treatment
Reprinted from: *Metals* **2019**, *9*, 104, doi:10.3390/met9010104 51

Kaitian Zhang, Heng Cui, Rudong Wang and Yang Liu
Mixing Phenomenon and Flow Field in Ladle of RH Process
Reprinted from: *Metals* **2019**, *9*, 886, doi:10.3390/met9080886 63

Branislav Buľko, Ivan Priesol, Peter Demeter, Peter Gašparovič, Dana Baricová and Martina Hrubovčáková
Geometric Modification of the Tundish Impact Point
Reprinted from: *Metals* **2018**, *8*, 944, doi:10.3390/met8110944 75

Branislav Buľko, Marek Molnár, Peter Demeter, Dana Baricová, Alena Pribulová and Peter Futáš
Study of the Influence of Intermix Conditions on Steel Cleanliness
Reprinted from: *Metals* **2018**, *8*, 852, doi:10.3390/met8100852 87

Feng Wang, Daoxu Liu, Wei Liu, Shufeng Yang and Jingshe Li
Reoxidation of Al-Killed Steel by Cr_2O_3 from Tundish Cover Flux
Reprinted from: *Metals* **2019**, *9*, 554, doi:10.3390/met9050554 97

Peiyuan Ni, Mikael Ersson, Lage T. I. Jonsson, Ting-An Zhang and Pär Göran Jönsson
Effect of Immersion Depth of a Swirling Flow Tundish SEN on Multiphase Flow and Heat Transfer in Mold
Reprinted from: *Metals* **2018**, *8*, 910, doi:10.3390/met8110910 105

Yandong Li, Tongsheng Zhang, Chengjun Liu and Maofa Jiang
Thermodynamic and Experimental Studies on Al Addition of 253MA Steel
Reprinted from: *Metals* **2019**, *9*, 433, doi:10.3390/met9040433 . . . . **119**

Kaitian Zhang, Jianhua Liu and Heng Cui
Investigation on the Slag-Steel Reaction of Mold Fluxes Used for Casting Al-TRIP Steel
Reprinted from: *Metals* **2019**, *9*, 398, doi:10.3390/met9040398 . . . . **127**

Tomas Mauder and Josef Stetina
High Quality Steel Casting by Using Advanced Mathematical Methods
Reprinted from: *Metals* **2018**, *8*, 1019, doi:10.3390/met8121019 . . . . **139**

Xianguang Zhang, Wen Yang, Haikun Xu and Lifeng Zhang
Effect of Cooling Rate on the Formation of Nonmetallic Inclusions in X80 Pipeline Steel
Reprinted from: *Metals* **2019**, *9*, 392, doi:10.3390/met9040392 . . . . **153**

Yanshen Han, Xingyu Wang, Jiangshan Zhang, Fanzheng Zeng, Jun Chen, Min Guan and Qing Liu
Comparison of Transverse Uniform and Non-Uniform Secondary Cooling Strategies on Heat Transfer and Solidification Structure of Continuous-Casting Billet
Reprinted from: *Metals* **2019**, *9*, 543, doi:10.3390/met9050543 . . . . **165**

Chenhui Wu, Cheng Ji and Miaoyong Zhu
Deformation Behavior of Internal Porosity in Continuous Casting Wide-Thick Slab during Heavy Reduction
Reprinted from: *Metals* **2019**, *9*, 128, doi:10.3390/met9020128 . . . . **181**

About the Editors

Karel Gryc (Assoc. Prof.), Associate Professor of Metallurgical Technology, earned his Ph.D. in Metallurgical Technology from the VSB - Technical University of Ostrava (VSB-TUO) in 2008. He has been a faculty member at VSB-TUO since 2003 where he gradually taught 16 courses, e.g., Iron and Steelmaking, Metallurgical Technologies, Cleanness and Usable Properties of Steel, Modelling and Visualization of Metallurgical Processes and Theory of Steelmaking Processes. His teaching has been at the undergraduate, graduate and postgraduate levels. Dr. Gryc has been involved in the Czech Metallurgical Society, the Czech Foundry Society and the Czech Society for New Materials and Technologies. Dr. Gryc has published over fifty papers related to steels, metal casting, metal simulation, solidification, and refining. His current work is focused on optimizing the production and quality of castings based on material and dimensional analyses and simulations.

Jan Falkus (Assoc. Prof.), Associate Professor, was born in 1957. In 1981, he graduated from the AGH University of Science and Technology in Kraków. From 1981 to 1983, he worked as a junior assistant at the Institute of Ferrous Metallurgy at the University of Mining and Metallurgy, and from 1983 as an assistant at the same institute. In 1990, he obtained a doctoral degree. From 1990, he was a second degree assistant. In 1999, he gained a habilitation colloquium at the Institute of Metallurgy and Materials Science. Since 2002, he became associate professor, and since 2012, he has been the head of the Department of Ferrous Metallurgy. His main research topics are: process metallurgy in steel production; studies of ferroalloy and cold models, mathematical modeling of processes. He is the author and co-author of over 100 publications on this research topic.

Preface to "Refining and Casting of Steel"

Steel has become the most requested material all over the world during the rapid technological evolution of recent centuries. As our civilization grows and its technological development becomes connected with more demanding processes, it is more and more challenging to fit the required physical and mechanical properties for steel in its huge portfolio of grades for each steel producer. It is necessary to improve the refining and casting processes continuously to meet customer requirements and to lower the production costs to remain competitive. New challenges related to both the precise design of steel properties and reduction in production costs are combined with paying special attention to environmental protection. These contradictory demands are the theme of this book. An Editorial briefly comments on fifteen contributions from experts in the refining and casting of steel which are published in this book, outlining recent and original advances in the fields of crucial multiphase interactions in the steel making furnaces: the basic oxygen furnace (BOF) and electric arc furnace (EAF). Next, the research focuses on the matters related to the secondary refining processes; fluid turbulence near to the metal–slag interface in the ladles, the evolution of inclusions for Ti-bearing steels treated with calcium, and the influence of aluminum on flux flow in the ladle during RH operations. Other papers are devoted to a geometry of impact area in the tundish, cleanliness consequences of an intermixing of different steel grades in the tundish, and the reoxidation of steel from tundish cover flux. Three papers are based on studies regarding the processes taking place in the mold from the viewpoint of heat transfer and multiphase flow influenced by SEN immersion depth, nozzle clogging, and the mold's flux interaction with steel. Finally, the authors contributed to this Special Issue by focusing their attention on the complex phenomenon connected with steel cooling, its solidification and quality.

Karel Gryc, Jan Falkus
Editors

Editorial

Refining and Casting of Steel

Karel Gryc [1,*] and Jan Falkus [2]

[1] Environmental Research Department, Faculty of Technology, Institute of Technology and Business in České Budějovice, Orkužní 517/10, 370 01 České Budějovice, Czech Republic
[2] Department of Ferrous Metallurgy, Faculty of Metals Engineering and Industrial Computer Science, AGH University of Science and Technology, Mickiewicza Av. 30, 30-059 Kraków, Poland; jfalkus@agh.edu.pl
* Correspondence: gryc@mail.vstecb.cz; Tel.: +420-777-18-78-98

Received: 1 February 2020; Accepted: 21 February 2020; Published: 24 February 2020

1. Introduction and Scope

Steel was the most requested material all over the world during the past fast technically evolving centuries. As our civilization grows and its technological development is connected with more demanding processes, it is more and more challenging to fit required physical and mechanical properties for steel in its huge portfolio of grades for each steel producer. It is necessary to improve the refining and casting processes continuously to meet customer requirements and lower the production costs to remain competitive.

New challenges relating to both the precise design of steel properties and reduction of production costs are combined with paying special attention to environmental protection. These contradictory demands are the theme of this Special Issue.

2. Contribution to the Special Issue

The papers covered by this Special Issue are state-of-the-art in the fields of crucial multiphase interactions in the steel making furnaces: basic oxygen furnace (BOF) [1] and electric arc furnace (EAF) [2]. Next, the research focuses on the matters related to the secondary refining processes; fluid's turbulences with nearby metal-slag interface in the ladles [3], the evolution of inclusions for Ti-bearing steels treated by calcium [4], and an influence of aluminum on flux flow in the ladle during Ruhrstahl Heraeus (RH) degassing operations [5]. Other papers are devoted to a geometry of impact area in the tundish [6], cleanliness consequences of an intermixing of the different steel grades in the tundish [7], and the reoxidation of steel from tundish cover flux [8]. Three papers are based on the studies, regarding the processes taking place in the mold from the viewpoint of heat transfer and multiphase flow influenced by deep submerged entry nozzle (SEN) immersion [9], nozzle clogging [10], and mold flux interaction with steel [11]. Finally, the authors contributed to this Special Issue by also focusing their attention on the complex phenomenon connected to steel cooling, its solidification and quality [12–15]. All published contributions and their briefly commented upon abstracts are organized based on the technological flow valid for modern steel production in the next subchapters.

2.1. Studies of the Slags in Steel-Making Furnaces

Baricová et al. [1] present the results of their investigation of changes in the chemical and mineralogical composition of slag during steel production in a BOF. This process was monitored using the slag samples that were collected during the period when oxygen blowing into a BOF was interrupted. Besides other findings, the results show that the structure of the slag related to its chemical composition was more significantly affected by the added slag-forming additives. The changes in the slag structure reflected the process of the gradual dissolution of lime in the slag. While in the primary slag, the predominating structure components were SiO_2 and lime, after the eighth minute of oxygen blowing, their contents significantly decreased and the predominating component was dicalcium silicate. After minute 24 of oxygen blowing, the final structure of the slag was formed, with the major component being tricalcium silicate, containing also dicalcium silicate, RO-phase, and calcium ferrites [1].

Assis et al. [2] investigated phosphorus equilibria between molten Fe–P alloys and $CaO-SiO_2-Al_2O_3-P_2O_5-FeO-MgO_{saturated}$ slags based on the trend of using up to 100% direct reduced iron (DRI) in the batch. The results indicate that there is a significant decrease in the phosphorus partition coefficient (LP) as alumina in the slag increases. The observed effect of alumina on the phosphorus partition is probably caused by the decrease in the activities of iron oxide and calcium oxide. Finally, an equilibrium correlation for phosphorus partition as a function of slag composition and temperature has been developed. It includes the effect of alumina and silica and is suitable for both oxygen and electric steelmaking-type slags [2].

2.2. Refining the Steel in Ladles

There are plenty of methods for how to study the phenomenon occurring during these refining processes. The kinetics of interaction at the metal-slag interface and also the possibility of a slag entrapment are strongly influenced by the character of fluid flow in the ladle. Three-phase interactions (metal-slag-argon) in ladle stirring operations have strong effects on the metal-slag mass transfer processes, Calderón-Hurtado et al. [3]. Specifically, the thickness of the slag controls the fluid turbulence to an extent that, once trespassing a critical thickness, increases of stirring strength no longer affects the flow. To analyze these conditions, a physical model considering the three phases was built to study liquid turbulence in the proximities of the metal-slag interface. A velocity probe placed close to the interface permitted the continuous monitoring and statistical analyses of any turbulence. The slag eye opening was found to be strongly dependent on the stirring conditions, and the mixing times decreased with thin slag thicknesses. Slag entrainment was enhanced with thick slag layers and high flow rates of the gas phase. A multiphase model was developed to simulate these results and was found to be a good agreement between experimental and numerical results [3].

Next, Li et al. [4] investigated the influence of aluminum on the evolution of the inclusion in Ti-bearing steel treated by Ca. Experimental simulations of steelmaking with different amounts of aluminum were achieved in the tube furnace at 1873 K, and field scanning electron microscopy and energy dispersive X-ray spectroscopy (FE-SEM and EDX) were employed to explore the characteristics of the inclusions in Ti-bearing steel during the calcium treatment process. It was found that morphologies, chemical compositions, and the size distribution of the inclusions were obviously different before and after calcium treatment. The calcium addition needed to be carefully considered regarding the mass fraction of aluminum with the purpose of modifying the solid inclusions to liquid phases. The thermodynamic analysis of inclusion formation in the Al–Ti–Ca–O system at 1873 K was conducted, as well as transformation behaviors of inclusions, including all types of solid inclusions and liquid phases during solidification. The thermodynamic equilibrium calculations are in good agreement with experimental data, which can be used to estimate inclusion formation in Ti-bearing steel [4].

Most of the current high-quality steels are treated under vacuum. Zhang et al. [5] have used adopted particle image velocimetry (PIV) with a 1:4 scaled water model. The results of mixing simulation experiments indicated that the mixing time decreased with the increase of the gas blowing rate. However, with the increase of snorkel immersion depth (SID), the mixing time presented a decreasing rend firstly and then increased. The measurement of flow fields of RH ladle by the PIV system can explain the phenomenon above. According to the characteristics of the flow field in RH ladle, the flow field can be divided into the mixing layer, the transition layer, and the inactive layer. On the one hand, the stirring powers in the RH ladle and vacuum chamber both increase with a higher gas blowing rate, leading to the decrease of mixing time. On the other hand, when SID increases from 400 mm to 480 mm, the gas blowing depth increases results in the mixing power, and the mixing time decreases at the beginning. Due to too much molten steel in the vacuum chamber and the expanding of the inactive layer in RH ladle however, the utilization rate of the gas driving force begins to decrease. Therefore, the mixing time starts to increase with the increase of SID [5].

2.3. Design, Intermix and Reoxidation Issues in the Tundish

Continuous casting operations start by pouring steel from the ladle to the tundish. Buľko et al. [6] reported the geometric modification of the tundish impact point. This paper compares the standard impact pad with the "spheric" spherical impact pad using computational fluid dynamics (CFD) tools and physical modelling. The evaluation criteria are residence time and flow in the tundish at three different casting speeds. Compared to the standard impact pad, based on our measurements of residence time distribution (RTD) curves using the water model, the "spheric" impact pad shortened the minimum residence times at casting speeds of 0.8 and 1.2 m·min^{-1}, which is on the level of 71% and 76% of the standard impact pad times under identical conditions. On the other hand, the "spheric" impact pad produced a 6% longer residence time than the standard impact at casting speed 1.6 m·min^{-1}. From a visual comparison of flow in the tundish, we can observe that the "spheric" impact pad produces a better flow pattern than the standard impact pad. It has no tendency to shortcut the flow at lower casting speeds. Moreover, dead zone areas are eliminated using the "spheric" impact pad. We can predict that using this impact pad in practice will have a positive influence on steel cleanliness, due to more dynamic steel flow at the steel-slag interface. Furthermore, the slag "eye" phenomenon can be reduced when the "spheric" impact pad is used, compared to using impact pads with a significant piston flow pattern [6].

Next, the study of authors Buľko et al. [7] reflects the fact that modern steel plants today produce a large portfolio of various steel grades, many for end-users demanding high quality. In order to utilize the maximum productivity of the continuous-casting machine, it is sometimes necessary to cast steel grades with different chemical compositions in one sequence. It is important, therefore, to know the possibilities of a specific continuous-casting machine to make the Intermix connections as short as possible. Any interference with established procedures may, however, have a negative impact on the cleanliness of the cast steel. Using physical and numerical simulation tools, it was found that reducing the steel level in the tundish during the exchange of ladles makes it possible to shorten the transition zone. Based on research results [7], it can be concluded that the ultra-low tundish (ULT) practice produces savings in terms of shorter transition zones, thereby increasing steel yields in the continuous-casting process. Under the given conditions, the ULT practice enabled a 20% shortening of the transition zone. The area of higher inclusion concentration during the ULT practice was located in the transition part of the slab, and therefore, had no negative effect on the final quality of the cast steel [7].

Wang et al. [8] studied aluminum-killed steel's reoxidation by Cr_2O_3 from the tundish cover flux. Reoxidation has long been a problem when casting ultra-low oxygen liquid steel. An experimental study of the reoxidation phenomenon caused by Cr_2O_3-bearing cover flux of Al-killed steel is presented here. $MgO-CaO-SiO_2-Al_2O_3-Cr_2O_3$ tundish cover fluxes with various Cr_2O_3 contents were used to study the effects of Cr_2O_3 on total oxygen content (T[O]) and the alumina and silicone loss of Al-killed steel at 1923 K (1650 °C). It was found that Cr_2O_3 can be reduced by Al to cause reoxidation, and the reaction occurs mainly within two to three min after the addition of the tundish cover flux with 5% and 10% Cr_2O_3 concentration. T[O] and Al loss increase with higher Cr_2O_3 concentration flux. Two controlled experiments were also carried out to investigate the oxygen transported to the steel by the decomposition of Cr_2O_3. It was calculated that when Al is present in steel, more than 90% of the reoxidation of Cr_2O_3 is caused by Al, and the rest is caused by decomposition [8].

2.4. Multiphase Flow and Multiphase Interactions in the Mould

The mold is the last steel production node where a liquid steel interacts with its surroundings. The effect of the immersion depth of a new swirling flow tundish SEN (Submerged Entry Nozzle) on the multiphase flow and heat transfer in mold was studied using numerical simulation by Ni et al. [9].

The RSM (Reynolds Stress Model) and the VOF (Volume of Fluid) model were used to solve the steel and slag flow phenomena. The results show that SEN immersion depth can significantly influence the steel flow near the meniscus. Specifically, an increase in the SEN immersion depth decreases the interfacial velocity, and this reduces the risk for the slag entrainment. The calculated Weber Number decreases from 0.8 to 0.2 when the SEN immersion depth increases from 15 cm to 25 cm. With a large SEN immersion depth, the steel flow velocity near the solidification front, which is below the mold level of the SEN outlet, increased. The temperature distribution has a similar distribution characteristic for different SEN immersion depths. The high temperature region is located near the solidification front. The temperature near the meniscus was slightly decreased when the SEN immersion depth was increased, due to an increased steel moving distance from the SEN outlet to the meniscus [9].

To solve the nozzle clogging issue in the continuous casting process of 253MA steel, a method of modifying solid inclusions to liquid phases is proposed by Li et al. [10]. The CALculation of PHAse Diagrams (CALPHAD) technique was employed to predict the liquid region of the Al_2O_3-SiO_2-Ce_2O_3 system. Then, a thermodynamic package based on the extracted data during the phase diagram optimization process was developed. This package was then used to compute the appropriate aluminum addition, which was 0.01% in 253MA steel. The Si-Al alloy was chosen as the deoxidant according to the thermodynamic analysis. The solid inclusions were ultimately modified to liquid phases at 1500 °C, when cerium was added through the equilibrium experiments in a $MoSi_2$ tube furnace [10].

Research related to the interaction of selected mold fluxes with aluminum-TRIP steel grades was conducted by Zhang et al. [11] The composition and property variations of two slags during a slag-steel reaction were analyzed. Accordingly, the crystalline morphologies of slag were discussed and the solid layer lubrication performance was evaluated by Jackson α factors. In addition, a simple kinetics equilibrium model was established to analyze the factors which affected SiO_2 consumption. The results revealed that slag-steel reacted rapidly in the first 20 min, resulting in a variation of viscosity and the melting temperature of slags. The slag-steel reaction also affected the crystal morphology significantly. Slag was precipitated as crystals with a higher melting temperature, a higher Jackson α factor, and a rougher boundary with the consumption of SiO_2 and the generation of Al_2O_3. In other words, although generated Al_2O_3 acted as a network modifier to decrease the viscosity of the liquid slag layer adjacent slab shell, the consumption of SiO_2 led to the deterioration of the lubrication performance in the solid slag layer adjacent copper, which was detrimental to the quality control for high Al-TRIP steel. Finally, a kinetics equilibrium model indicated that it is possible to reduce a slag-steel reaction by adjusting factors, such as the diffusion coefficient k, c_{SiO2}, ρ_f and L_f, during the actual continuous casting process [11].

2.5. Cooling, Solidification and Deformation of Steel During Continuous Casting

The main concept of Mauder's et al. work [12] is to utilize advanced numerical modelling techniques with a self-regulation algorithm in order to reach optimal casting conditions for real-time casting control. A fully 3-D macro-solidification model for the continuous casting (CC) process and an original fuzzy logic regulator are combined. The fuzzy logic (FL) regulator reacts to signals from two data inputs, the temperature field and the historical steel quality database. FL adjusts the cooling intensity as a function of casting speed and pouring temperature. This approach was originally designed for the special high-quality high-additive steel grades such as higher strength grades, steel for acidic environments, steel for the offshore technology and so forth. However, the mentioned approach can also be used for any arbitrary low-carbon steel grades. The usability and results of this approach are demonstrated for steel grade S355, were the real historical data from the quality database contains approximately 2000 heats. The presented original solution, together with the large steel quality databases, can be used as an independent CC prediction control system [12].

Cooling rate effect on formation of non-metallic inclusions in X80 pipeline steel is the topic of the research presented by Zhang et al. [13]. Non-metallic inclusions have a strong influence on the hydrogen-induced cracking (HIC) and sulfide stress cracking (SSC) in pipeline steels, which should be well controlled to improve the steel resistance to HIC and SSC. The effects of cooling rate on the formation of non-metallic inclusions have been studied both experimentally and thermodynamically. It was found that the increasing cooling rate increased the number density and decreased the size of the inclusions, while the inverse results were obtained by decreasing the cooling rate. Furthermore, as the cooling rate decreased from 10 to 0.035 K/s, the inclusions were changed from Al_2O_3-CaO to Al_2O_3-CaO-MgO-CaS. At a high cooling rate, the reaction time is short and the inclusions cannot be completely transformed, which should be mainly formed at high temperatures. Meanwhile, at a low cooling rate, the inclusions can be gradually transformed and tend to follow the equilibrium compositions [13].

Han et al. [14] reported results from the comparison of different secondary cooling strategies from the viewpoint of heat transfer and the solidification structure of billets. Water flux distribution largely influences the heat transfer and solidification of continuously cast steel billets. In this paper, a secondary cooling strategy of transverse non-uniform water flux (i.e., higher flux density on billet center), was established and compared with the uniform cooling strategy using mathematical modelling. Specifically, a heat transfer model and a cellular automaton finite element coupling model were established to simulate the continuous casting of the C80D steel billet. The water flux was measured using different nozzle configurations to assist the modelling. The mathematical results were validated by comparing the surface temperature and the solidification structure. It is shown that the non-uniform cooling strategy enables the increase of corner temperature and reduction in surface temperature difference, while a higher reheating rate is found on the surface center of the billet. Moreover, the non-uniform cooling strategy can enhance the cooling effect and refine the solidification structure. Accordingly, the liquid pool length is shortened, and the equiaxed crystal density is increased along with the decreased equiaxed crystal ratio. The uniform cooling strategy contributes to reducing internal cracks of billet, and the non-uniform one is beneficial for surface quality and central segregation. For C80D steel, the non-uniform cooling strategy outperforms the uniform one [14].

Based on all of the presented research areas, heavy reduction and its deformation behavior of internal porosity, investigated by Wu et al. [15], should be mentioned as the last but also most valuable topic included in our Special Issue. Heavy reduction (HR) is a novel technology that could effectively improve the internal porosities and other internal quality problems in continuously cast steel, during which a large reduction deformation is implemented at and after the strand solidification end. In the present paper, non-uniform solidification of the wide-thick slab was calculated with a two-dimensional (2D) heat transfer model. Based on the predicted temperature distribution at the solidification end of the casting strand, a three-dimensional (3D) thermal-mechanical coupled model was developed for investigating the deformation behavior of the internal porosities in a wide-thick slab during HR. An Arrhenius-type constitutive model for the studied steel grade was derived based on the measured true stress-strain with single-pass thermo-simulation compression experiments and applied to the 3D thermal-mechanical coupled model for improving the calculation accuracy. With the developed 3D thermal-mechanical coupled model, deformation behavior of the two artificial porosities located at the slab center of 1/2 width and 1/8 width during HR was investigated under different conditions of HR deformation, HR start position, and HR reduction mode. Based on the calculated porosity closure degree (η_s), and the corresponding equivalent strain (ε_{eq}), under different HR conditions, a prediction model that describes the quantitative relationship between η_s and ε_{eq} was derived for directly and accurately evaluating the process effect of HR on improving the internal porosities in a wide-thick slab [15].

3. Conclusions

The Special Issue, "Refining and Casting of Steel" and its research articles represent interesting examples from the most crucial challenges for maximizing the productivity and quality of continuously cast steel grades. The guest editors suppose that these papers should be inspiring for the scholars, researchers and technologists who are actively involved in this field. We hope that the presented articles will help them during new research studies, debates, and discussions.

Acknowledgments: The guest editors would like to thank all who contributed directly and indirectly to the successful development of this Special Issue. The guest editors thank all the scholars and authors who submitted their manuscripts and were willing to publish their research activities in this Special issue. Special mention and sincere thanks to the reviewers who agreed to review the articles and provide feedback to improve the quality of the manuscripts. Credits should also be given to the editors and to Managing Editor Natalie Sun and also all the staff of the Metals Editorial Office for their contribution and support in the publication process of this issue.

Conflicts of Interest: The authors declare no conflict of interest.

References

1. Baricová, D.; Pribulová, A.; Futáš, P.; Buľko, B.; Demeter, P. Change of the Chemical and Mineralogical Composition of the Slag during Oxygen Blowing in the Oxygen Converter Process. *Metals* **2018**, *8*, 844. [CrossRef]
2. Assis, A.N.; Tayeb, M.A.; Sridhar, S.; Fruehan, R.J. Phosphorus Equilibrium Between Liquid Iron and CaO-SiO_2-MgO-$Al2O_3$-FeO-$P2O_5$ Slags: EAF Slags, the Effect of Alumina and New Correlation. *Metals* **2019**, *9*, 116. [CrossRef]
3. Calderón-Hurtado, F.A.; Dávila, R.M.; Chattopadhyay, K.; García-Hernández, S. Fluid Flow Turbulence in the Proximities of the Metal-Slag Interface in Ladle Stirring Operations. *Metals* **2019**, *9*, 192. [CrossRef]
4. Li, Y.; Zhang, T.; Duan, H. Influence of Al on Evolution of the Inclusions in Ti-Bearing Steel with Ca Treatment. *Metals* **2019**, *9*, 104. [CrossRef]
5. Zhang, K.; Cui, H.; Wang, R.; Liu, Y. Mixing Phenomenon and Flux Field in Ladle of RH Process. *Metals* **2019**, *9*, 886. [CrossRef]
6. Buľko, B.; Priesol, I.; Demeter, P.; Gašparovič, P.; Baricová, D.; Hrubovčáková, M. Geometric Modification of the Tundish Impact Point. *Metals* **2018**, *8*, 944. [CrossRef]
7. Buľko, B.; Molnár, M.; Demeter, P.; Baricová, D.; Pribulová, A.; Futáš, P. Study of the Influence of Intermix Conditions of Steel Cleanliness. *Metals* **2018**, *8*, 852. [CrossRef]
8. Wang, F.; Liu, D.; Liu, W.; Yang, S.; Li, J. Reoxidation of Al-Killed Steel by Cr_2O_3 from Tundish Cover Flux. *Metals* **2019**, *9*, 554. [CrossRef]
9. Ni, P.; Ersson, M.; Jonsson, L.T.I.; Zhang, T.A.; Jönsson, P.G. Effect of Immersion Depth of a Swirling Flow Tundish SEN on Multiphase Flow and Heat Transfer in Mold. *Metals* **2018**, *8*, 910. [CrossRef]
10. Li, Y.; Zhang, T.; Liu, C.; Jiang, M. Thermodynamic and Experimental Studies on Al Addition of 253MA Steel. *Metals* **2019**, *9*, 433. [CrossRef]
11. Zhang, K.; Liu, J.; Cui, H. Investigation on the Slag-Steel Reaction of Mold Fluxes Used for Casting Al-TRIP Steel. *Metals* **2019**, *9*, 398. [CrossRef]
12. Mauder, T.; Stetina, J. High Quality Steel Casting by Using Advanced Mathematical Methods. *Metals* **2018**, *8*, 1019. [CrossRef]
13. Zhang, X.; Yang, W.; Xu, H.; Zhang, L. Effect of Cooling Rate on the Formation of Nonmetallic Inclusions in X80 Pipeline Steel. *Metals* **2019**, *9*, 392. [CrossRef]
14. Han, Y.; Wang, X.; Zhang, J.; Zeng, F.; Chen, J.; Guan, M.; Liu, Q. Comparison of Transverse Uniform and Non-Uniform Secondary Cooling Strategies on Heat Transfer and Solidification Structure of Continuous-Casting Billet. *Metals* **2019**, *9*, 543. [CrossRef]
15. Wu, C.; Ji, C.; Zhu, M. Deformation Behavior of Internal Porosity in Continuous Casting Wide-Thick Slab during Heavy Reduction. *Metals* **2019**, *9*, 128. [CrossRef]

© 2020 by the authors. Licensee MDPI, Basel, Switzerland. This article is an open access article distributed under the terms and conditions of the Creative Commons Attribution (CC BY) license (http://creativecommons.org/licenses/by/4.0/).

Article

Change of the Chemical and Mineralogical Composition of the Slag during Oxygen Blowing in the Oxygen Converter Process

Dana Baricová *, Alena Pribulová, Peter Futáš, Branislav Buľko and Peter Demeter

Technical University of Kosice, Faculty of Materials, Metallurgy and Recycling, Institute of Metallurgy, Park Komenskeho 14, 04001 Kosice, Slovak; alena.pribulova@tuke.sk (A.P.); peter.futas@tuke.sk (P.F.); branislav.bulko@tuke.sk (B.B.); peter.demeter@tuke.sk (P.D.)
* Correspondence: dana.baricova@tuke.sk; Tel.: +421-55-602-2755

Received: 28 September 2018; Accepted: 16 October 2018; Published: 18 October 2018

Abstract: The article presents the results of the investigation of changes in the chemical and mineralogical composition of slag during steel production in a blown oxygen converter. This process was monitored using the slag samples that were collected during the period when oxygen blowing into an oxygen converter was interrupted. The slag samples were collected after 150 s (2.5 min), then after 5, 8, 11, and 24 min of oxygen blowing, and in minute 27 when oxygen blowing was terminated. The sampling was carried out within five consecutive melting processes. The article presents and documents the changes in the contents of CaO, CaO (free), Fe (total), FeO, SiO_2, and in the basicity of the slag during oxygen blowing. It also provides the characteristics of individual structural components formed during oxygen blowing and a detailed description of the lime assimilation process, including the formation of the final structure of the slag, consisting of dicalcium silicate ($2CaO \cdot SiO_2$), tricalcium silicate ($3CaO \cdot SiO_2$), RO-phase, and calcium ferrites ($2CaO \cdot Fe_2O_3$). The results of the investigation of the changes in the chemical composition of the slag during oxygen blowing in an oxygen converter were compared with the changes in the structural composition of the slag.

Keywords: oxygen converter slag; chemical composition of slag; structural composition of slag

1. Introduction

Metallurgical slags play an important role in the processes of steel production and treatment. The knowledge of the formation and development of chemical and mineralogical properties of steel slags is essential. Steel production in oxygen converters is primarily characterised by a high rate of individual operations and this must be considered when determining the slag production rate. Therefore, it is important to know the key factors influencing the rate of slag formation. Additionally, slags must have the required physical and chemical properties to fulfil its purpose during the steel production. And, last but not least, the final chemical and mineralogical composition of slag has a significant impact on its final properties and, hence, its further utilization. Until now, very few scientific papers have dealt with these issues at such a complex level. For the first time, the chemical and mineralogical compositions of steelmaking slags were directly linked.

The main factors that determine the rate of slag formation are the contents of oxides of iron and manganese in the slag, the temperature of the steel, and the conditions under which the lime is added [1]. When the slag formation is delayed, corrosion of the furnace lining increases, oxidised impurities are not completely absorbed, and lime utilization is low. The slag formation process is a key factor affecting the quality of the produced steel and the output of the production process as such.

The important roles of slags during the steel production are described by the author [2]. Steelmaking slag acts as a sink for impurities during refining of steel, and it controls oxidizing and reducing potential of the bath during refining through FeO content. Slag prevents the passage of nitrogen and hydrogen from the atmosphere to the liquid steel in the bath, provides protection to the liquid steel from re-oxidation, and insulates the liquid bath and reduces thermal losses.

In order to control the slag regimen in an oxygen converter, it is important to understand the process of formation of the structure and composition of slag during the steel production. This process must also be studied in relation to the developmental trends in the field of steel production and a growing number of variables.

The importance of converter slag as dephosphorization and desulphuration media has ceased with implementation of special pre- and post-converter treatments. The transfer from top blowing to combined blowing converters has decreased the amount of slagged iron that, together with the tendency towards lower silicon and manganese contents, has decreased slag to steel ratio. Still, the control of slag forming and behaviour during the blow is of great importance [3].

The generated compositions and amounts of slag will depend on the process in which the slag is formed and on the raw materials used in the process, e.g., iron input and fluxing ingredients, such as burnt lime, governed by the proportions of materials required for an effective process [4].

Slag in a BOF (basic oxygen furnace) is heterogeneous and always contains some entrained gas bubbles and solid material (either un-dissolved or precipitated) [5].

Knowledge of the chemical, mineralogical, and morphological properties of steel slags is essential because their cementitious and mechanical properties, which play a key role in their utilization, are closely linked to these properties [6].

The authors of paper [7] reported physical and chemical characteristics of the converter steel slag, as well as its comprehensive utilization by means of internal and external recycling. The major structural phases of the converter steel slag mainly include dicalcium silicate ($2CaO \cdot SiO_2$), tricalcium silicate ($3CaO \cdot SiO_2$), RO-phase (a solid solution of CaO-FeO-MnO-MgO), and calcium ferrite ($2CaO \cdot Fe_2O_3$). There are great differences in Vickers hardness between individual mineral phases of the converter steel slag.

At present, after the separation of the metallic fraction that is recycled directly in a metallurgical plant, the processing of the converted slag may be divided into three main trends. Within the first one, the converter slag is used in the building industry [8–11]. According to the second trend, steel slag is used in agriculture [12]. The third trend shows that the steel converter slag is added directly into a blast-furnace, agglomeration, or steel charge [13,14].

The purpose of the present article was to analyse the process of slag formation and development in a top-blown oxygen converter on the basis of the changes in the chemical and mineralogical composition. This process may be best examined if the slag samples are collected in the period when the oxygen blowing is interrupted. The slag samples were collected after 150 s (2.5 min), then after 5, 8, 11, and 24 min of oxygen blowing, and after the oxygen blowing was terminated in minute 27. For this purpose, the sampling was carried out during five consecutive melting processes (A, B, C, D, E) in a top-blown, 190-tonne oxygen converter. The total melting time was 46 min while the blowing time was 27 min in all the melting processes.

2. Materials and Methods

The charge used in the oxygen converter (VSZ a.s., Kosice, Slovakia) consisted of pig iron and steel scrap, in approximately identical quantities in all melting processes. The slag-forming additives were lime and dolomitic lime. In the case of Melting Processes D and E, the quantities of the added slag-forming additives were identical. In Melting Process C, the quantities of lime were approximately 25% higher. In Melting Process B, the percentage of slag-forming additives was the highest, and in addition to lime and dolomitic lime, magnesite was used as well. In Melting Process A, a portion of the lime was replaced with the demetallized converter slag. The slag-forming additives were added into the oxygen converter gradually. The first dose was always added in the beginning of oxygen blowing and the second dose before minute 2 of oxygen blowing. In Melting Process B, it was necessary to add an extra dose of slag-forming additives due to the unsatisfactory result of the pre-test chemical analysis. The quantities of the charged raw materials used in the steel production in individual melting processes and the doses of slag-forming additives are presented in Table 1.

Table 1. Quantities of the charged raw materials and doses of the slag-forming additives.

Sample	Pig Iron (t)	Scrap (t)	Lime (kg)	Dolomitic Lime (kg)	Demetallized Slag (kg)	Magnesite (kg)
Melt A	150.1	40.0	5500 1000	2000 -	1 000 -	- -
Melt B	149.0	40.0	9000 1500 1300	1500 1000 3000	- - -	985 - 990
Melt C	152.9	36.0	6750 4500	1000 1000	- -	- -
Melt D	152.5	42.0	5000 3100	2000 -	- -	- -
Melt E	153.0	42.0	5000 3100	1000 1000	- -	- -

The chemical analysis of the collected slag samples was carried out in the accredited laboratory by X-ray quantometer RIGKU (VSZ a.s., Kosice, Slovakia), and was supplemented with the identification of the slag basicity (B = CaO/SiO_2), as well as the analysis of free lime, as presented in Table 2. The content of free lime was determined applying the method of chemical analysis. This method uses the selective dissolution of free lime in the mixture of sugar and alcohol, and the subsequent determination through the titration with hydrochloric acid. The described methodology is typically used when performing an analysis of free lime directly in a metallurgical plant. However, this methodology is not standardised. Additionally, slag samples were treated applying a standard method, that is, grinding and polishing, in order to prepare metallographic scratch patterns, which were subsequently examined using a NEOPHOT 32 optical microscope (Technical university of Kosice, Kosice, Slovakia). Thus, individual structural components and their morphology were observed and photographically documented.

The performed analyses also included the EDX microscopy analysis that was aimed at qualitative as well as quantitative confirmation of the percentages of individual structural phases present in the slag. This EDX analysis was carried out using a Joel JSM-7000F scanning electron microscope (SAV, Kosice, Slovakia). This analysis was applied to a complete series of slag samples collected from the Melting Process D. The purpose thereof was to predict individual structural components of the slag. The analysis and the microscopic observation facilitated acquiring the best possible understanding of the development of the slag structure in the convertor.

Table 2. Chemical analysis of slag samples.

Sample	Time (min)	Fe(total) (%)	FeO (%)	SiO$_2$ (%)	CaO (%)	MgO (%)	MnO (%)	P$_2$O$_5$ (%)	S (%)	B	CaO (free) (%)
A2.5	2.5	22.00	20.80	19.06	27.06	4.43	6.49	1.47	0.04	1.42	7.42
A5	5.0	23.12	23.91	15.34	35.67	4.94	6.41	1.94	0.03	2.33	4.63
A8	8.0	20.43	16.09	16.98	42.15	7.05	4.36	1.64	0.04	2.48	4.93
A11	11.0	8.60	10.92	18.36	47.95	7.46	4.10	0.98	0.03	2.61	2.97
A24	24.0	12.29	10.78	13.26	49.40	6.65	3.11	0.96	0.03	3.73	1.83
A27	27.0	15.00	18.13	12.24	50.20	5.89	5.62	1.21	0.09	4.10	0.95
B2.5	2.5	23.80	25.87	19.10	33.37	4.03	7.06	1.40	0.04	1.75	8.39
B5	5.0	14.52	14.80	18.27	40.66	4.78	7.39	1.59	0.03	2.23	6.38
B8	8.0	11.28	11.77	18.53	43.82	4.85	9.39	1.34	0.04	2.36	5.45
B11	11.0	14.31	14.87	10.64	55.21	5.56	4.67	1.32	0.03	5.19	6.59
B24	24.0	15.77	15.52	12.20	47.26	6.03	5.78	1.32	0.04	3.87	4.81
B27	27.0	19.06	23.15	10.06	51.76	5.08	4.24	1.04	0.13	5.15	1.77
C2.5	2.5	20.22	24.14	18.06	35.05	4.63	8.09	1.00	0.03	1.94	7.86
C5	5.0	17.10	18.68	16.25	43.52	6.65	8.00	1.30	0.03	2.68	6.22
C8	8.0	6.48	8.05	15.60	57.76	5.44	4.55	1.09	0.04	3.70	5.42
C11	11.0	7.80	3.38	15.24	63.37	4.84	3.34	1.14	0.04	4.16	4.42
C24	24.0	12.71	4.31	14.16	47.53	5.44	3.86	1.00	0.08	3.36	1.84
C27	27.0	15.01	15.09	11.82	50.98	5.87	4.66	1.36	0.10	4.31	0.89
D2.5	2.5	18.54	16.09	19.30	37.11	5.44	4.71	1.12	0.00	1.92	10.97
D5	5.0	15.40	15.66	16.32	44.03	8.26	3.90	0.75	0.03	2.70	9.02
D8	8.0	18.54	16.14	14.00	38.41	5.64	3.93	0.87	0.00	2.74	7.35
D11	11.0	18.38	17.04	10.76	52.99	7.86	3.66	1.19	0.08	4.92	7.80
D24	24.0	18.87	20.69	9.80	50.47	7.96	3.73	1.39	0.07	5.15	2.10
D27	27.0	19.84	21.13	10.83	48.62	6.75	4.21	1.28	0.12	4.49	0.58
E2.5	2.5	27.96	27.39	18.19	33.37	4.86	3.12	0.50	0.00	1.83	8.97
E5	5.0	26.37	30.46	13.73	37.23	4.66	6.23	2.00	0.03	2.71	5.79
E8	8.0	19.10	22.80	14.97	39.54	4.86	5.77	2.01	0.03	2.64	6.30
E11	11.0	14.32	10.63	12.12	48.03	4.86	2.17	0.99	0.00	3.96	6.98
E24	24.0	16.23	20.19	9.97	42.90	4.06	4.29	1.37	0.08	4.30	2.06
E27	27.0	21.44	22.13	10.24	46.55	4.34	5.79	1.28	0.12	4.55	0.72

The experimental melting processes also included the sampling during the period when oxygen blowing was interrupted, always after minute 8 and 24, and in minute 27 when oxygen blowing was terminated. Also, the temperature and activity of oxygen in the metal were measured. The chemical composition of the metal, identified in the accredited laboratory by an ARL 8800 spectrometer, the temperature of the metal, and oxygen activity by a CELOX immersion probe are listed in Table 3.

Table 3. Chemical composition of metal samples, temperature and activity of oxygen in the metal.

Sample	Time (min)	C (%)	Mn (%)	Si (%)	P (%)	S (%)	T$_{steel}$ (°C)	aO (ppm)
A$_M$0 pig iron after desulphurization		4.530	0.540	0.790	0.068	0.011	-	-
A$_M$8	8	2.520	0.090	0.010	0.020	0.015	1418	10.0
A$_M$24	24	0.036	0.114	0.010	0.004	0.019	1645	882.7
A$_M$27	27	0.031	0.069	0.005	0.007	0.016	1665	989.5
B$_M$0 pig iron after desulphurization		4.470	0.510	0.890	0.081	0.012	-	-
B$_M$8	8	2.240	0.140	0.010	0.033	0.033	1425	10.8
B$_M$24	24	0.040	0.120	0.010	0.007	0.018	1630	669.4
B$_M$27	27	0.020	0.050	0.005	0.004	0.017	1651	957.6
C$_M$0 pig iron after desulphurization		4.600	0.610	0.760	0.066	0.014	-	-
C$_M$8	8	2.490	0.090	0.010	0.019	0.016	1480	17.8
C$_M$24	24	0.041	0.162	0.010	0.013	0.015	1667	798.5
C$_M$27	27	0.036	0.135	0.005	0.010	0.014	1659	931.2
D$_M$0 pig iron after desulphurization		4.400	0.480	0.790	0.078	0.012	-	-
D$_M$8	8	2.370	0.100	0.010	0.036	0.015	1391	19.2
D$_M$24	24	0.055	0.119	0.010	0.010	0.020	1659	759.7
D$_M$27	27	0.033	0.084	0.010	0.009	0.017	1664	983.2
E$_M$0 pig iron after desulphurization		4.600	0.470	0.810	0.069	0.009	-	-
E$_M$8	8	2.290	0.090	0.010	0.018	0.022	1375	18.5
E$_M$24	24	0.053	0.097	0.010	0.006	0.023	1590	618.8
E$_M$27	27	0.028	0.066	0.005	0.006	0.017	1655	1191.0

3. Results

3.1. Changes in the Chemical Composition of Slag during Oxygen Blowing

In the beginning of oxygen blowing, the steel scrap represented the lowest layer inside the converter and it was covered with liquid raw iron with the temperature of 1350 °C. Subsequently, the first dose of slag-forming additives was added and the oxygen jet was triggered. The second dose was added gradually, before minute 2 of oxygen blowing. The melting temperature of the pure lime is 2570 °C. The dissolution as such began after the termination of the passive period that consisted of heating the melt frozen on the lime parts. It was, therefore, necessary, in the beginning of oxygen blowing, to increase the temperature of the bath, as soon as possible, particularly in the upper section of the converter where the parts of undissolved lime, covered with the frozen melt, were floating on the level of the molten metal. Therefore, in the beginning of oxygen blowing, the so-called "soft blowing" is always applied. In this case, the oxygen jet was placed above the bath level, and the oxygen flow rate was reduced. In the beginning of oxygen blowing, oxygen from the oxygen jet in the impact zone was dissolved. The bath temperature rapidly increased. During oxygen blowing, the temperature near the oxygen jet reached over 2000 °C. Following the melting of the slag phase, the oxygen blowing mode changed into the so-called "hard blowing". The oxygen jet was shifted lower towards the bath level and intensity of oxygen blowing was increased while the entire content of the bath was intensively stirred and oxidised.

The changes in the chemical composition of the slag and in the slag basicity during oxygen blowing in the top-blown oxygen converter for individual melting processes are listed in Table 2. The beginning of blowing was accompanied with intensive oxidation of the upper section of the bath. Temperatures gradually increased due to the running exothermal reactions. Oxygen began to dissolve in the metal bath and this was followed by the oxidation of iron and other additive components. The first to oxidise was silicon due to its strongest affinity for oxygen. As shown in Table 3, the average value of the silicon content in raw iron was 0.808%. After 8 min of oxygen blowing, the silicon contents decreased, in all the melting processes, down to 0.010%. As a result of a low percentage of dissolved CaO and a high concentration of SiO_2 in the slag, very aggressive acidic slag was formed in the beginning of oxygen blowing. The slag basicity after 2.5 min of oxygen blowing ranged from 1.42 to 1.94. During oxygen blowing, there was an increase in the quantity of slag from the oxides that are transferred into the slag and the slag-forming additives were gradually dissolved. The basicity of the slag increased. At the end of the refining process, the basicity values ranged between 4.1 and 4.55. In Melting Process B, the final basicity value was 5.15; this was caused by adding another dose of slag-forming additives that were added due to the unsatisfactory chemical composition of the metal at the pre-test. With the increasing percentage of the dissolved lime during oxygen blowing, the content of CaO (free) proportionally decreased. Nevertheless, this process is imperfect and at the end of oxygen blowing the final slag contains undissolved lime parts in the quantities representing 0.58% to 0.95%. In Melting Process B, the final content of free lime reaches the value of as much as 1.77%, as a result of adding the third dose of slag-forming additives. In the details of the study, it is possible to find unassimilated pieces of lime in the all samples. For example, Figure 1 shown free lime in the slag within Melting Process B, which is visible even to the naked eye.

Figure 1. Free lime in the slag at the end of oxygen blowing in the melting process B, sample B27.

The content of free lime in steel slag is the key factor that hinders potential applications of metallurgical slag in the building industry. If the free lime is not stabilised, it is hydrated at sites where it comes into contact with the surrounding atmosphere and its volume increases, thus causing the volumetric instability and the consequent decomposition of the slag, according to Equation (1):

$$\mathrm{CaO}\left(1\,\mathrm{cm}^3\right) \overset{H_2O}{\rightarrow} \underset{\text{portlandite}}{\mathrm{Ca(OH)}_2}\left(1.92\,\mathrm{cm}^3\right) \overset{CO_2}{\rightarrow} \underset{\text{calcite}}{\mathrm{CaCO}_3}\left(2.14\,\mathrm{cm}^3\right) + H_2O \qquad (1)$$

The content of SiO_2 in the slag reached the highest values in the beginning of oxygen blowing (18.06% to 19.30%). Due to the fact that silicon has the strongest affinity for oxygen, the majority portion of silicon oxidises within 8 min of oxygen blowing. As a result of the gradual dissolution of lime, transfer of other oxidised admixtures into the metal, and an increase in the total quantity of slag, the content of SiO_2 in the slag gradually decreased during oxygen blowing. At the end of oxygen blowing, the content of SiO_2 in the slag represented 9.8% to 12.24%.

In the beginning of oxygen blowing, the content of total iron in the slag was rather high, ranging from 18.54% to 27.96%. This was caused by the fact that the surface of the bath was highly oxidised due to the "soft blowing", which resulted in oxidation of Fe from the metal bath into FeO and its transfer into the slag. Very intensive mixing of the surface parts of the bath also resulted in a higher content of metallic Fe in the slag in form of drops; this corresponds to the high content of Fe (total). During oxygen blowing, when the operation mode was switched to "hard blowing" and after increasing the quantity of the slag, the percentage of Fe (total) decreased as a result of the reverse reduction of FeO from the slag into the bath. The lowest values of Fe (total) and FeO contents in the slag were observed approximately between minutes 8 and 11 of oxygen blowing. At the end of oxygen blowing, when the temperature of the metal bath was high and the admixture components were oxidised, the oxidation of Fe into FeO ran again in the bath and resulted in the transfer thereof into the slag. At the end of the refining process, Fe (total) ranged from 15.00% to 21.44%.

3.2. Changes in the Mineralogical Composition of Slag during Oxygen Blowing

The slag samples collected 2.5 min after the beginning of oxygen blowing reflected the status immediately after the slag-forming additives were added into the converter. Majority of slags were of the glossy character that was caused by the presence of silicon oxide in the slag. In the diffractogram of the X-ray diffraction analysis, the dominating structural patterns were the lines of calcium oxide or calcium hydroxide that were formed as a result of CaO hydration in the sample. Sporadically, monticellite $2(FeO,MnO,MgO,CaO)\cdot SiO_2$ and dicalcium silicate $2CaO\cdot SiO_2$ precipitated. Figure 2 presents a lime part in the primary slag, detected 2.5 min after the beginning of oxygen blowing. The figure clearly shows the beginning of the infiltration of a lime part by the slag, as well as mostly glossy character of the slag surrounding the lime part, and the beginning of the formation of dicalcium silicate precipitation front around the lime part.

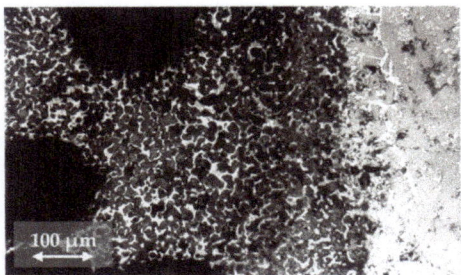

Figure 2. A lime part in the glassy silicate slag after 2.5 min of oxygen blowing, in the melting process D, sample D2.5.

In the slag samples collected after minute 5 of oxygen blowing, the basic structural component was again the free lime, in form of calcium hydroxide at the time of the X-ray diffraction analysis. In addition to the free lime, the structure also comprised of dicalcium silicate $2CaO \cdot SiO_2$ and, in a small amount, also calcium ferrites $CaO \cdot Fe_2O_3$. Lime parts lay inside the glassy silicate matrix. This stage was characterised by the formation of a dicalcium silicate margin around the lime part and a high degree of infiltration of the lime part by the slag, including the initial formation of calcium silicates and RO-phase that consisted of free oxides (Fe,Mn,Ca,Mg)O inside the lime part. Figure 3 clearly shows the formation of dicalcium silicate margin around the lime part in the glassy silicate slag.

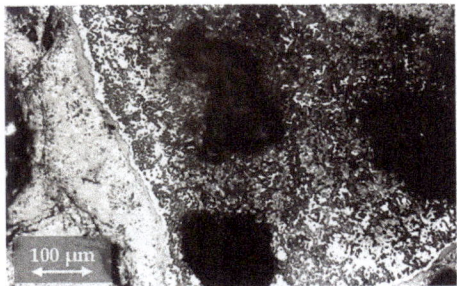

Figure 3. A lime part in the glassy silicate slag and the beginning of the formation of a dicalcium silicate margin after minute 5 of the refining process, melting process E, sample E5.

In the slag samples collected after minute 8 of oxygen blowing, there was a significant decrease in the number of pieces of unassimilated lime, or more precisely, reduction of their sizes. Dicalcium silicate became the basic structural component. As compared to the previous group of slags, collected after minute 5 of blowing, there was a slight increase in the content of calcium ferrite $CaO \cdot Fe_2O_3$. The structure also contained the primary silicate slag that is characterised by its glassy character. On the basis of the performed analyses we can state that approximately in minute 8 of the refining process the final structure of the slag began to form. Figure 4 shows the lime part in the stage preceding the total assimilation.

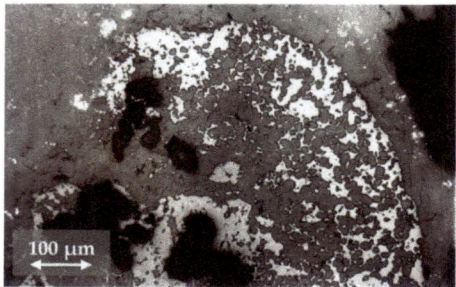

Figure 4. Decomposition of the lime part in the slag after minute 8 of oxygen blowing. Sample D8, Melting Process D.

The slag samples collected after minute 11 of oxygen blowing contained only the residues of unassimilated lime. The fact that even more calcium oxide was transferred into the slag from the assimilated lime resulted in the transformation of a part of dicalcium silicate into tricalcium silicate, Equation (2):

$$2CaO \cdot SiO_2 + CaO \rightarrow 3CaO \cdot SiO_2 \qquad (2)$$

The morphology of calcium silicates was different at various structure sites; some sites were present with globular particles, at some sites the predominating components were acicular formations, as shown in Figure 5.

Figure 5. Needles of tricalcium silicate and globular particles of dicalcium silicate in the slag after minute 11 of oxygen blowing, in the melting process B, sample B11.

The structure of slag samples collected after minute 24 was rather homogenous as to their morphology. The slag contained tricalcium silicate, dicalcium silicate, RO-phase, and calcium ferrites. The product of dephosphorisation, $3CaO \cdot P_2O_5$, was bound to dicalcium silicate; the product of desulphurisation, CaS, was bound to calcium ferrites. The structure still contained a certain portion of unassimilated lime, probably originating in the last dose of slag-forming additives. However, it was not in the form of pure calcium oxide, but the solid solution of CaO-FeO containing 10% FeO. The formation of this solid solution was accompanied with the oxidation of iron in the last stages of the refining process. A typical appearance of such structure is presented in Figure 6. It contains needles of tricalcium silicate, dicalcium silicate, as well as RO-phase and calcium ferrites located between the arms of the needles.

Through a visual assessment it was possible to observe dicalcium silicate in irregular formations, representing globular, acicular and irregular formations. As the lime gradually dissolved, the slag became even more intensively saturated with the lime. Subsequently, dicalcium silicate transformed into tricalcium silicate that was predominantly separated out in the acicular form.

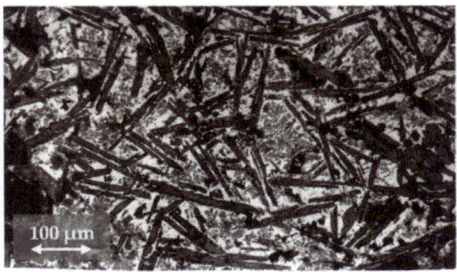

Figure 6. Needles of tricalcium silicate and globular particles of dicalcium silicate, RO-phase, and calcium ferrites in the slag sample collected after minute 24 of oxygen blowing, in the melting process D, sample D24.

In the slag samples collected after the oxygen blowing was terminated, there was no change in their structural composition, as compared to the samples collected after minute 24 of oxygen blowing. The structure of the slag was very homogenous, see Figure 7. Due to intensive mixing in the bath as well as further dissolution of lime in the slag structure, the majority component was tricalcium silicate. The percentage of dicalcium silicate, separated out in the slag mainly in form of irregular formations, was gradually decreasing. The percentage of tricalcium silicate, which tends to separate out in acicular formations, was increasing. Due to intensive mixing, the homogeneity of the slag was increasing with longer refining times.

Predominating components were tricalcium silicate, dicalcium silicate, RO-phase and calcium ferrites. The structure of the slag also contained a small amount of unassimilated lime. The presence of dicalcium silicate, calcium ferrites in the final slag in Melting Process D is presented in Figure 8.

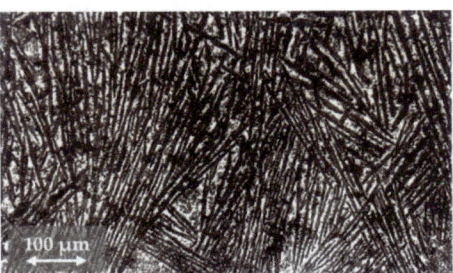

Figure 7. Final structure of the slag collected after termination of oxygen blowing, in the melting process D, sample D27.

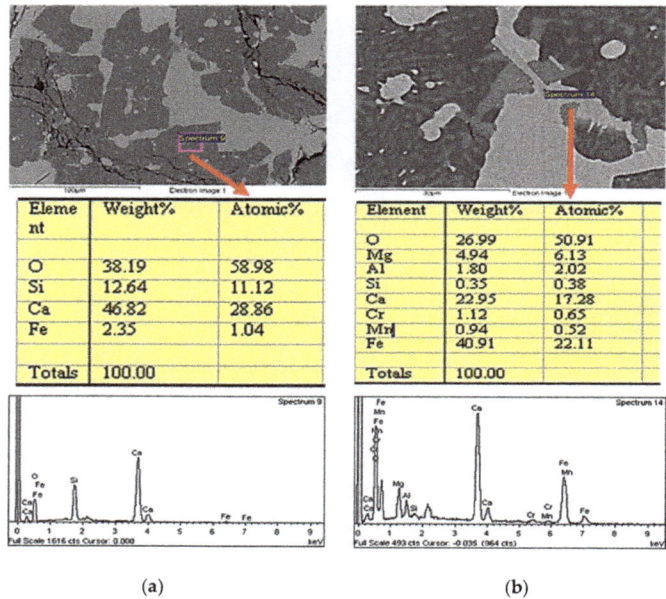

Figure 8. Structure of the slag in Melting Processes D and EDX—the analysis proving the presence of dicalcium silicate and calcium ferrites in the final slag in Melting Process D: (**a**) dicalcium silicate, and (**b**) calcium ferrites.

4. Discussion

As mentioned above, five consecutive melting processes were analysed. In all cases, the same grade of steel was produced. The charge used in the oxygen converter consisted of pig iron and steel scrap in approximately identical quantities in all melting processes, where lime and dolomitic lime were used as the slag-forming additives. In Melting Process C, the quantities of lime were approximately 25% higher. In Melting Process B, it was necessary to add an extra dose of slag-forming additives due to the unsatisfactory result of the pre-test chemical analysis. In Melting Process A, a portion of the lime was replaced with the demetallized converter slag. The amounts and types of the added slag-forming additives did not have any significant effect on the formation of slag, or on its chemical and mineralogical composition.

The rate of slag formation is determined by the rate of dissolution of lime, the melting temperature of which is, as mentioned above, 2570 °C. During the production of steel in an oxygen converter, such temperature could not be achieved in the entire melt content. Dissolution of CaO was therefore rather slow. After the oxygen jet was triggered, oxidation of the bath began. The first to oxidise was silicon; it reacted with CaO under concurrent formation of $2CaO \cdot SiO_2$ while its melting temperature is 2130 °C) [15]. The slag formation was very positively affected by mixed oxides that were formed in the $FeO-MnO-CaO$, $CaO-Fe_2O_3$, and $CaO-P_2O_5$ systems that have lower melting temperatures than CaO. An example of dissolving CaO in the primary slag is shown in Figure 9.

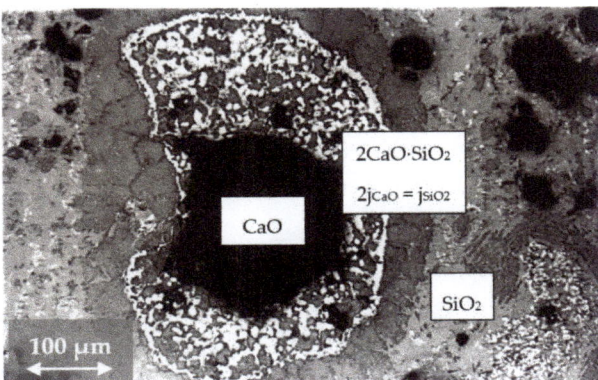

Figure 9. A small piece of undissolved lime in the slag from Melting Process C, sample C3, collected after minute 8 of oxygen blowing.

There was a gradient of CaO activity from the lime surface towards the slag; in the opposite direction, there was a gradient of SiO_2 activity. Activity of CaO was decreasing towards the slag; similarly, activity of SiO_2 was decreasing towards the lime part. Therefore, there must be a certain distance from the lime part at which the values of diffuse solutions of CaO ($2j_{CaO}$) and diffuse solutions SiO_2 (j_{SiO2}) are in the ratio of $2j_{CaO} = j_{SiO2}$. At this point, there is a dicalcium silicate precipitation front into which equivalent amounts of CaO and SiO_2 constantly diffuse and in which new amounts of dicalcium silicate constantly precipitate until the precipitation surface is fully filled. No other precipitation was running because CaO and SiO_2 were spatially separated. Another process of lime dissolution consisted in mechanical decomposition of a dicalcium silicate layer which was significantly facilitated by "soft blowing". Potential subsequent processes may include recurrent diffusion of more CaO from the lime, formation of another dicalcium silicate layer, its decomposition, or even complete dissolution of the lime part. Once this process is completed, few undissolved crystals remain.

In the first part of the refining process, following the addition of first portions of slag-forming additives, majority of the slag was of glassy character due to the predominating content of silicon oxide in the slag. From this glassy phase, monticellite $2(FeO,MnO,MgO,CaO) \cdot SiO_2$ or the mixture of $2FeO \cdot SiO_2$, $2MnO \cdot SiO_2$, and $2CaO \cdot SiO_2$ crystals precipitated, to a larger or smaller extent. There were also small amounts of mixed oxide and RO-phase $(Fe,Mn,Ca,Mg)O$. The monticellite glassy slag was only little active in the beginning of blowing as a result of a low content of the RO-phase. The purpose of the slag mode was to transfer this low-activity monticellite slag into dicalcium silicate, or tricalcium silicate, containing the RO-phase. The transformation of monticellite into dicalcium silicate ($2CaO \cdot SiO_2$), or tricalcium silicate ($3CaO \cdot SiO_2$), ran as described by Equations (3) and (4) [16]:

$$2(FeO,MnO,MgO,CaO) \cdot SiO_2 + CaO \rightarrow 2CaO \cdot SiO_2 + RO\text{-phase} \quad (3)$$

$$2CaO \cdot SiO_2 + CaO \rightarrow 3CaO \cdot SiO_2 \quad (4)$$

Approximately in the middle of the refining process, homogenous glass ceased to exist, the predominating structural components were monticellite and RO-phase, under the significant presence of dicalcium silicate, the slag contained emulsified small particles of raw iron. In the third fourth of the refining process, dicalcium silicate or tricalcium silicate precipitated, depending on the following addition of lime into the slag. Calcium ferrites began to precipitate and metallic iron and free lime were present. At the end of oxygen blowing, the main components of the slag structure were dicalcium silicate, tricalcium silicate and RO-phase. There was a significant increase in the contents of calcium ferrites and the slag also contained metallic iron and free lime. Metallic iron occurred in two forms, as granules or as facets. Facets were formed during the cooling of residual melt, enriched with CaO and FeO, through the following mechanism (5):

$$(CaO\text{-}FeO)_{liquid} \rightarrow calcium\ ferrite + Fe(metal) \quad (5)$$

Figure 10 presents a part of the CaO-SiO$_2$-FeO ternary scheme [17] where the arrows indicate the slag development in an oxygen converter during oxygen blowing from the primary slag to the final slag for all of the analysed melting processes.

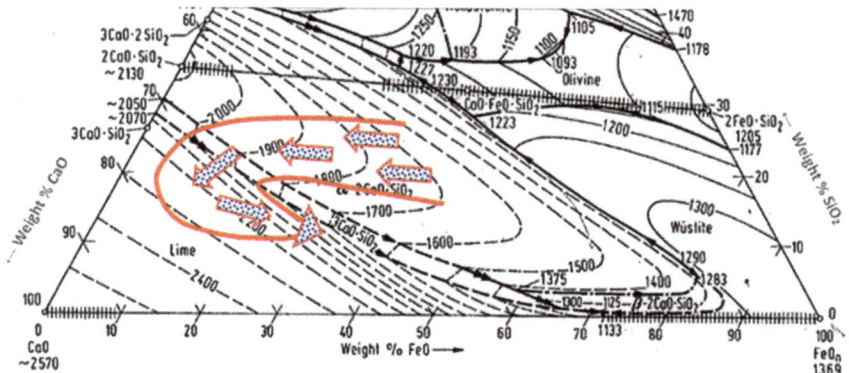

Figure 10. Slag development in an oxygen converter during oxygen blowing, from the primary to the final slag, for all the analysed melting processes highlighted in a part of the CaO-SiO$_2$-FeO ternary scheme, adapted from [17].

The change in the mineralogical composition of the steel slag fully corresponded with the change in their chemical composition. Figure 11 presents the change in the chemical composition of the slag formed in Melting Process D during oxygen blowing in the oxygen converter and the corresponding change in the slag structure.

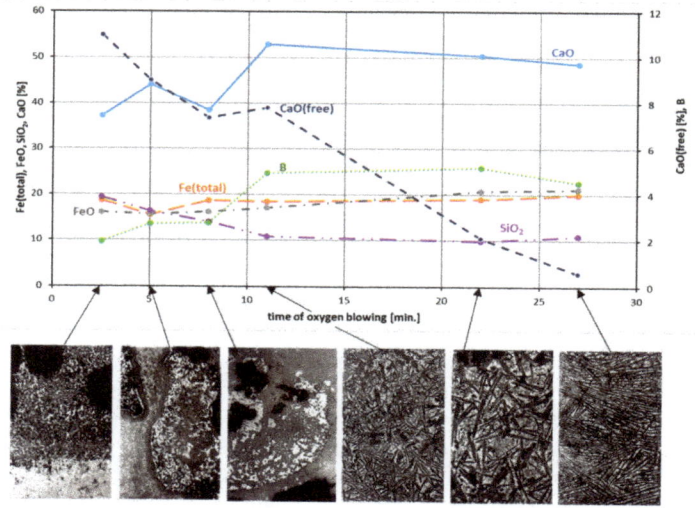

Figure 11. Change in the chemical composition of the slag formed in Melting Process D during oxygen blowing in the oxygen converter and the corresponding change in the slag structure.

5. Conclusions

The amount of the oxygen converter slag ranges from 120 to 150 kg per tonne of the produced crude steel. The amount of the produced slag primarily depends on the type of technology, the ratio of scrap to pig iron in the batch, the grade of the steel being produced, the initial chemical composition of pig iron, and the amount of slag additives added. At present, efforts are exerted to reduce the amount of slag; this may be achieved through the pre-treatment of pig iron, secondary metallurgy of the molten steel, and careful selection of batch materials. Despite all the efforts, the amount of the produced slag represents about 13% of the total weight of the batch. The knowledge of the creation and development of chemical and mineralogical properties of steel slags is essential. It is important to know the key factors influencing the rate of slag formation. Steelmaking slag must have the required physical and chemical properties to fulfil its purpose during the steel production, and the final chemical and mineralogical composition of slag has a significant impact on its final properties and hence its further utilization.

The chemical composition of slag was very closely related to the chemical composition of the metal; it corresponded to chemical processes running while oxygen was blown into the metal. The amount of the added slag-forming additives, and demetallized slag did not have any significant effect on the formation of slag, chemical, and mineralogical composition. The final content of lime in the slag ranged from 46 to 50%, but the process of lime dissolution was imperfect and after oxygen blowing terminated, the final contents of CaO (free) in the final slags was ranging from 0.58 to 1.77%. The SiO_2 content in the primary slag was 18–20%; in minute 8, it slightly increased and then gradually decreased down to the level of approximately 10% in all slags. The amount of FeO in the slag is largely dependent on the blowing mode and on the position of the oxygen lance.

At the end of blowing, the content of FeO in the slag was approximately 15% to 22%. The development of changes in the content of total iron in the slag in all cases corresponded to the development of changes in the content of FeO in the slag.

The structure of the slag related to its chemical composition and was more significantly affected by the added slag-forming additives. The changes in the slag structure reflected the process of gradual dissolution of lime in the slag. While in the primary slag the predominating structure components were SiO_2 and lime, after minute 8 of oxygen blowing their contents significantly decreased and the predominating component was dicalcium silicate. After minute 24 of oxygen blowing, the final structure of the slag was formed, with the major component being tricalcium silicate, containing also dicalcium silicate, RO-phase, and calcium ferrites.

Author Contributions: D.B. and A.P. performed experimental analysis, wrote the manuscript, funding, and also evaluated results of change of chemical and mineralogical composition; P.D. designed the experiments performed; B.B. provided paper revision; and P.F. performed a graphical evaluation of experiments.

Acknowledgments: This work was supported by the Scientific Grant Agency of The Ministry of Education of the Slovak Republic nos. VEGA 1/0703/16, VEGA1/0073/17, and VEGA 1/0868/17.

Conflicts of Interest: The authors declare no conflict of interest.

References

1. Yugov, I.P. Mechanism and kinetics of optimized slag formation in an oxygen converter. *Metallurgist* **2005**, *49*, 307–310. [CrossRef]
2. Role of Slag in Converter Steelmaking. Available online: http://ispatguru.com/role-of-slag-in-converter-steelmaking/ (accessed on 3 July 2018).
3. Jalkanen, H.; Holappa, L. On the role of slag in the oxygen converter process. In Proceedings of the VII International Conference on Molten Slags Fluxes and Salts, Johannesburg, South Africa, 25–28 January 2004; pp. 71–76.
4. Analysis of Metallurgical Processes and Slag Utilisation in an Integrated Steel Plant Producing Advanced High Strength Steels. Available online: https://www.diva-portal.org/smash/get/diva2:1013704/FULLTEXT01.pdf (accessed on 3 August 2018).

5. Deo, B.; Overbosch, A.; Snoeijer, B.; Das, D.; Srinivas, K. Control of slag formation, foaming, flopping, and chaos in BOF. *Trans. Indian Inst. Met.* **2013**, *66*, 543–554. [CrossRef]
6. Yildirim, I.Z.; Prezzi, M. Chemical, mineralogical, and morphological properties of steel slag. *Adv. Civ. Eng.* **2011**, *2011*. [CrossRef]
7. Zhao, J.; Yan, P.; Wang, D. Research on mineral characteristics of converter steel slag and its comprehensive utilization of internal and external recycle. *J. Clean. Prod.* **2017**, *156*, 50–61. [CrossRef]
8. Jiang, Y.; Ling, T.C.; Shi, C.; Pan, S.Y. Characteristics of steel slags and their use in cement and concrete. *Resour. Conserv. Recycl.* **2018**, *136*, 187–197. [CrossRef]
9. Wan, J.; Wu, S.; Xiao, Y.; Chen, Z.; Zhang, D. Study on the effective composition of steel slag for asphalt mixture induction heating purpose. *Constr. Build. Mater.* **2018**, *178*, 542–550. [CrossRef]
10. Baricová, D.; Pribulová, A.; Demeter, P.; Buľko, B.; Rosová, A. Utilizing of the metallurgical slag for production of cementless concrete mixtures. *Metalurgija* **2012**, *51*, 465–468.
11. Morata, M.; Saborido, C.; Fontserè, V. Slag aggregates for railway track bed layers: Monitoring and maintenance. In Proceedings of the 15th International Conference on Railway Engineering Design and Operation, Madrid, UK, 19–21 July 2016; pp. 283–294, ISSN 1743-3509 (on-line). [CrossRef]
12. Rex, M. The use of BF, converter and ladle slags in European agriculture-benefits or risks? Slag-Providing solutions for global construction and other markets. In Proceedings of the 4th European Slag Conference, Oulu, Finland, 20–21 June 2005; pp. 51–55, ISSN 1617-5867.
13. Diao, J.; Zhou, W. System assessment of recycling of steel slag in converter steelmaking. *J. Clean. Prod.* **2016**, *125*, 159–167. [CrossRef]
14. Mihok, Ľ.; Fedičová, D. Recycling of demetallized steelmaking slag into charge of basic oxygen converter. *Metalurgija* **2000**, *39*, 93–99.
15. Thermochemical and Mineralogical Tables for Geochemical Modeling. Available online: http://thermoddem.brgm.fr/species/larnitegamma (accessed on 12 October 2018).
16. Kijac, J. Vysokoteplotné procesy výroby ocele I. In *High temperature processes in steelmaking I*, 1st ed.; Hutnícka fakulta TU v Košiciach: Košice, Slovakia, 2006; ISBN 80-8073-515-8.
17. Allibert, M. *Slag atlas*, 2nd ed.; Verein Deutscher Eisenhuttenleute: Dusseldorf, Germany, 1995; p. 126. ISBN 3-514-00457-9.

© 2018 by the authors. Licensee MDPI, Basel, Switzerland. This article is an open access article distributed under the terms and conditions of the Creative Commons Attribution (CC BY) license (http://creativecommons.org/licenses/by/4.0/).

Article

Phosphorus Equilibrium Between Liquid Iron and CaO-SiO$_2$-MgO-Al$_2$O$_3$-FeO-P$_2$O$_5$ Slags: EAF Slags, the Effect of Alumina and New Correlation

Andre N. Assis [1,*], Mohammed A. Tayeb [2], Seetharaman Sridhar [3] and Richard J. Fruehan [4]

1. Vallourec Tubes, 2669 Martin Luther King Jr. Blvd, Youngstown, OH 44510, USA
2. SABIC, Jeddah, 23955, Saudi Arabia; Tayebmab@sabic.com
3. Metallurgical and Materials Engineering, Colorado School of Mines, 1500 Illinois St, Golden, CO 80401, USA; sseetharaman@mines.edu
4. Materials Science and Engineering, Carnegie Mellon University, 5000 Forbes Ave, Pittsburgh, PA 15213, USA; fruehan@cmu.edu
* Correspondence: anogueira.assis@gmail.com

Received: 4 December 2018; Accepted: 15 January 2019; Published: 23 January 2019

Abstract: The increased use of electric arc furnace (EAF) steelmaking using up to 100% direct reduced iron (DRI) has prompted an interest in better control of phosphorus since iron ore and, consequently, DRI have higher phosphorus and silica compared to scrap. There is limited work reported on slag chemistries corresponding to that in the EAF when DRI is used. In the current research, phosphorus equilibria between molten Fe–P alloys and CaO-SiO$_2$-Al$_2$O$_3$-P$_2$O$_5$-FeO-MgO$_{saturated}$ slags were investigated. The results indicate that there is a significant decrease in the phosphorus partition coefficient (L_P) as alumina in the slag increases. The observed effect of alumina on the phosphorus partition is probably caused by the decrease in the activities of iron oxide and calcium oxide. Finally, an equilibrium correlation for phosphorus partition as a function of slag composition and temperature has been developed. It includes the effect of alumina and silica and is suitable for both oxygen and electric steelmaking-type slags.

Keywords: dephosphorization; EAF; BOF; phosphorus equilibrium

1. Introduction

Phosphorus is typically an undesirable residual element that has to be controlled. The development of new steel grades for next-generation applications result in ever tightening ranges for many residuals. Also, the steel industry is under constant pressure to keep costs down and maintain profitable margins. These opposing trends pressure the steelmakers to further optimize their processes to obtain maximum quality while use less "noble" iron sources.

In EAF steelmaking, the use of DRI has been increasing with plants using charges of up to 100% DRI continuously. This source of iron has higher alumina and phosphorus contents than typical scrap and, thus, the impact of these elements must be evaluated with respect to dephosphorization.

The present work investigates the phosphorus equilibrium for slag chemistries relevant to EAF steelmaking, including the effect of alumina in the phosphorus partition. It is also a continuation of a previously published paper [1] that included the literature review on the topic, experimental setup, and Basic Oxygen Furnace (BOF) data.

1.1. The Dephosphorization Reaction

The equilibrium of phosphorus between liquid metal and slag has been extensively studied since the 1930s. Assis et at. [1] recently published an extensive literature review on many of the works that proposed correlations to predict the phosphorus equilibrium partition between metal and slag. In general, it is well documented that dephosphorization increases with increasing slag basicity and decreases with increasing temperatures. Basu and coworkers [2,3] also noted that dephosphorization initially increases with increasing FeO in the slag but then decreases after a certain point. They proposed that this behavior is a function of the slag basicity and temperature.

When slag species are considered, the phosphorus oxidation reaction (dephosphorization) can be expressed in an ionic form:

$$\underline{P} + 2.5(FeO) + 1.5\left(O^{2-}\right) = \left(PO_4^{3-}\right) + 2.5Fe \tag{1}$$

where the underlined species is dissolved in liquid iron. The capacity of the slag to hold the phosphorus, a term called apparent phosphorus equilibrium constant is defined as

$$K_P = \frac{(\%P)_{slag}/[\%P]_{steel}}{(\%T.Fe)^{2.5}} \tag{2}$$

where $(\%P)_{slag}$ and $[\%P]_{steel}$ are the concentrations of phosphorus in the slag and the steel respectively and $(\%T.Fe)$ is the total iron in the slag.

The phosphorus partition coefficient, L_P, can be defined as

$$L_P = \frac{(\%P)_{slag}}{[\%P]_{steel}} \tag{3}$$

In the 1980s, Suito et al. [4–8] studied phosphorus partitioning on various slag systems at temperatures between 1823 K and 1923 K. In 1984, Suito et al. [5] proposed a correlation for K_P:

$$\log\left(\frac{L_P}{T.Fe^{2.5}}\right) = 0.0720[(\%CaO) + 0.3(\%MgO) + 0.6(\%P_2O_5) + 0.144(\%SiO_2) \\ + 0.6(\%MnO)] + \frac{11570}{T} - 10.52 \tag{4}$$

Assis et al. [1] noted that the form of Suito's correlation describes how slag composition changes the activity coefficients of O^{2-}, $PO_4{}^{3-}$, and Fe^{2+}. The coefficients are the first interaction coefficients of a Taylor series.

In 2000, Ide et al. [9] revised the MgO coefficient from 0.30 to 0.15. They [9] used CaO crucibles to vary the MgO content in the slag.

1.2. The Effect of Alumina on Dephosphorization

Alumina is present in steelmaking slags and usually ranges between 3 and 7 wt.%. Alumina comes from gangue in DRI and Al in the scrap as well as from flux additions. It has not been extensively studied in primary steelmaking, and most of the literature work with respect to alumina was conducted for hot metal or ladle dephosphorization [10–15].

In 1946 Balajiva and coauthors [12] noticed a decrease of phosphorus partition for experiments containing high alumina levels. Ladle dephosphorization for highly basic slags saturated with MgO was studied by Ishii and Fruehan [13]. They found that increasing (Al_2O_3/FeO) ratio reduced the slag phosphorus capacity. Similarly, Li and coauthors [14] reported a negative influence of alumina on the phosphorus capacity of ladle dephosphorization, especially for slags where the $Al_2O_3/(Al_2O_3 + SiO_2)$

ratio exceeded 0.66. The binary basicities were very high (above 5), the equilibration time was only 1 to 3 h and the slags contained Na$_2$O, which can improve phosphorus partition [8].

Miyata [15] reported the results of hot metal dephosphorization where 5 wt.% Al$_2$O$_3$ was added to the slag system CaO-SiO$_2$-FeO$_x$. This work recorded an improvement in dephosphorization, which claimed to be caused by the increase in slag fluidity. Mansour Al-Harbi [11] used the MTDATA thermodynamic package [10] and found that by increasing the alumina addition from 5 wt.% to 15 wt.%, the phosphorous content in the steel was reduced by almost 50% enhancing the phosphorus holding capacity of the slag.

2. Materials and Methods

A complete description of the experimental approach was described in detail by Assis et al. [1]. A brief description is provided below.

The current work concentrates on binary basicities (CaO/SiO_2) ranging from 1.0–2.7, FeO concentrations of up to 40 wt.%, and alumina of up to ~15 wt.% in the slag system CaO-SiO$_2$-MgO$_{Saturated}$-FeO-P$_2$O$_5$-Al$_2$O$_3$. Most of the experiments were performed at 1873K (1600 °C); some experiments were performed at 1823 K (1550 °C) and 1923 K (1650 °C). Equilibrium was attained from both sides, i.e., phosphorus was transferred from slag to metal and from metal to slag. To do this, experiments were mostly run in pairs; one with initial phosphorus partition, L_P, above equilibrium and the other with L_P below equilibrium.

2.1. Equilibration Time

A wide range of holding times has been reported in previous works [2–5,12,16] for the phosphorus reaction. It was demonstrated by Assis et al. [1] that equilibrium could be achieved in ~10 h for the present conditions. Thus, 10 h of holding time was used in the present work.

2.2. Master Fe–P Alloy

A master Fe–P alloy was prepared by melting electrolytic iron containing 99.97 wt.% Fe and 0.0004 wt.% P with iron phosphide (Fe3P) powder in an MgO crucible. The mixture was melted using a box furnace (CM 1712 GS FL, CM Furnaces Inc., Bloomfield, NJ, USA) heated at a rate of 2 K/min up to 1873 K (1600 °C) and held for 3 h. The furnace atmosphere was controlled by flowing high purity Argon gas. The furnace was cooled at a rate of 2 K/min to room temperature. The crucible was then crushed and the master alloy was polished to remove any traces of MgO from the surface. The composition of the master alloy was determined by inductively coupled plasma mass spectroscopy (ICP-MS). The final phosphorus content was 0.054 wt.%.

2.3. Iron Oxide

FeO was synthesized by mixing dried, reagent grade Fe$_2$O$_3$ with a slight excess of electrolytic iron power with a ratio of approximately 0.98. It was melted in an MgO crucible at 1273 K (1000 °C) under argon atmosphere. The temperature of 1273 K was chosen to not allow MgO dissolution to take place. The FeO-Fe mixture was then crushed and pulverized; any excess iron was removed by magnetic separation. The resulting powder was characterized with X-ray diffraction and confirmed that FeO had been produced.

2.4. Master Slag

A master slag was prepared with a targeted binary basicity of 1.0 and MgO content of 18 wt.%. CaO, SiO$_2$, and MgO were dried in air at 1173 K (900 °C) for 8 h in a resistance box furnace. The products were mixed and melted for 4 h at 1873 K (1600 °C) under argon atmosphere in a graphite crucible. The liquid slag was then quenched in cold water by opening the furnace and pouring the contents of the crucible into the water. The solid slag was pulverized using a puck mill (Shatterbox 8530,

SPEX SamplePrep LLC, Metuchen, NJ, USA) and decarburized in an MgO crucible for 10 h at a temperature of 1473 K (1200 °C) in air atmosphere. The master slag was used to facilitate melting during the equilibrium experiments. The individual slag composition for each experiment was adjusted accordingly by adding other oxides to the master slag.

2.5. Experimental Procedure

For each experiment, approximately 12 g of Fe–P alloy and 6 g of premelted mixed slag were allowed to equilibrate for 10 h in high-density MgO crucibles. Ultra-high-purity argon was used for the atmosphere to prevent the oxidation of the melt. The oxygen partial pressure content was measured to be between 10^{-5} to 10^{-6} atm. The present work has studied only two-phase equilibrium between liquid metal and slag. Three-phase equilibrium takes much longer to be achieved due to poor contact between all three phases or due to the lower density of the gas phase limiting the amount of mass transferred per unit time. Phosphorus was added in the form of Fe_3P into the metal and $Ca_2P_2O_7$ in the slag. All experiments were carried out in a horizontal tube furnace (CM 1730-12 HTF, CM Furnaces Inc., Bloomfield, NJ, USA) with ten $MoSi_2$ heating elements. The furnace was calibrated by inserting a type B thermocouple through the alumina reaction tube and measuring the temperature at intervals of 2 cm. In order to achieve 1873 K (1600 °C) in the furnace hot zone, the controller of the furnace needed to be set to 1918 K (1654 °C). The furnace calibration was always rechecked upon replacement of the alumina reaction tube.

The furnace of heated at a rate of 2 K/min until the desired equilibrium temperature was achieved. The temperature was held for 10 h. The furnace was then cooled to 1573 K (1300 °C) at a rate of −21 K/min to promote rapid solidification of the samples and held for 10 min. The cooling would then continue at a rate of −2 K/min until room temperature was achieved.

The metal analysis for phosphorus was mostly carried out by ICP-MS at West Penn Testing Group (New Kensington, PA, USA) with a reported reproducible limit detection of 20 ppm; most of the experiments exceeded that level. A few metal samples were also analyzed for phosphorus by optical emission spectroscopy (OES) at Clark Testing (Jefferson Hills, PA, USA) and glow discharge mass spectrometry (GDMS) at EAG Laboratories (San Diego, CA, USA). Some metal samples were analyzed for total oxygen content using LECO combustion analysis. The slag samples were analyzed for CaO, SiO_2, MgO, Al_2O_3, FeO, and P_2O_5 using energy dispersive X-ray fluorescence (ED-XRF) at West Penn Testing (New Kensington, PA, USA).

3. Results

Initial and final compositions of the metal and slag for the current equilibrium experiments are listed in Tables 1 and 2, respectively. Sample coding consists of two capital letters followed by two numbers, e.g., MS05. The letters M and S stand for metal and slag respectively. The two digits refer to the order of the experiments. When it starts with M then S (MSXX), it means the experiment was designed to transfer phosphorus from metal to slag, i.e., dephosphorization, similarly, SMXX are experiments where rephosphorization occurred. As previously mentioned, a two-phase slag-metal equilibrium was established which is faster than a three-phase equilibrium with gaseous species. Randomly selected experiments were repeated to ascertain the reproducibility of the experimental results. The reproducibility in the estimation of phosphorus partition was within 6 per cent.

Table 1. Initial chemical compositions, temperature, and holding time for current equilibrium experiments; square brackets denote concentration in the metal.

Exp. ID	T K	Ht [1] h	B2 [2]	B3 [3]	CaO wt.%	SiO_2 wt.%	MgO wt.%	FeO wt.%	Al_2O_3 wt.%	P_2O_5 wt.%	[P] wt.%
MS01	1873	10	1.00	0.98	30.28	30.23	10.02	27.72	0.58	0.93	0.0456
SM01	1873	10	0.99	0.99	30.19	30.38	10.10	27.80	0.27	1.01	0.0008
MS02	1873	10	1.50	1.49	38.02	25.38	8.79	26.13	0.21	0.87	0.0426

Table 1. Cont.

Exp. ID	T K	Ht [1] h	B2 [2]	B3 [3]	CaO wt.%	SiO$_2$ wt.%	MgO wt.%	FeO wt.%	Al$_2$O$_3$ wt.%	P$_2$O$_5$ wt.%	[P] wt.%
SM02	1873	10	1.50	1.49	38.03	25.32	8.78	26.07	0.21	1.00	0.0018
MS03	1873	10	2.07	2.06	42.97	20.76	5.36	29.50	0.13	0.91	0.0425
SM03	1873	10	2.07	2.06	43.04	20.77	5.36	29.34	0.13	1.00	0.0011
MS04	1873	10	1.03	1.02	25.42	24.79	13.84	34.15	0.24	0.87	0.0426
SM04	1873	10	1.03	1.02	25.52	24.77	13.85	33.94	0.24	0.99	0.0043
MS05	1873	10	1.03	1.02	23.44	22.74	12.91	39.19	0.23	0.87	0.0426
SM05	1873	10	1.03	1.02	23.52	22.76	12.87	39.00	0.23	0.99	0.0047
MS06	1873	10	1.50	1.49	30.84	20.54	9.19	37.72	0.22	0.87	0.0426
SM06	1873	10	1.51	1.49	30.89	20.50	9.19	37.55	0.22	1.04	0.0032
MS07	1873	10	2.09	2.07	43.88	21.02	5.42	28.32	0.13	0.86	0.0425
MS08	1873	10	1.01	0.86	29.98	29.83	16.57	17.47	4.90	0.93	0.0523
SM08	1873	10	1.01	0.86	29.92	29.66	16.55	17.54	4.95	1.06	0.0019
MS09	1873	10	1.01	0.75	28.11	27.95	16.45	16.88	9.40	0.89	0.0523
SM09	1873	10	1.01	0.75	28.15	27.90	16.43	16.76	9.45	1.01	0.0019
MS10	1873	10	1.00	0.67	28.31	28.17	17.28	10.94	14.12	0.88	0.0523
SM10	1873	10	1.01	0.67	28.39	28.14	17.29	10.77	14.09	1.02	0.0025
MS11	1873	10	2.01	1.21	33.42	16.66	8.91	28.94	11.01	0.88	0.0523
SM11	1873	10	2.01	1.21	33.39	16.64	8.88	28.92	10.95	1.03	0.0023
MS12	1873	10	2.10	1.57	37.86	18.02	7.14	29.81	6.13	0.87	0.0523
SM12	1873	10	2.10	1.57	37.85	18.01	7.13	29.70	6.11	1.02	0.0015
MS14	1873	10	1.00	0.93	31.08	31.00	15.53	18.61	2.59	0.87	0.0523
SM14	1873	10	1.01	0.93	31.08	30.84	15.58	18.51	2.59	1.07	0.0020
MS15	1873	10	3.51	2.52	49.17	14.03	3.75	26.04	5.50	1.42	0.0523
SM15	1873	10	3.51	2.52	49.19	14.02	3.69	25.87	5.49	1.64	0.0009
MS16	1873	10	1.74	1.40	37.61	21.57	8.63	25.73	5.33	0.92	0.0523
SM16	1873	10	1.74	1.40	37.71	21.63	8.66	25.32	5.38	1.09	0.0010
MS17	1873	10	1.69	1.69	41.73	24.62	7.37	25.04	0.11	0.94	0.0523
SM17	1873	10	1.68	1.68	41.53	24.65	7.34	25.16	0.11	1.03	0.0015
MS20	1873	10	3.79	1.90	37.59	9.91	7.02	34.48	9.91	0.98	0.0551
MS21	1873	10	2.00	1.00	27.98	13.99	13.12	29.72	14.02	1.01	0.0542
MS22	1823	10	1.00	0.90	29.31	29.21	14.68	22.37	3.26	0.85	0.0521
SM22	1823	10	1.02	0.91	29.21	28.77	14.85	22.41	3.16	1.27	0.0018
MS23	1873	4	1.75	1.40	37.62	21.49	8.60	25.73	5.45	0.90	0.0550
SM23	1873	4	1.76	1.40	37.65	21.45	8.63	25.52	5.46	1.08	0.0012
MS24	1923	10	1.00	0.92	31.20	31.12	15.54	18.29	2.63	0.87	0.0546
SM24	1923	10	1.01	0.93	31.25	31.03	15.53	18.13	2.66	1.05	0.0019
MS25	1923	10	2.68	1.44	39.80	14.88	8.10	23.41	12.68	0.97	0.0541
MS26	1923	10	2.00	1.00	27.95	13.96	12.95	30.02	14.09	0.88	0.0537
MS27	1823	10	2.67	1.46	39.91	14.96	8.02	23.58	12.45	0.93	0.0538
MS28	1823	10	2.00	1.00	28.08	14.01	12.90	29.92	14.05	0.89	0.0538
MS29	1873	10	1.01	1.00	32.05	31.89	16.96	17.53	0.22	1.00	0.0541
SM29	1873	10	1.01	1.00	32.03	31.83	17.01	17.51	0.22	1.04	0.0015

[1] Ht = holding time; [2] B2 = CaO / SiO$_2$; [3] B3 = CaO / (SiO$_2$ + Al$_2$O$_3$).

Table 2. Equilibrium chemical compositions (final reported chemical analysis); square brackets denote concentration in the metal. Values were normalized to 100 pct.

Exp. ID	B2 [1]	B3 [2]	CaO wt.%	SiO$_2$ wt.%	MgO wt.%	FeO wt.%	Al$_2$O$_3$ wt.%	P$_2$O$_5$ wt.%	[P] wt.%	[O] wt.%
MS01	1.00	0.98	30.03*	29.98*	15.17	22.83	0.80	1.18	0.0141	NA [3]
SM01	0.99	0.99	29.95*	30.13*	15.44	22.87	0.48	1.13	0.0131	NA [3]
MS02	1.50	1.49	38.75*	25.87*	10.64	23.57	0.00*	1.17	0.0029	NA [3]
SM02	1.50	1.49	38.52*	25.65*	10.20	24.58	0.00*	1.05	0.0030	NA [3]
MS03	2.07	2.06	44.29*	21.40*	6.49	26.54	0.00*	1.29	0.0024	NA [3]
SM03	2.07	2.06	43.70*	21.09*	7.03	27.06	0.00*	1.13	0.0024	NA [3]

Table 2. Cont.

Exp. ID	B2 [1]	B3 [2]	CaO wt.%	SiO$_2$ wt.%	MgO wt.%	FeO wt.%	Al$_2$O$_3$ wt.%	P$_2$O$_5$ wt.%	[P] wt.%	[O] wt.%
MS04	1.03	1.02	26.69*	26.02*	16.35	29.73	0.00*	1.22	0.0100	NA [3]
SM04	1.03	1.02	26.24*	25.48*	17.45	29.73	0.00*	1.09	0.0070	NA [3]
MS05	1.03	1.02	25.40*	24.65*	15.56	33.09	0.00*	1.29	0.0090	NA [3]
SM05	1.03	1.02	24.78*	23.98*	16.40	33.73	0.00*	1.11	0.0080	NA [3]
MS06	1.50	1.49	33.14*	22.07*	10.37	33.08	0.00*	1.33	0.0024	NA [3]
SM06	1.51	1.49	32.65*	21.67*	10.58	33.85	0.00*	1.25	0.0024	NA [3]
MS07	2.09	2.07	43.72*	20.94*	7.69	26.45	0.00*	1.21	0.0024	NA [3]
MS08	1.01	0.86	30.63*	30.47*	17.44	15.71	4.66	1.08	0.0309	NA [3]
SM08	1.01	0.86	30.70*	30.43*	17.84	14.79	5.09	1.15	0.0262	NA [3]
MS09	1.01	0.75	27.96*	27.80*	18.64	15.21	9.39	0.99	0.0343	NA [3]
SM09	1.01	0.75	28.16*	27.91*	18.79	15.07	9.13	0.94	0.0315	NA [3]
MS10	1.00	0.67	28.00*	27.86*	19.72	10.27	13.55	0.61	0.0855	NA [3]
SM10	1.01	0.67	28.38*	28.12*	19.57	10.21	13.13	0.59	0.0764	NA [3]
MS11	2.01	1.21	35.06*	17.48*	10.64	24.24	11.32	1.25	0.0060	NA [3]
SM11	2.01	1.21	35.07*	17.48*	10.40	24.65	11.22	1.18	0.0061	NA [3]
MS12	2.10	1.57	39.74*	18.91*	8.95	25.16	6.19	1.04	0.0033	NA [3]
SM12	2.10	1.57	39.34*	18.72*	8.68	26.07	6.16	1.03	0.0032	NA [3]
MS14	1.00	0.93	33.59	31.67	16.37	14.63	2.74	1.01	0.0240	NA [3]
SM14	1.01	0.93	33.36	31.28	16.00	15.47	2.89	1.00	0.0200	NA [3]
MS15	3.51	2.52	49.69*	14.17*	3.79	25.37	5.46	1.53	0.0008	0.035
SM15	3.51	2.52	49.54*	14.12*	3.78	25.55	5.44	1.57	0.0007	0.035
MS16	1.74	1.40	38.92*	22.32*	8.95	23.09	5.57	1.15	0.0027	0.072
SM16	1.74	1.40	38.82*	22.27*	8.90	23.34	5.54	1.13	0.0028	0.070
MS17	1.69	1.69	43.18*	25.47*	7.63	22.62	0.00*	1.10	0.0013	0.086
SM17	1.68	1.68	42.30*	25.11*	7.51	24.01	0.00*	1.07	0.0018	0.084
MS20	3.79	1.90	38.51*	10.15*	9.01	31.54	9.33	1.45	0.0012	NA [3]
MS21	2.00	1.00	28.58*	14.29*	16.04	26.83	12.86	1.40	0.0100	NA [3]
MS22	1.00	0.90	33.06	30.47	15.62	16.72	2.86	1.28	0.0106	NA [3]
SM22	1.02	0.91	33.05	30.45	15.58	16.71	2.88	1.33	0.0103	NA [3]
MS23	1.75	1.40	37.14*	21.21*	9.43	26.11	4.85	1.26	0.0036	NA [3]
SM23	1.76	1.40	37.01*	21.09*	9.72	26.47	4.60	1.11	0.0026	NA [3]
MS24	1.00	0.92	31.71*	31.63*	18.70	14.06	2.77	1.13	0.0290	NA [3]
SM24	1.01	0.93	32.02*	31.80*	18.91	13.67	2.52	1.08	0.0275	NA [3]
MS25	2.68	1.44	40.68*	15.20*	11.06	19.52	12.08	1.45	0.0035	NA [3]
MS26	2.00	1.00	29.01*	14.49*	16.14	25.62	13.46	1.29	0.0092	NA [3]
MS27	2.67	1.46	39.31*	14.73*	8.05	26.47	10.15	1.28	0.0023	NA [3]
MS28	2.00	1.00	28.43*	14.19*	13.17	31.16	11.94	1.11	0.0086	NA [3]
MS29	1.06	1.04	34.34	32.54	18.54	12.81	0.4	1.37	0.0248	NA [3]
SM29	1.05	1.04	34.74	32.96	17.95	12.79	0.36	1.21	0.0193	NA [3]

[1] B2 = CaO / SiO$_2$; [2] B3 = CaO/(SiO$_2$ + Al$_2$O$_3$); [3] NA = Not Analyzed; * Estimated values.

The initial MgO saturation content in all slags used were estimated using FactSage. However, there was a consistent increase of the MgO concentration (MgO pick up) in all slag samples through the course of the experiments. The authors have recently published a paper about FactSage's tendency to slightly underestimate MgO saturation levels elsewhere [17].

As seen in Table 2, CaO and SiO$_2$ were estimated for most of the experiments. Ideally, the initial and final weights of CaO and SiO$_2$ in the slag should be equal and preserve a constant *CaO/SiO$_2$* ratio during the phosphorus reaction. Therefore, the expected change in their relative concentrations is small. To test this approach, several samples were analyzed for CaO and SiO$_2$ in order to evaluate the differences between the initial and final *CaO/SiO$_2$* ratios. The difference in the ratio was found to be less than ±0.08. This may have been due to hydration of the stored powders. The authors therefore have assumed that the initial and final *CaO/SiO$_2$* ratios to be equal and their contents were determined by mass balances.

4. Discussion

4.1. The Effect of Alumina on the Phosphorus Partition

There is limited amount of work describing the effect of alumina on the phosphorus partition. A series of equilibrium experiments were conducted to study the effect of alumina on the phosphorus partition. From the data in Figure 1, it is apparent that at constant CaO/SiO_2 ratio and FeO content, alumina decreases the phosphorus partition. At alumina concentration of 9–11 wt.% and CaO/SiO_2 ratio of 1.0 and 2.0, the phosphorus partition was lowered by more than half compared to slags with no alumina.

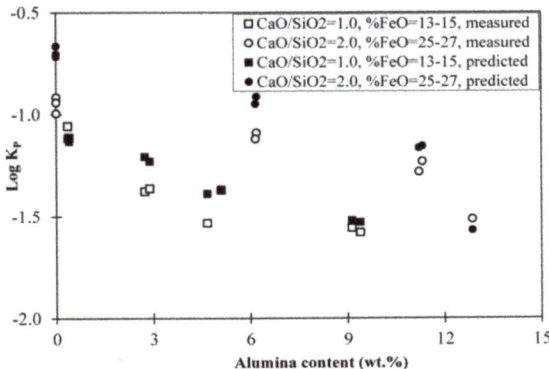

Figure 1. log K_P versus slag alumina content for binary basicities of 1.0 and 2.0 at FeO compositions of 13–15 and 25–27 wt.%, respectively, at 1873 K (1600 °C). Solid markers predicted by Equation (5).

Experiment 15 was carried out with a highly basic slag (CaO/SiO_2 = 3.5) and Al_2O_3 concentration of 5 wt.% in order to assess if the effect of alumina on dephosphorization persists at higher basicity. It was compared with one of the experiments from Assis et al. [1] that used similar slag and metal compositions. The result confirms the trend observed at lower basicity that as the alumina is added, the phosphorus partition is lowered.

The results discussed above are in agreement with some of the ones reported by Selin [18] and shown in Table 3. The experiments by Selin [18] labeled A21, E2, and C4 a binary basicity of about 1.6 with increasing alumina content. The effect of alumina on L_P is evident and L_P decreased from 129 to 38.5 when comparing experiments A21 and C4.

Table 3. Selected results from Selin and the present work.

Exp. ID	B2 [1]	B3 [2]	CaO wt.%	SiO$_2$ wt.%	MgO wt.%	FeO wt.%	Al$_2$O$_3$ wt.%	P$_2$O$_5$ wt.%	[P] wt.%	[O] ppm	L$_P$
C4 [3]	1.60	1.15	35.3	22.0	11.5	20.2	8.78	0.93	0.0105	980	38.5
E2 [3]	1.60	1.36	38.2	23.9	10.2	22.4	4.23	0.91	0.0065	1094	61
A21 [3]	1.65	1.65	40.8	24.7	8.4	24.3	0.0	0.98	0.0033	1185	129
MS16	1.74	1.40	38.92*	22.32*	8.95	23.09	5.57	1.15	0.0027	720	186
SM16	1.74	1.40	38.82*	22.27*	8.90	23.34	5.54	1.13	0.0028	700	176
MS17	1.69	1.69	43.18*	25.47*	7.63	22.62	0.00*	1.10	0.0013	860	370
SM17	1.68	1.68	42.30*	25.11*	7.51	24.01	0.00*	1.07	0.0018	840	258

[1] B2 = CaO / SiO$_2$; [2] B3 = CaO / (SiO$_2$ + Al$_2$O$_3$); [3] data from Selin [18]; * Estimated values.

From the above discussion, alumina levels expected in EAF and BOF steelmaking are detrimental to the dephosphorization ability of the slag. It is more significant in the slag chemistries prevalent in the EAF compared to those in the BOF. Further investigations must be made to clarify whether this effect is due to changes in the activity of iron oxide (oxygen potential), oxygen anions (slag structure), the activity of phosphate ion in the slag, or a combination of two or more of these.

4.2. Comprehensive Equation for Phosphorus Equilibrium

The purpose of this section is to present the method that was used to develop a comprehensive equation for phosphorus partition for both EAF- and BOF-type slags. The adopted methodology is different from previous treatments but its final form is similar to the one proposed by Suito et al. [4,5]: it assumes that the log K_P is a linear function of the slag components and reciprocal temperature. The present data was combined with those from Assis et al. [1] and Suito et al. [4,5]. The experimental approach for the work of Assis et al. [1] and the current research was similar and thus it is reasonable to combine both data sets in the assessment of phosphorus partition. Literature data were also examined and evaluated considering two important aspects: the attainment of equilibrium (the equilibration time) and the presence of any solid phases in the slag.

It was demonstrated that holding times of four hours yielded compositions very close to equilibrium and 10 h ensured attainment of equilibrium. The holding time in the experiments conducted by Suito and coworkers [4,5] varied between three hours for non-SiO_2-containing slags and five h for slags with high SiO_2 and low FeO. The authors therefore assumed Suito's experiments were close or at equilibrium.

The second point to discuss is the presence of solid phases (undissolved oxides) in the slag which will affect the "apparent" slag composition making it either more or less demanding with respect to dephosphorization. The slags in the present experiments were assured to be completely liquid. Literature data was initially examined by FactSage [19] version 6.4 (ThermFact Inc., Montréal, Canada and GTT–Technologies, Herzogenrath, Germany) using the CON1 database, which was later integrated into the FToxid database and contains a more accurate description of slags containing P_2O_5 to predict the amount of solid phases in equilibrium with the liquid slag. This approach provided a better estimate of the actual liquid slag composition. The calculations made with FactSage 6.4 using CON1 were compared with FactSage 7.2 (ThermFact Inc., Montréal, Canada and GTT–Technologies, Herzogenrath, Germany) using FToxid and were found to be in close agreement. In FactSage 7.2 Equilib module, the same calculations can be carried by selecting the databases FToxid, FTmisc, and FactPS. The solution phases to be selected are FTmisc-FeLQ, FToxid-SLAGA, FToxid-MeO_A, FToxid-aC2Sa, FToxid-bC2Sa, FToxid-C2SP, FToxid-C3Pr, FToxid-C3Pa, FToxid-C3Pb, FToxid-M3Pa, FToxid-CMPc, and FToxid-M2Pa. A FactSage macro was used to expedite the calculations.

The fraction of liquid slag in the experiments conducted by Suito et al. [4,5] ranged between 81 and 99% according to the FactSage calculations. On the other hand, while evaluating some other literature data [2,3], significant concentrations of solids (C_2S and monoxide) were present in the slags with a calculated liquid fraction as low as 27%. The large amount of solids makes it difficult to evaluate the phosphorus partition in the remaining liquid.

An additional set of data, which was used in formulating the new equilibrium correlation, is that of Ide and Fruehan [9]. They used calcia crucibles and CaO saturated slags with a binary basicity of 2.9, and varied the concentration of MgO between 2.9 and 10.3 wt.%. The liquid fraction of the slags was above 80% according to the FactSage estimations. The use of calcia crucibles enables the MgO content to be varied without the constraints of maintaining a MgO saturated slag and provides a potentially better estimation of the MgO coefficient.

The steps to find the new correlation:

1. The MgO coefficient was fixed as reported by Ide and Fruehan [9]. The coefficient is $0.072 \times 0.15 = 0.0108$.
2. Evaluate log K_P -0.0108(%MgO) as a function of *wt.%CaO, wt.%SiO$_2$* and *wt.%P$_2$O$_5$* using the data from Suito et al. [4,5], Assis et al. [1], and the present data that did not contain Al_2O_3.

The resulting coefficients are 0.073 for CaO, 0.0105 for SiO$_2$, and 0.070 for P$_2$O$_5$. All p-values were below 0.05. The resulting equation is

$$\log K_P - 0.0108(\%MgO) = 0.073(\%CaO) + 0.0105(\%SiO_2) + 0.070(\%P_2O_5) - 4.3$$

3. Fix the CaO, MgO, SiO$_2$, and P$_2$O$_5$ coefficients as done above for MgO and evaluate the Al$_2$O$_3$ coefficient using the present alumina-containing data. The resulting coefficient is 0.0160 for Al2O3. The p-value was below 0.05. The resulting equation is

$$\log K_P - 0.0108(\%MgO) - 0.073(\%CaO) - 0.0105(\%SiO_2) - 0.070(\%P_2O_5)$$
$$= 0.160(\%Al_2O_3) - 4.34$$

The temperature dependence and intercept constant were then evaluated using the data from the present work and Suito et al. [4,5] for 1823–1923 K (1550–1650 °C). The new intercept is of −10.46 with a 95% confidence interval of 0.1. The temperature dependence is the same as reported by Suito et al. [4,5]. The final equation obtained is

$$\log\left(\frac{L_P}{T.Fe^{2.5}}\right) = 0.073[(\%CaO) + 0.148(\%MgO) + 0.96(\%P_2O_5) + 0.144(\%SiO_2)$$
$$+ 0.22(\%Al_2O_3)] + \frac{11570}{T} - 10.46 \pm 0.1 \quad (5)$$

It is interesting to note that both CaO coefficient of 0.073 and the constant of −10.46 are in good agreement to the original correlation reported by Suito et al. [5] (0.072 and −10.52, respectively). Equation (5) is the same as reported by Assis et al. [1] but with the newly calculated Al$_2$O$_3$ coefficient included. Figure 2 compares the measured log K_P and newly calculated apparent phosphorus equilibrium constants by Equation (5). The new correlation represents well different sets of laboratory data including the current results and those of Assis et al. [1], Ide et al. [9], and Suito et al. [4,5]. The R^2 of 0.95 is high, and because the equation was fitted using multiple sets of data, it is capable of predicting the phosphorus partition over a wider range of slag compositions than previously developed correlations. Additionally, the mean absolute error is 0.152. A comparison of the measured L_P with the predicted L_P by Equation (5) is shown in Figure 3.

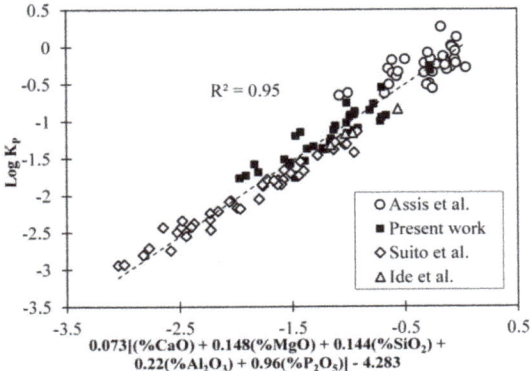

Figure 2. Apparent equilibrium constant versus Equation (5) at 1873 K (1600 °C). Data sets from Suito et al. [4,5], Ide et al. [9], Assis et al. [1], and the present work.

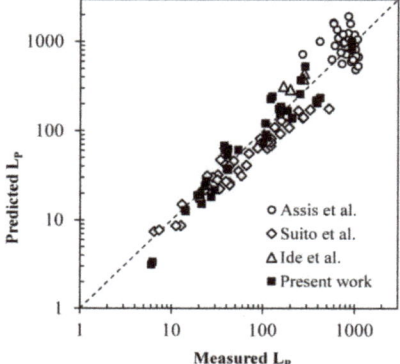

Figure 3. Measured phosphorus partition versus predicted phosphorus partition by Equation (5). Data sets from Suito et al. [4,5], Ide et al. [9], Assis et al. [1], and the present work.

In the new correlation, the coefficients of alumina and silica are small but positive. However, one must not confuse positive coefficients as enhancers of dephosphorization. Increasing silica and/or alumina would dilute CaO, which has the highest coefficient, thus decreasing K_P. Figure 4 demonstrates the change of log K_P when silica or alumina increases while the concentrations of the remaining components are adjusted proportionally. It can be seen that increasing either SiO_2 or Al_2O_3 decreases K_P.

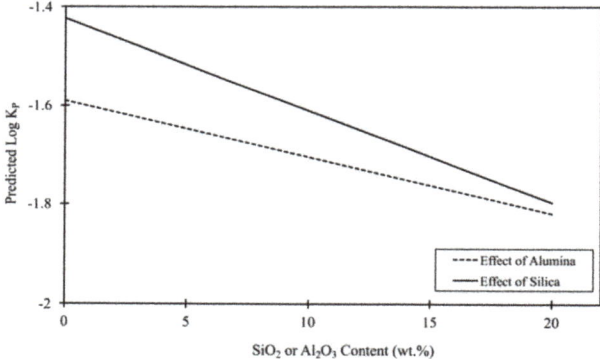

Figure 4. Calculated log K_P versus silica and alumina concentration as predicted by Equation (5).

The steelmaker can use Equation (5) to easily predict the equilibrium phosphorus partition in their operations by simply comparing the actual L_P measured using physical steel and slag samples with the equilibrium L_P predicted by Equation (5). The departure from equilibrium is related to the driving force for dephosphorization and can be a helpful indicator of the kinetics inside the furnace. As such, Equation (4) can be integrated into kinetic models and mass transfer equations to simulate furnace behavior. This however is not covered in this publication.

Equation (5) is a statistical correlation and is only valid within the range in which it was originally evaluated. This range is shown in Table 4.

Table 4. Range of slag compositions valid for Equation (5).

Variable	Min	Max
%CaO	54.01	10.94
%SiO$_2$	33.75	0.48
%T.Fe	59.43	7.93
%MgO	21.94	3.78
%P$_2$O$_5$	0.51	5.46
%Al$_2$O$_3$	0	21.05
Temperature	1823	1923

5. Conclusions

This paper presented the continuation of the work described by Assis et al. [1]. It presented new sets of phosphorus equilibrium data using EAF-type slags with and without alumina, demonstrated the negative effect of alumina on L_P, and proposed a new improved correlation for predicting the phosphorus equilibrium partition. This new correlation includes the effect of alumina and can be used for both BOF and EAF-type slags to predict the equilibrium phosphorus partition. The negative effect of alumina in the correlation occurs due to the dilution of CaO in the slag. However, further investigations are needed to understand how alumina affects slag structure, the activity of iron oxide and the activity of the phosphate ion in the slag. Equation (5) is easy to use and enables the steelmaker to evaluate furnace performance with respect to dephosphorization. In a future publication, we will evaluate the departure from equilibrium in an extensive collection of plant data and discuss which types of furnaces operate closer to equilibrium.

Author Contributions: Conceptualization: A.N.A. and M.A.T.; Methodology: A.N.A.; Validation: R.J.F. and S.S.; Formal Analysis: M.A.T.; Investigation: M.A.T.; Writing—Original Draft Preparation: M.A.T.; writing—Review and Editing: A.N.A.; Visualization: A.N.A. and M.A.T.; Supervision: R.J.F. and S.S.; Project Administration: R.J.F.; Funding Acquisition: R.J.F. and S.S.

Funding: This research was funded by the Center for Iron and Steelmaking Research (CISR) at Carnegie Mellon University.

Acknowledgments: The authors would like to thank the industrial members of the Center for Iron and Steelmaking Research (CISR) at Carnegie Mellon University

Conflicts of Interest: The authors declare no conflicts of interest.

References

1. Assis, A.N.; Tayeb, M.A.; Sridhar, S.; Fruehan, R.J. Phosphorus Equilibrium Between Liquid Iron and CaO-SiO2-MgO-Al2O3-FeO-P2O5 Slag Part 1: Literature Review, Methodology, and BOF Slags. *Metal. Mater. Trans. B* **2015**, *46*, 2255–2263. [CrossRef]
2. Basu, S.; Lahiri, A.K.; Seetharaman, S. Phosphorus partition between liquid steel and CaO-SiO2-P2O5-MgO slag containing low FeO. *Metal. Mater. Trans. B* **2007**, *38*, 357–366. [CrossRef]
3. Basu, S.; Lahiri, A.K.; Seetharaman, S. Phosphorus Partition between Liquid Steel and CaO-SiO 2-FeO x-P 2 O 5-MgO Slag Containing 15 to 25 Pct FeO. *Metal. Mater. Trans. B* **2007**, *38*, 623–630. [CrossRef]
4. Suito, H.; Inoue, R.; Takada, M. Phosphorus distribution between liquid iron and MgO saturated slags of the system CaO-MgO-FeOx-SiO2. *Tetsu-to-Hagané* **1981**, *67*, 2645–2654. [CrossRef]
5. Suito, H.; Inoue, R. Phosphorus Distribution between MgO-saturated CaO-FetO-SiO2-P2O5-MnO Slags and Liquid Iron. *Trans. Iron Steel Inst. Jpn.* **1984**, *24*, 40–46. [CrossRef]
6. Suito, H.; Inoue, R. Effect of Calcium Fluoride on Phosphorus Distribution between MgO-saturated Slags of the System CaO-MgO-FeOx-SiO2 and Liquid Iron. *Tetsu-to-Hagané* **1982**, *68*, 1541–1550. [CrossRef]
7. Suito, H.; Inoue, R. Manganese equilibrium between molten iron and MgO-saturated CaO-FetO-SiO2-MnO-slags. *Trans. Iron Steel Inst. Jpn.* **1984**, *24*, 257–265. [CrossRef]
8. Suito, H.; Inoue, R. Effects of Na2O and BaO additions on phosphorus distribution between CaO-MgO-FetO-SiO2-slags and liquid iron. *Trans. Iron Steel Inst. Jpn.* **1984**, *24*, 47–53. [CrossRef]

9. Ide, K.; Fruehan, R.J. Evaluation of phosphorus reaction equilibrium in steelmaking. *Iron Steelmak.* **2000**, *27*, 65–70.
10. Davies, R.H.; Dinsdale, A.T.; Gisby, J.A.; Robinson, J.A.J.; Martin, S.M. MTDATA-thermodynamic and phase equilibrium software from the national physical laboratory. *Calphad* **2002**, *26*, 229–271. [CrossRef]
11. AlHarbi, M. *Design a New Slag Chemistry to Improve De-Phosphorization Level (>70%)*; SABIC: Saudi Arabia, 2011.
12. Balajiva, K.; Quarrell, A.G.; Vajragupta, P. A laboratory investigation of the phosphorus reaction in the basic steelmaking process. *J. Iron Steel Inst.* **1946**, *153*, P115.
13. Ishii, H.; Fruehan, R.J. Dephosphorization equilibria between liquid iron and highly basic CaO-based slags saturated with MgO. *Iron Steelmak. (USA)* **1997**, *24*, 47–54.
14. Li, G.; Hamano, T.; Tsukihashi, F. The effect of Na2O and Al2O3 on dephosphorization of molten steel by high basicity MgO saturated CaO-FeOx-SiO2 slag. *ISIJ Int.* **2005**, *45*, 12–18. [CrossRef]
15. Miyata, M. Hot Metal Dephosphorization with CaO-SiO2-Al2O3-FetO Flux. *Curr. Adv. Mater. Proces.* **1995**, *8*, 101.
16. Balajiva, K.; Vajragupta, P. The Effect of Temperature on the Phosphorus Reaction in the Basic Steelmaking Process. *J. Iron Steel Inst.* **1946**, *155*, 563–567.
17. Tayeb, M.A.; Assis, A.N.; Sridhar, S.; Fruehan, R.J. MgO solubility in steelmaking slags. *Metal. Mater. Trans. B* **2015**, *46*, 1112–1114. [CrossRef]
18. Selin, R. The role of Phosphorus, Vanadium and slag forming oxides in direct reduction based steelmaking. *R. Inst. Technol.* **1987**.
19. Bale, C.W.; Bélisle, E.; Chartrand, P.; Decterov, S.A.; Eriksson, G.; Gheribi, A.E.; Hack, K.; Jung, I.-H.; Kang, Y.-B.; Melançon, J. FactSage thermochemical software and databases, 2010–2016. *Calphad* **2016**, *54*, 35–53. [CrossRef]

© 2019 by the authors. Licensee MDPI, Basel, Switzerland. This article is an open access article distributed under the terms and conditions of the Creative Commons Attribution (CC BY) license (http://creativecommons.org/licenses/by/4.0/).

Article

Fluid Flow Turbulence in the Proximities of the Metal-Slag Interface in Ladle Stirring Operations

Fabian Andrés Calderón-Hurtado [1], Rodolfo Morales Dávila [1,2,*], Kinnor Chattopadhyay [2,3] and Saúl García-Hernández [4]

1. Department of Metallurgy and Materials Engineering, Instituto Politécnico Nacional-ESIQIE, Ed. 7, Colonia Zacatenco, C.P. 07738 Mexico City, Mexico; fabandrescal@gmail.com
2. Department of Materials Science and Engineering, University of Toronto, 184 College Street, Suite 140, Toronto, ON M5S 3E4, Canada; kinnor.chattopadhyay@utoronto.ca
3. Department of Mechanical and Industrial Engineering, University of Toronto, 184 College Street, Suite 140, Toronto, ON M5S 3E4, Canada
4. Instituto Tecnológico de Morelia, Metallurgy Graduate Center, Av. Tecnológico No. 1500, Morelia, C.P. 58120 Michoacán, Mexico; iq_sagahz@hotmail.com
* Correspondence: rmorales@ipn.mx; Tel.: +52-1-55-5454-8322

Received: 13 December 2018; Accepted: 3 February 2019; Published: 6 February 2019

Abstract: Three-phase interactions (metal-slag-argon) in ladle stirring operations have strong effects on the metal-slag mass transfer processes. Specifically, the thickness of the slag controls the fluid turbulence to an extent that once trespassing a critical thickness, increases of stirring strength no longer effect the flow. To analyze these conditions, a physical model considering the three phases was built to study liquid turbulence in the proximities of the metal-slag interface. A velocity probe placed close to the interface permitted the continuous monitoring and statistical analyses of any turbulence. The slag eye opening was found to be strongly dependent on the stirring conditions, and the mixing times decreased with thin slag thicknesses. Slag entrainment was enhanced with thick slag layers and high flow rates of the gas phase. A multiphase model was developed to simulate these results and was found to be a good agreement between experimental and numerical results.

Keywords: ladle stirring; turbulence; slag; interface; refining; mixing time; slag opening

---

1. Introduction

Stirring of liquid steel in ladles through argon bubbling has various functions, like thermal and chemical homogenization, speeding metal–slag mass transfer rates (melt desulfurization), supposedly floatation of inclusions, and, when necessary, steel cool down to cast at the desired temperature. Unfortunately, however, this operation is not free from serious drawbacks. Among these we have the opening of a slag eye through which oxygen and nitrogen can be absorbed, possible slag entrainment if the stirring intensity of the bath is high, and enhancement of melt-refractory and slag-refractory reactions which will degrade steel cleanliness. Another important operational factor, which depends on the steel tapping practice, oxygen and sulfur contents of steel at tapping, and amount and type of additions, is the thickness of the slag layer. The thickness of this upper phase definitively influences the stirring conditions of the bath for a given energy input. It is natural to think that, among the various phases involved in this process, there must be a narrow window of opportunity in the process where one can get the best contact. Thermally and chemically homogeneous baths, reasonable desulfurization rates lasting a span time from 5 to 8 min, low refractory wearing-rates, and capture of inclusions that reach the metal-slag interface especially during the rising-time period are the goals of this process. In such a complex-multiphase system, reaction-thermodynamics and fluid flow phenomena interact intimately in a way difficult to understand even today. The present work was focused on fluid flow and, specifically, on the turbulence of the liquid metal, which is close to the metal–slag interface since this region is critical for floating inclusions.

Fluid flow of liquid steel in ladles has been studied from many points of view, using physical and mathematical models. The structure of the gas-liquid plume was studied using a mechanistic approach by Krishnapisharody and Irons [1,2], establishing models to estimate the size of the slag eye opening (SEO) area and the height of the spout region as the function of bath height and gas flow rate. The same authors developed correlations linking averaged velocities of the liquid and gas phases and gas volume fraction along the plume height as functions of gas flow rate and bath height [3]. Spout height was defined through a dimensionless variable involving gas flow rate [4] presenting a unified theory for two-phase flows dynamics in the plume [5]. Mixing times are considered useful information in estimating the thermal and chemical homogenization speediness of steel in ladles under some given flow rate of gas. Various authors have proposed simple engineering correlations to estimate the mixing time for gas-liquid systems [6]. For example, Mazumdar and Guthrie [7–9] estimated this parameter and the plume velocity (it was the averaged two-phase velocity in the plume) through the ladle dimensions and the flow rate of the bubbling gas. Table 1 shows these correlations for two-phase, gas-liquid flows, according to other researches [6–11]. However, in actual steel ladle systems, the influence of the layer on the mixing time has indicated the mixing times are larger than for two-phase flows [12–14]. All these studies reported that the thickness of the slag weakens the mixing kinetics in steel; it can even be said that the thickness of the slag is more important than its physical properties in this regard [15,16]. It must be said that low viscosity slags enhance mass transfer between both phases, and tough slag entrainment in the melt can be brought on by rises of gas flow rate [17]. The correlations so far reported to date, were used to calculate mixing times with an upper slag and are presented in Table 2 [10,11,16,18–21]. The slag eye opening area is another important parameter of the process since, depending on its size, since steel can be less or more contaminated by the atmospheric air. Its measurement, through infrared video cameras, is also important to determine the actual flow rates of argon, since the injection through the porous plug is, most of the time, inaccurate due to the partial obstruction of the plug surface by debris of metal and slag or leaks. One of the first works attempting to study the process conditions on the SEO area was that of Yonezawa and Schwerdtfeger [22] and Subagyo et al. [23]. The operating process parameters factors affecting the size of the SEO are summarized in Table 3 [24,25]. It is worthy to mention that all correlations should be taken with caution, as we employed a scaled down model under room temperature. There are other works related to the present topic by Liu et al. [26], who combined the Large Eddy Simulation, Volume of Fluid Model (VOF), and Discrete Phase Model, to simulate the slag eye opening area, but these authors did not provide information about the influence of the slag thickness on the mixing time nor on the structure of the flow. Kumar et al. [27] employed a similar approach to study the influence of the gas flow rate on the slag eye opening area for a fixed slag thickness but did not go further on the flow structure and turbulence level at the metal-slag interface. Therefore, the focus of this work was first to analyze what of all available correlations in the literature were the most appropriate to estimate the mixing time and the slag eye opening area. Second, for first time in the open literature, the level of turbulence at the metal-slag interface for some given operating conditions was introduced. Finally, the combined effects of slag thickness and gas flow rate on mixing time, slag eye opening area, and flow structure were reported as useful tools for steelmakers.

Table 1. Mixing time correlations for baths without an upper layer.

Mixing Time		Energy	Units of Energy	Reference
$\tau_m = 100\dot{\varepsilon}^{-0.42}$	(1)	$\dot{\varepsilon} = \frac{\rho_L g Q H_L}{\rho_L \pi R_2 H_L}$	W/ton	[10]
$\tau_m = 125\dot{\varepsilon}^{-0.289}$	(2)	$\dot{\varepsilon} = \frac{\rho_L g Q H_L}{\rho_L \pi R_2 H_L}$	W/ton	[10]
$\tau_m = 37\varepsilon_m^{-0.33} H_L^{-1} R^{1.66}$	(3)	$\dot{\varepsilon} = \frac{\rho_L g Q H_L}{\rho_L \pi R_2 H_L}$	W/ton	[11]
$\tau_m = 1200 Q_g^{-0.47} D^{1.97} H_L^{-1} v_L^{0.47}$	(4)	-	-	[6]

ρ_L = liquid density, g = gravity constant, Q, Q_g = gas flow rate, H_L = bath height, R = ladle radius, D = ladle diameter, v_L = kinematic viscosity of the liquid, and τ_m = mixing time.

Table 2. Mixing time correlations for a denser phase with a lighter top layer [16].

Correlation	N	r/R	θ	Top Layer	Reference
$\tau_m = 100\dot{\varepsilon}^{-0.42}$	1	0, 0.5	-	Polystyrene balls	[10]
$\tau_m = 125\dot{\varepsilon}^{-0.289}$	1	0	-	-	[11]
$\tau_m = 1910 Q_g^{-0.217} D^{1.49} H_L^{-1.0} v_L^{0.37} \left[\frac{\rho_L - \rho_s}{\rho_L}\right]^{0.243}$	1	0	-	Silicon oil and pentane	[18]
$\tau_m = 60.2 Q_g^{-0.33} R^2 H_L^{-1.0} h_s^{0.6} \left[\frac{\sigma_s}{\mu_s}\right]^{-0.022}$	2	0.5	180	Petroleum ether, mustard oil and benzene	[19]
$\tau_m = 152 Q_g^{-0.33} R^{2.33} H_L^{-1.0} \left(\frac{h_s}{H_L}\right)^{0.3} v_s^{0.033} \left[\frac{\rho_L - \rho_s}{\rho_L}\right]^{-0.044}$	1, 2	0.5, 0	180	Petroleum ether, mustard oil and benzene and silicon oil	[20]
$\tau_m = 2.33 \dot{\varepsilon}_{ms}^{-0.34} H_{eff}^{-1.0}$	1	0	-	Kerosene and silicon oil	[16]
$\tau_m = 9.83 N^{0.1025} \dot{\varepsilon}^{-0.364} \left(\frac{r}{R}\right)^{-0.0051} \left[\frac{h_s}{H_L}\right]^{0.004}$	1,2 3	0.33 0.5 0.67 0.80	120 180	Engine oil blue, engine oil red and soybean oil	[21]

τ_m = mixing time (s), Q, Q_g = gas flow rate, D and L = diameter and height of the container, respectively, v_s = the kinematic viscosity of the lighter phase, v = the kinematic viscosity of the denser phase, ρ_L and ρ_s = the densities of the denser and lighter phase, respectively, μ_s = the dynamic viscosity of the lighter phase, σ_s = the surface tension of the lighter phase, $\dot{\varepsilon}$ = the specific potential energy input in W/ton, except for Khajavi and Barati [16] in W/kg, h_s = the thickness of the lighter phase, H_L = bath height, N = the number of nozzles, θ = their separation angle, and H_{eff} = effective height, including the denser and lighter phases as defined by Khajavi and Barati [15]. The units of all other variables are expressed in SI units. This table was complemented with the last row in the present work.

Table 3. Correlations to estimate the size of the open steel eye during stirring operations in ladles.

Ref.	Correlation	N	r/R	Systems	Constraints
[22]	$\log\left(\dfrac{A_{es}}{h_s H}\right) = -0.69897 + 0.90032 \log\left(\dfrac{Q^2}{g h_s^5}\right) - 0.14578 \left[\log\left(\dfrac{Q^2}{g h_s^5}\right)\right]^2 + 0.0156 \left[\log\left(\dfrac{Q^2}{g h_s^5}\right)\right]^3$ (1)	1	0	Mercury–oil, Liquid steel–slag	$\Phi_{orifice} = 0.5$ mm, $0.01 \leq \dfrac{Q^2}{g h_s^5} \leq 10{,}000$. Other diameters of the orifice give different correlations
[23]	$\dfrac{A_{es}}{(h_s + H)^2} = 0.02 \mp 0.002 \left(\dfrac{Q^2}{g h_s^5}\right)^{0.375 \mp 0.0136}$ (2)	–	0, 1	Mercury–oil, Liquid steel–slag	It is a modification of the precedent correlation
[24]	$A_e^* = -0.76(Q^*)^{0.4} + 7.15(1-\rho^*)^{-1/2} (Q^*)^{0.73} (h^*)^{-1/2}$, $Q^* = \dfrac{Q}{g^{0.5} H^{2.5}}$, $A_e^* = \dfrac{A_e}{H^2}$, $A_p^* = \dfrac{A_p}{H^2} = 1.41 (Q^*)^{0.4}$, $\rho^* = \dfrac{\rho_s}{\rho_L}$, (3) $h^* = \dfrac{h_s}{H}$	1	various	Water–paraffin, water–motor oil, CaCl$_2$–paraffin oil, Hg–silicon oil, water–silicon oil and steel–slag	Assumed to be for general application of various systems and different orifice positions
[25]	$\dfrac{A_p}{h_s H} = 3.25 \left(\dfrac{U_p^2}{g h_s}\right)^{1.28} \left(\dfrac{\rho_L}{\Delta \rho}\right)^{0.55} \left(\dfrac{v_s}{h_s U_p}\right)^{-0.05}$ $U_p = 17.4 Q^{0.244} H^{-0.08} \left(\dfrac{\rho_g}{\rho_L}\right)^{0.0218}$ (4)	1	0, 0.5	Water–petroleum-ether, water–coconut oil, water–mustard oil	Applicable for $\varepsilon \sim 0.01$ W/kg, $0.75 < H/D < 1.5$, $\nu \sim 10^{-6}\,\text{m}^2/\text{s}$, and $0.006 < h_s < 0.05$ for centric position of the orifice

All correlations are given in SI units. Q = gas flow rate, h_s = slag thickness, H = the height of the denser phase, U_p = plume velocity, A_p = the plume area, A_e, A_{es} = the area of the eye opening, v_s = kinematic viscosity of the lighter phase, ρ_s and ρ_L = densities of the lighter and denser phases, respectively, and g = gravity constant, $\Phi_{orifice}$ = diameter of the orifice. The plume velocity was calculated using the equation of Yonezawa and Schwerdtfeger [22].

2. Experimental Setup

The experimental setup consisted of a 1/3 scale ladle, made of transparent plastic by a billet company located in Mexico and equipped with a bottom plug to stir steel with argon. The dimensions of the model and the positions of the plug and the slide gate are reported in Figure 1a,b. The ladle was filled with tap water through a top pipe until the operational scaled-height corresponded to the plant. The injection of argon was modeled using air supplied by an air compressor and injected through an orifice in the ladle bottom, with a pressure of 2.5 kg/cm^2. The gas flow rate was measured through a mass flowmeter located between the compressor and the injection point. The lighter phase was food-oil and a layer of this material was conformed before starting the experiment to obtain the desired thickness. For visualization records of the experiments, two video cameras were placed, one on the top of the bath and another one facing the wall of a flat chamber attached outside the ladle wall filled with water to avoid optical distortions from the curvature of the vessel. The camera on the top recorded the images of the bath surface without and with an oil layer. The video recordings were decomposed into images of the eye opening with a frequency of 2 s^{-1}. Quantitative measurements of these areas were performed, using a previously calibrated image analyzer software [28] and following a similar procedure as reported by Peranandhanthan and Mazumdar [25]. The scale down criterion to 1/3 was carried out with the equation $Q_m = f^{2.5} Q_p$, derived from the Froude number [29], where Q_p and Q_m are the gas flow rate in the prototype and the model, respectively, and f is the scale factor.

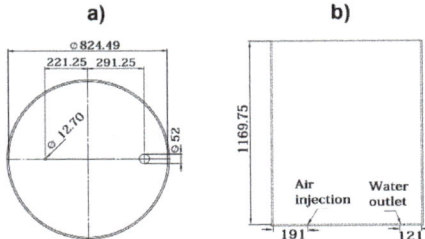

Figure 1. (a) Dimensions of the model (mm). (b) Position of the nozzle.

Twenty cubic centimeters of an aqueous solution of food red-colorant was employed as tracer, injected 100 mm below the bath surface near the geometric center of the ladle. A peristaltic pump was used for extracting samples from the bath, which were fed into a colorimeter cell to obtain the instantaneous concentration of the tracer. The analog signals of the colorimeter were converted into digital signals through a data acquisition card in a PC permitting real-time plotting of the tracer concentration versus time. Fluid flow turbulence was captured through a 10 million Hz ultrasonic transducer, or probe, immersed in the bath 20 mm below its surface and located in the ladle wall just opposite the wall which was nearest to the injection orifice. This probe measured the horizontal velocities in the bath at this plane, as well as all turbulent variables associated with the flow. Figure 2 shows a scheme of the experimental setup described here. The physical properties of the three phases—water, oil, and air—, as well as other details of the experimentation are shown in Table 4.

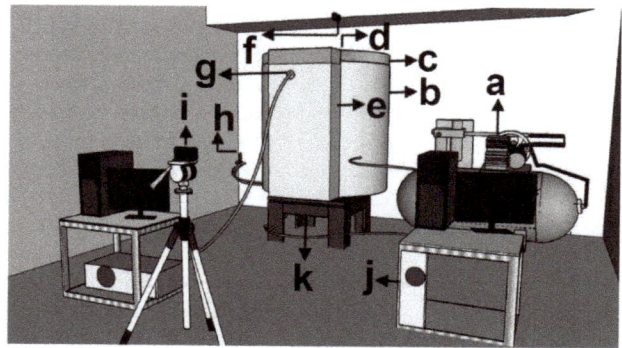

Figure 2. Experimental set up. (**a**) Air compressor. (**b**) Water. (**c**) Oil layer. (**d**) Tracer injection. (**e**) Flat chamber. (**f**) Upper camera. (**g**) Ultrasonic transducer. (**h**) Mass flowmeter. (**i**) Frontal camera. (**j**) Colorimeter cell. (**k**) Air injection.

Table 4. Experimental conditions and physical properties of the multiphase model.

Flow Rates Of Gas	m^3/s Model	5.33 × 10^{-5}	1.07 × 10^{-4}	2.14 × 10^{-4}	4.28 × 10^{-4}	5.50 × 10^{-4}
	l/min Ladle	52	100	200	400	500
Physical Properties of Fluids (293 K)						
	Density Kg/m^3	Viscosity Pa-s	Surface Tension N/m		Interfacial Tension N/m	
Water	1000	0.001003	0.073		0.0565	
Oil	913	0.060	0.040			
Air	1.24	1.8 × 10^{-5}				

Other features: Nozzle diameter = 6 mm, bath height = 0.90 m, and scale up criterion = the Fr number.

3. The Multiphase Model

To simulate the interaction among the multiphase system, the VOF was applied [30]. This model uses a common pressure-velocity field by solving a single set of momentum transfer equations and uses as a phase indicator, for including the presence of interfaces, the volume fraction of a phase by the solving the corresponding advection equation. The equation of the phase indicator is,

$$\frac{\partial \alpha_i}{\partial t} + (u_m \cdot \nabla)\alpha_i = 0, \tag{1}$$

where the unit value of α_i corresponds to a cell full of fluid 1, while a zero value indicates that the cell contains no fluid 1, u_m is the mix velocity. To avoid numerical diffusion the equation should be solved using second order explicit discretization equation in time and space, updating the indicator through the velocity field [31]. The pressure-velocity field is simulated by resolving the continuity and Navier-Stokes equations,

$$\nabla \cdot u_k = 0 \tag{2}$$

$$\frac{\partial u_k}{\partial t} + u_k \cdot \nabla u_k = -\frac{1}{\rho_k} \nabla p_k + \nu_k \nabla^2 u_k, \tag{3}$$

where the u_k is the Reynolds Averaged Navier-Stokes (RANS) velocity of the turbulent flow, and p_k and ν_k are the pressure and velocity in the direction k. The interface boundary conditions or momentum jump conditions are expressed as,

$$\sum_{k=1}^{2} T_k n_k = 2\sigma_I H_I n_I, \tag{4}$$

where T_k is the total stress interfacial tensor, n_k is the normal vector to the interfacial surface, and σ_I, H_I and n_I are the surface tension, the radius of curvature, and the normal vector to the interface which was simulated through the Continuous Surface Model of Brackbill [32]. For the present case, the physical properties of the multiphase system, (including the food-oil [33]), were calculated as,

$$\rho_m = \rho_w \alpha_w + \rho_o \alpha_o + \rho_a \alpha_a \tag{5}$$

$$\mu_m = \mu_w \alpha_w + \mu_o \alpha_o + \mu_a \alpha_a. \tag{6}$$

The constraint for the volume fraction was, $\alpha_w + \alpha_o + \alpha_a = 1$ (7)

where the sub-indexes w, o, and a hold for water, oil, and air, respectively. The k-ε model [34,35] was used to simulate the turbulence of the flow combined with Equations (2) and (3) to obtain the pressure-velocity field, which was employed to update the advection equation of the indicator. The model, which was based in the turbulent viscosity hypothesis [35], required the solution of two other equations for the turbulent kinetic energy and its dissipation rate,

$$\frac{\partial(\rho_m k)}{\partial t} + \frac{\partial(\rho_m k u_i)}{\partial x_i} = \frac{\partial}{\partial x_j}\left[\left(\mu_m + \frac{\mu_m^t}{\sigma_k}\right)\frac{\partial k}{\partial x_j}\right] + C_{1\varepsilon}\frac{\varepsilon}{k} + G_k + G_b - \rho_m \varepsilon \tag{8}$$

$$\frac{\partial(\rho_m \varepsilon)}{\partial t} + \frac{\partial(\rho_m \varepsilon)}{\partial x_j} = \frac{\partial}{\partial x_j}\left[\left(\mu_m + \frac{\mu_m^t}{\sigma_\varepsilon}\right)\frac{\partial}{\partial x_j}\right] + C_{1\varepsilon}\frac{\varepsilon}{k}(G_k + C_{3\varepsilon}G_b) - C_{2\varepsilon}\rho_m \frac{\varepsilon^2}{k}. \tag{9}$$

G_k was the generation of kinetic energy due to the interaction between the gradients of the mean velocity and the Reynolds stresses (energy extracted from the mean flow):

$$G_k = -\rho_m u_i' u_j' \frac{\partial u_k}{\partial x_j}. \tag{10}$$

And G_b is the energy generated by buoyancy forces, given by Equation (11):

$$G_b = -g_i \frac{\mu_m^t}{Pr_t} \frac{\partial \rho_m}{\partial x_i}. \tag{11}$$

The scalars k and ε are used to calculate the turbulent viscosity through Equation (12),

$$\mu_m^t = \rho_m C_\mu \frac{k^2}{\varepsilon}, \tag{12}$$

where $C_\mu = 0.09$, $C_{1\varepsilon} = 1.44$, $C_{2\varepsilon} = 1.92$, $\sigma_k = 1.0$, $\sigma_\varepsilon = 1.3$, and $C_{3\varepsilon} = 1.0$.

The solution of the governing equations with the boundary conditions and all source terms were obtained through the commercial package ANSYS [36]. In all solid surfaces, a no-slip boundary condition was applied, and the wall log-law was used to link the outer grid with the computational elements in the boundary layer. In the nozzle, an entry gas-velocity boundary condition was applied and, in the top surface of the bath, a pressure one was applied. The calculation domain was divided by 2,000,000 hexahedral-100% low skewedness-structured cells, using a second order discretization scheme and a segregated-explicit coupled computing procedure. The Pressure Implicit of Splitting Operations (PISO) algorithm was employed for the pressure-velocity coupling [36]. Testing was not carried out regarding the independence of the numerical results, it was assumed that this requirement was accomplished. The calculations were conducted under transient conditions. The model was running for 300 s, ensuring a stable flow condition by reviewing the data file every 1 s. A criterion for convergence was fixed when the sum of all residuals for the dependent variables were less than 10^{-4} with unbalances less than 1%. It was assumed that these settings ensured independence of the numerical results from the mesh size, although analysis in this regard was not actually carried out.

4. Results and Discussion

4.1. Flow Parameters

The mixing time was considered when 95% of the global tracer concentration was achieved for the two-phase flow, calculated using the experimental conditions of Table 4 and the correlations in Table 1. These calculated mixing times were compared with the corresponding experimental mixing times, and the two values plotted in Figure 3a. Here it was evident that correlations 1 and 3 in Table 1 were the most approximated to the experimental measurements. Since the presence of the upper phase affected the fluid dynamics of the lower phase, the same procedure was applied for calculating the mixing times of the three-phase system (two immiscible liquids and gas phase), using the correlations of Table 2 and comparing these values with the experimental measurements. The results are presented in Figure 3b, where, as it is evident, five predicted values are very approximated to the experimental ones. Carefully observing these two correlations for the two-phase and three-phase flows, given their highest accuracy in predicting the experimental mixing times, the following observations could be established:

1. For two-phase flows, mixing time was basically dependent on the stirring energy, the bath height, and on the geometry of the ladle.
2. Tall baths favored smaller values of the mixing time.
3. For three-phase flows, thicker slags increased the mixing time.
4. The mixing time was basically independent from the physical properties of the upper phase, such as density and viscosity, and depends more on its thickness.

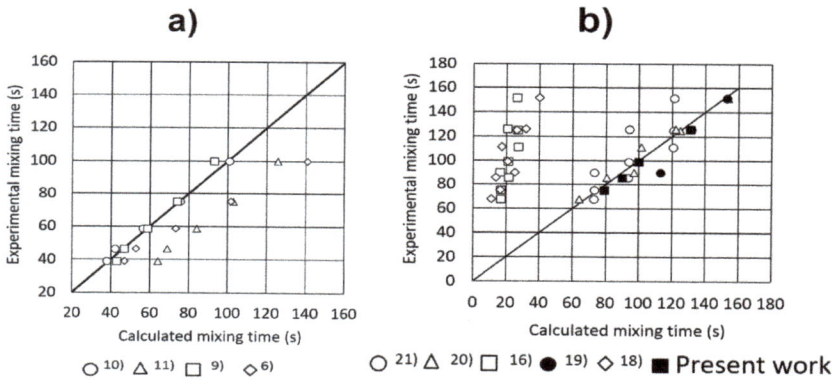

Figure 3. Experimental versus calculated mixing times. (**a**) Bath without an upper layer. (**b**) Bath with an upper layer.

The later observation was the most oriented to discussion of what someone would intuitively think of as the physical properties of the slag or upper phase would be determinant. However, it was evident from the presented results and those of other researchers that the control of mixing was essentially the thickness due to the dissipation rate of kinetic energy when the gas–liquid plume strains the interface. That is, once the liquid, driven by the buoyancy forces imparted by the bubbles, reached the interface, the upper phase was displaced, receding toward the ladle wall and forming the spout where the bubbles burst, consuming energy in the process.

The experimental SEO areas along the experimental time are shown in Figure 4, and these areas suffered natural variations due to the turbulence of the flow but tended to reach constant values for any given operating condition.

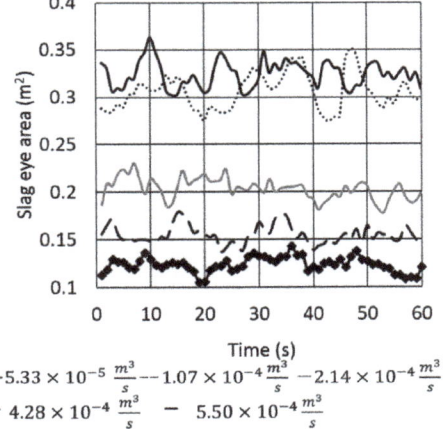

Figure 4. Effects of gas flow rate for an upper layer thickness of 0.02 m on the variations of the eye opening area with time.

Figure 5 shows the SEO areas simulated with the mathematical model and compared with corresponding photos of the experimental SEO areas. As seen, there was a very good qualitative agreement between the mathematical simulations and the experimental SEO areas; even the shapes of the SEOs were very similar.

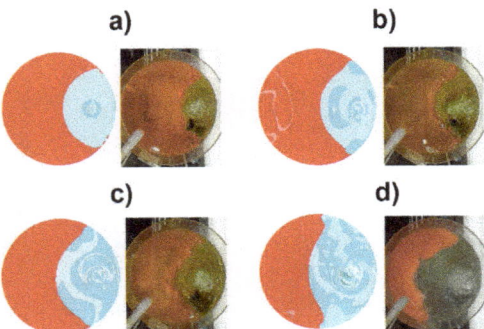

Figure 5. Comparison between experimental and numerical views of the slag eye opening area for an oil thickness 0.02 m. (**a**) Flow 5.33×10^{-5} m^3/s. (**b**) 1.07×10^{-4} m^3/s. (**c**) 2.14×10^{-4} m^3/s. (**d**) 4.28×10^{-4} m^3/s.

To test the quantitative capacity of the mathematical model in predicting the experimental SEO areas, the images of the digital images and the areas of the corresponding photos in Figure 5 were measured with the image analyzer. The results of the digital or numerical SEO areas were compared with the measured SEO areas in Figure 6 and, as is seen, the agreement is excellent, granting the reliability of the mathematical model to study the dynamics of the three-phase flow system. In the same Figure, the SEO areas calculated with the correlations of Table 3, could be compared with the experimental SEOs, and it was evident that the best correlation was the number 4. Therefore, this correlation is recommended as a simple engineering tool to estimate the SEO for operating steel ladles. Accordingly, it could be said that the present mathematical model is useful in testing the reliability of engineering correlations for calculating mixing time and SEO areas.

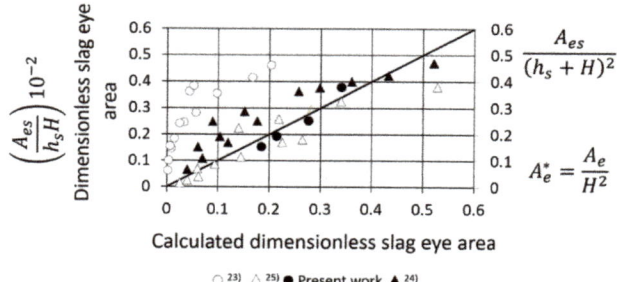

Figure 6. Slag eye opening area.

4.2. Flow Structure

The presence of a slag or an upper layer decreased the turbulence of the denser liquid, as is seen in Figure 7a,b for the velocity field in a vertical plane passing through the axis of the injector without and with a 0.01 m thick layer, respectively. Without an upper layer, the recirculating flow was large enough to include the full plane. However, even a thin upper layer restricted the recirculating flow forming a free shear boundary layer in the upper right extreme, near the ladle wall. Figure 7c,d shows the velocity fields in perpendicular planes to those shown in Figure 7a,b.

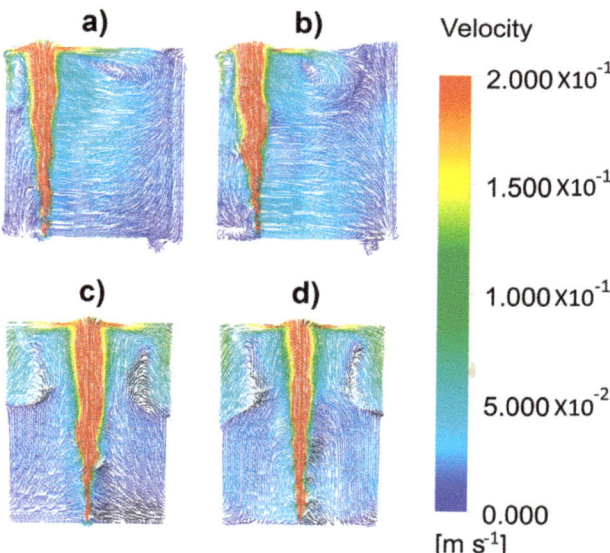

Figure 7. Simulated velocity fields of the liquid phase for a gas flow rate of 1.07×10^{-4} m^3/s and different thicknesses of the oil layer. Front view (**a**) 0 m and (**b**) 0.01 m. Side view (**c**) 0 m and (**d**) 0.01 m.

As seen, in these planes, more than half of these planes had small fluid velocities and the last third of the upper region was subjected to the effects of recirculating flows induced by the two-phase plume. Figure 8a,b correspond to Figure 7a,b with upper layer thicknesses of 0.02 and 0.04 m, respectively.

The thicker upper layers constrained more the upper recirculating flow, leaving larger regions of small liquid velocities. Once the plume reached the bath surface, it encountered the upper layer, which was displaced forming the slag eye opening and the spout. Figure 8c,d show the velocity fields in planes which are perpendicular to those shown in Figure 8a,b, respectively. The regions with small velocities increased considerably; with an upper layer of 0.02 m, there was still a recirculating flow in the denser flow made contact with the upper phase. However, when the upper layer thickness was 0.04 m, the contact between the denser and upper layer was constrained to a very limited area surrounding the spout. Under these conditions the exchange between both liquid phases was very poor and, from a practical point of view, very small refining capacity was left and the capture of inclusions was limited. Since we used a 1/3 scale model, a thick 0.12 m slag would keep stagnant practically all the melt for a flow a rate of 100 l/min. An option was to increase the flow rate of argon to intensify the contact between both liquids, but thick slags were easily entrained in the melt bulk [37], increasing the presence of inclusions.

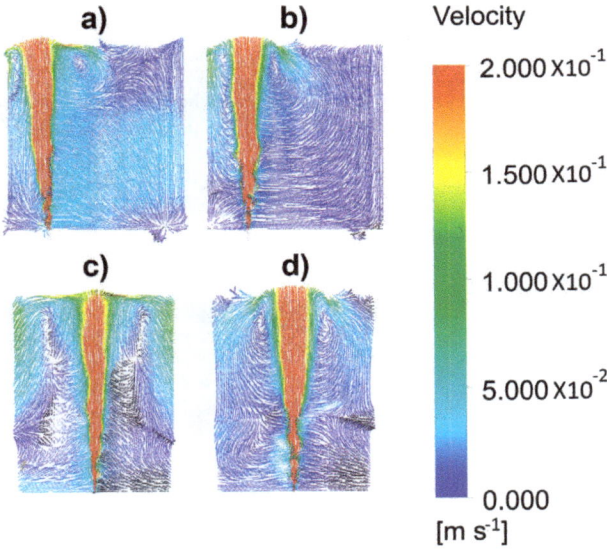

Figure 8. Simulated velocity fields of the liquid phase for a gas flow rate of 1.07×10^{-4} m^3/s and different thicknesses of the oil layer. Front view (**a**) 0.02 m and (**b**) 0.04 m. Side view (**c**) 0.02 m and (**d**) 0.04 m.

Figure 9a,b shows the velocity fields in horizontal planes at different bath heights, for a ladle without and with an upper layer 0.01 m thick. Without the upper layer, horizontal-rotating motions are observed, but even the presence of a thin upper layer changed the flow pattern to divided recirculating flows at each side of the plume. Thicker layers of 0.02 and 0.04 m in Figure 10a,b, respectively, intensified the split of the velocity fields at each side of the plume, though, with a 0.02 m thick layer, the motion in the liquid bulk was still appreciable. Hence, a layer of 0.06 m in the actual ladle was thin enough to maintain good stirring conditions and even increased the flow of rate of gas to intensify the mass transfer during the desulfurization process. This thickness represented 2.2% of the total bath height, and it would be recommendable since it kept a volume large enough to capture inclusions and to desulfurize steel. Thinner layers would not be enough to keep dissolved all the sulfur necessary for a given steel grade, and thicker slags would constrain the motion of steel, thereby decreasing its contact with the slag.

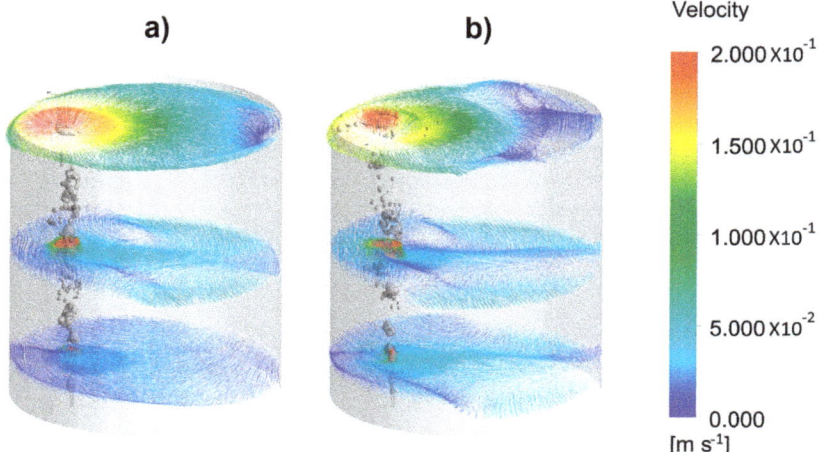

Figure 9. Simulated velocity fields of the liquid phase in different horizontal planes, at 100 mm, 450 mm, and 880 mm, for a gas flow rate of 1.07×10^{-4} m^3/s and different thicknesses of the oil layer. (**a**) 0 m and (**b**) 0.01 m.

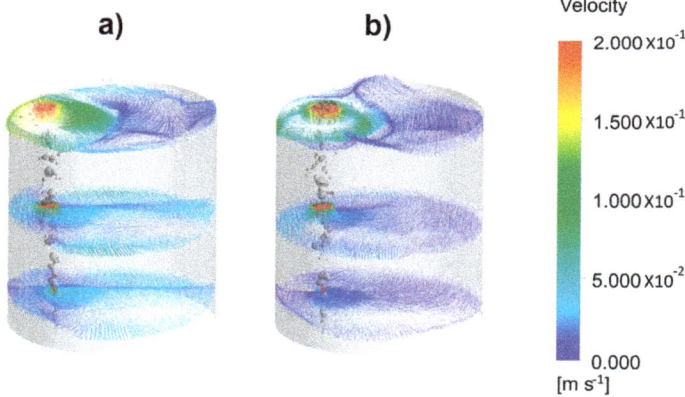

Figure 10. Simulated velocity fields of the liquid phase in different horizontal planes, at 100 mm, 450 mm, and 880 mm, for a gas flow rate of 1.07×10^{-4} m^3/s and different thicknesses of the oil layer. (**a**) 0.02 m and (**b**) 0.04 m.

The proximities of the metal-slag interface are the most important to capture inclusions by the slag phase. Therefore, it is important to understand the turbulence conditions in this region. Figure 11a–d show the measured horizontal-velocities with the ultrasound probe along the distance from the opposite wall of the plume to the wall behind the plume. The dotted line corresponds to the average velocities of the experimental measurements, and the interrupted lines correspond to the predictions of the mathematical model. When there was not an upper layer, as in Figure 11a, the velocity fluctuations near the wall were large and their amplitudes grew at a maximum at the location of the plume, due to the bursting effect of the bubbles. Even under the presence of a thin layer, as in Figure 11b, the velocity fluctuations were considerably damped and remained large in the region of the plume for the reason adduced above. As the layer got thicker, as in Figure 11c,d, the velocity fluctuations suffered further dampening to the point where there were practically no more, and leaving only those velocity spikes corresponding to the plume. Regarding the mathematical model, although its predictions do not match as well the averaged experimental velocities, they lay inside the magnitudes of the fluctuating velocities. The reasons of this mismatch are basically three, the first is that the VOF model solves only one set of equations for all three phases, and the second is that this model predicts averaged velocities of turbulent flows. The third reason is the impossibility to match the computing time with the experimental ones. All these factors impede a good matching between simulations and measurements. However, given these upheavals the mathematical model makes predictions of the trends of the velocity fields. Another possible source of error could be the nit independence of the numerical results from mesh size.

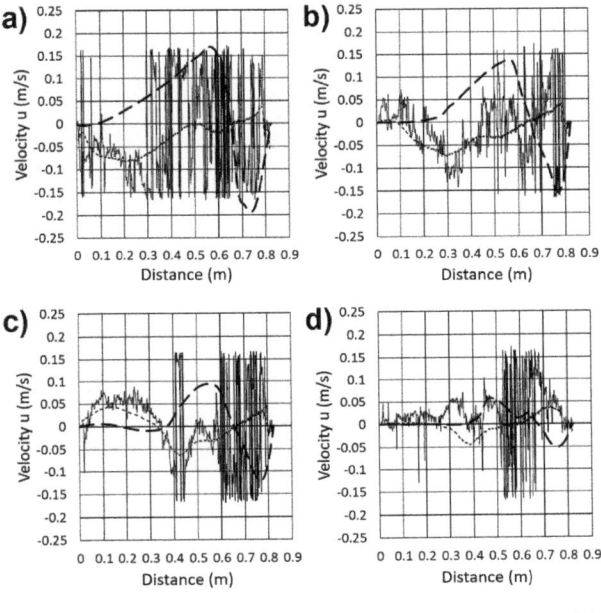

— Instantaneous velocities – – Numerical velocities with the VOF model
- - - Averaged experimental velocities

Figure 11. Measured versus simulated velocities 20 mm below the metal-slag interface for different thicknesses of the upper phase layer at a fixed gas flow rate of 1.07×10^{-4} m^3/s. (**a**) 0 m, (**b**) 0.01 m, (**c**) 0.02 m, and (**d**) 0.04 m.

Further information about the flow structure in proximities of the metal-slag interface c be seen through the fields of streamlines shown in Figure 12a–d corresponding to Figure 11a–d, respectively. With no upper layer, the stream lines initiating from the spout followed a regular irradiating pattern. With a thin layer, this pattern was slightly disrupted, and two regular vortexes were formed symmetrically and close to the opposite ladle wall. Thicker layers subdivided the flow in many local vortexes with different orientations, making the flow practically stagnant.

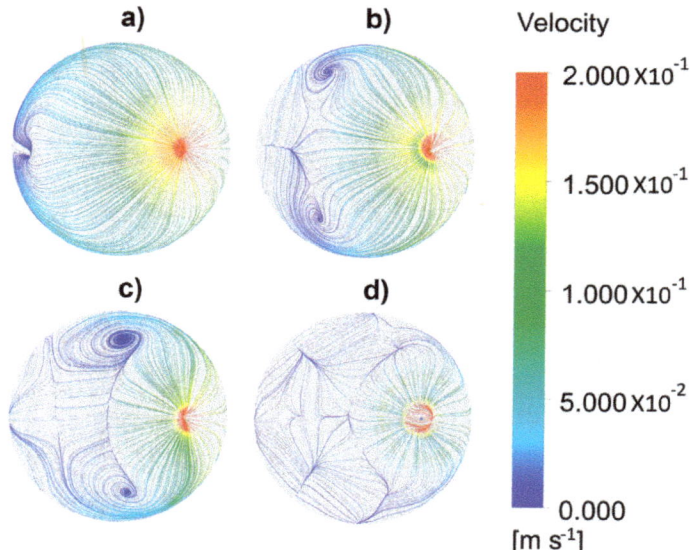

Figure 12. Streamlines of the flow for different thicknesses of the upper layer for a gas flow rate of 1.07×10^{-4} m^3/s. (**a**) 0 m, (**b**) 0.01 m, (**c**) 0.02 m, and (**d**) 0.04 m.

4.3. Closure

Despite the limitations of the model, made explicit before, it was evident that mixing times are dependent on stirring energy and bath geometry in two-phase flows. In three phase flows, mixing time depends on these factors and on the thickness of the slag phase, the physical properties of the upper layer play a secondary role. The SEO area in three-phase flows depends on the flow rate of the stirring gas, the physical properties of the upper layer and on the bath height. As the thickness of the upper layer increases, the mixing time increases and the bath approaches near-stagnant condition. The SEO area increases with the flow rate of gas and decreases with the slag thickness. Denser upper layers favor larger SEO areas and high viscosity layers decrease it.

There has been research claiming that bubbles play an important role to float out inclusions through mechanism of their adherence on their surfaces [38] and carrying them to the bath surface. Rather, it is the motion of steel, originated by the energy provided by the buoyant bubbles, that carries the inclusions in contact with the slag which, eventually, absorbs them. Hence, it was difficult to agree with this view, as their area volume ratio is very small, as can be seen in Figures 9 and 10. It could be criticized that these simulations do not reflect actual sizes in liquid steel. However, water models always report big bubbles sizes, that may or may not be disintegrated, and given the surface tension of liquid steel, argon bubbles in steel should be twice as large as those observed in the models. Instead, it is the flow near the proximities of the metal–slag-interface which control the floatation of inclusions. When this region is disrupted by strong forced convection flows, the efficiency of floatation decreases. It is only when this region comes to some velocity fields of small magnitude that the efficiency to float inclusions gets high. This explains the need of having what it is called "rinsing time", which refers to the last period after refining, when the stirring energy is decreased to allow floatation of inclusions in the Stokes regime. This operation is well-known by all steelmakers, and it is certainly a very necessary one to attain the best of steel cleanliness.

Author Contributions: Investigation, F.A.C.-H. and S.G.-H.; Methodology, R.M.D.; Supervision, K.C.

Funding: This research was funded by Consejo Nacional de Ciencia y Tecnología (CoNaCyT).

Acknowledgments: We give the thanks to IPN and (CoNaCyT) for the support of this work. One of us (R.M.) gives the thanks to University of Toronto for making his stay in Toronto possible.

Conflicts of Interest: The authors declare no conflict of interest.

Abbreviations

List of Symbols
A_e, A_{es}—Slag eye opening area
A_p—Area of the plume
u_i—Velocity in direction "i"
Q and Q_g—Gas flow rate
L—Ladle height
R—Ladle radius
U_p—Plume velocity
D—Ladle diameter
h_s—Thickness of the oil or slag layer
H_{eff}—Effective bath height
N—Number of plugs or orifices in the ladle bottom
r—Radial position
T_k—Interfacial stress tensor
n_k—Normal vector to the interfacial surface
H, H_L—Bath height
k—Turbulent kinetic energy
G_k and G_b—the generation terms of kinetic energy by the mean flow and by the buoyancy, respectively.
g_i—Gravity constant
Pr—Prandtl number

Greek Letters
α_i Volume fraction of phase "i"
τ_m Mixing time
$\dot{\varepsilon}$—Specific potential energy input
ρ_i Density of phase "i"
ε—Dissipation rate of turbulent kinetic energy
ν_i Kinematic viscosity of phase "i"
σ_i Surface tension
μ_s—Dynamic viscosity of the slag
$\Delta\rho \frac{\rho_s}{\rho_L}$

Sub-Indexes
m.—mixture
s.—slag
o.—oil
w.—water

References

1. Krishnapisharody, K.; Irons, G.A. A Model for Slag Eyes in Steel Refining Ladles Covered with Thick Slag. *Metall. Mater. Trans. B* **2015**, *46*, 191–198. [CrossRef]
2. Krishnapisharody, K.; Irons, G.A. Modeling of Slag Eye Formation over a Metal Bath Due to Gas Bubbling. *Metall. Mater. Trans. B* **2006**, *37*, 763–772. [CrossRef]
3. Krishnapisharody, K.; Irons, G.A. A Study of Spouts on Bath Surfaces from Gas Bubbling: Part II. Elucidation of Plume Dynamics. *Metall. Mater. Trans. B* **2007**, *38*, 377–388. [CrossRef]
4. Krishnapisharody, K.; Irons, G.A. A Study of Spouts on Bath Surfaces from Gas Bubbling: Part I. Experimental Investigation. *Metall. Mater. Trans. B* **2007**, *38*, 367–375. [CrossRef]
5. Krishnapisharody, K.; Irons, G.A. A Unified Approach to the Fluid Dynamics of Gas–Liquid Plumes in Ladle Metallurgy. *ISIJ Int.* **2010**, *50*, 1413–1421. [CrossRef]

6. Iguchi, M.; Nakamura, K.; Tsujino, R. Mixing Time and Fluid Flow Phenomena in Liquids of Varying Kinematic Viscosities Agitated by Bottom Gas Injection. *Metall. Mater. Trans. B* **1998**, *29*, 569. [CrossRef]
7. Mazumdar, D.; Guthrie, R.I.L. Discussion of "mixing time and fluid flow phenomena in liquids of varying kinematic viscosities agitated by bottom gas injection". *Metall. Mater. Trans. B* **1999**, *30*, 349–351. [CrossRef]
8. Mazumdar, D.; Guthrie, R.I.L. Hydrodynamic modeling of some gas injection procedures in ladle metallurgy operations. *Metall. Mater. Trans. B* **1985**, *16*, 83–90. [CrossRef]
9. Mazumdar, D.; Guthrie, R.I.L. Mixing models for gas stirred metallurgical reactors. *Metall. Mater. Trans. B* **1986**, *17*, 725–733. [CrossRef]
10. Haida, O.; Emi, T.; Yamada, S.; Sudo, F. Scaninjet II, Part I. In Proceedings of the 2nd International Conference on Injection Metallurgy, MEFOS-Jerkentoret, Lulea, Sweden, 12–13 June 1980.
11. Ying, Q.; Yun, L.; Liu, L. Scaninjet III, Part I. In Proceedings of the 2nd International Conference on Injection Metallurgy, MEFOS-Jernkontoret, Lulea, Sweden, 15–17 June 1983.
12. Li, B.; Yin, H.; Zhou, C.Q.; Tsukihashi, F. Modeling of Three-phase Flows and Behavior of Slag/Steel Interface in an Argon Gas Stirred Ladle. *ISIJ Int.* **2008**, *48*, 1704–1711. [CrossRef]
13. Liu, Z.; Li, L.; Li, B. ISIJ International, Modeling of Gas-Steel-Slag Three-Phase Flow in Ladle Metallurgy: Part I. *Phys. Model.* **2017**, *57*, 1971.
14. Singh, U.; Anapagaddi, R.; Mangal, S.; Padmanabhan, K.A.; Singh, A.K. Multiphase Modeling of Bottom-Stirred Ladle for Prediction of Slag–Steel Interface and Estimation of Desulfurization Behavior. *Metall. Mater. Trans. B* **2016**, *47*, 1804–1816. [CrossRef]
15. Mazumdar, D.; Guthrie, R.I.L. Modeling Energy Dissipation in Slag-Covered Steel Baths in Steelmaking Ladles. *Metall. Mater. Trans. B* **2010**, *41*, 976–989. [CrossRef]
16. Khajavi, L.T.; Barati, M. Liquid Mixing in Thick-Slag-Covered Metallurgical Baths—Blending of Bath. *Metall. Mater. Trans. B* **2010**, *41*, 86–93. [CrossRef]
17. Jonsson, P.G.; Sichen, D. Viscosities of LF Slags and Their Impact on Ladle Refining. *ISIJ Int.* **1997**, *37*, 484–491. [CrossRef]
18. Yamashita, S.; Miyamoto, K.; Iguchi, M.; Zeze, M. Model Experiments on the Mixing Time in a Bottom Blown Bath Covered with Top Slag. *ISIJ Int.* **2003**, *43*, 1858–1860. [CrossRef]
19. Mazumdar, D.; Kumar, D.S. In *Proceedings: Oxygen in Steelmaking*; Irons, G., Sun, S., Eds.; MetSoc of CIM: Westmount, QC, Canada, 2004; pp. 311–323.
20. Patil, S.P.; Satish, D.; Peranandhanthan, M.; Mazumdar, D. Mixing Models for Slag Covered, Argon Stirred Ladles. *ISIJ Int.* **2010**, *50*, 1117–1124. [CrossRef]
21. Amaro-Villeda, A.M.; Ramirez-Argaez, M.A.; Conejo, A.N. Effect of Slag Properties on Mixing Phenomena in Gas-stirred Ladles by Physical Modeling. *ISIJ Int.* **2014**, *54*, 1–8. [CrossRef]
22. Yonezawa, K.; Schwerdtfeger, K. Spout eyes formed by an emerging gas plume at the surface of a slag-covered metal melt. *Metall. Trans.* **1999**, *30*, 411–418. [CrossRef]
23. Subagyo, G.A.; Brooks, G.A. Irons. Spout Eyes Area Correlation in Ladle Metallurgy. *ISIJ Int.* **2003**, *43*, 262–263. [CrossRef]
24. Krishnapisharody, K.; Irons, G.A. An Extended Model for Slag Eye Size in Ladle Metallurgy. *ISIJ Int.* **2008**, *48*, 1807–1809. [CrossRef]
25. Peranandhanthan, M.; Mazumdar, D. Modeling of Slag Eye Area in Argon Stirred Ladles. *ISIJ Int.* **2010**, *50*, 1622–1631. [CrossRef]
26. Liu, W.; Tang, H.; Yang, S.; Wang, M.; Li, J.; Li, Q.; Liu, J. Numerical Simulation of Slag Eye Formation and Slag Entrapment in a Bottom-Blown Argon-Stirred Ladle. *Metall. Trans. B* **2018**, *49*, 2681–2691. [CrossRef]
27. Ramasetti, E.; Visuri, V.-V.; Sulasalmi, P.; Mattila, R.; Fabritius, T. Modeling of the Effect of the Gas Flow Rate on the Fluid Flow and Open-Eye Formation in a Water Model of a Steelmaking Ladle. *Steel Res. Int.* **2018**, *90*. [CrossRef]
28. Image Processing and Analysis in Java. Consulted in October 2018. Available online: http://imagej.nih.gov/ij/index.html.ImageJ-1.49 (accessed on 5 October 2018).
29. Frank, M. *White, Fluid Mechanics*, 4th ed.; McGraw-Hill: New York, NY, USA, 1999; p. 294.
30. Hirt, C.W.; Nichols, B.D. Volume of fluid (VOF) method for the dynamics of free boundaries. *J. Comput. Phys.* **1981**, *39*, 201–225. [CrossRef]
31. Ferziger, J.H.; Peric, M. *Computational Methods for Fluid Dynamics*; Springer: Berlin, Germany; New York, NY, USA, 2002; pp. 72–75.

32. Brackbill, J.U.; Kothe, D.B.; Zemach, C. A continuum method for modeling surface tension. *J. Comput. Phys.* **1992**, *100*, 335–354. [CrossRef]
33. Sahasrabudhe, S.N.; Rodriguez-Martinez, V.; O'Meara, M.; Frakas, B.E. Density, viscosity, and surface tension of five vegetable oils at elevated temperatures: Measurement and modeling. *Int. J. Food Prop.* **2017**, *20*, S1965–S1981. [CrossRef]
34. Pope, S.B. *Turbulent Flows*; Cambridge Press: New York, NY, USA; Cambridge, UK; London, UK, 2000; pp. 387–457.
35. Wilcox, D.C. *Turbulence Modeling for CFD*; D.C.W. Industries Inc: La Cañada, CA, USA, 2000; pp. 103–218.
36. ANSYS-FLUENT. Available online: www.ansys.com/Products/Fluids/ANSYS-Fluent (accessed on 3 May 2018).
37. Han, Z.; Holappa, L. Mechanisms of Iron Entrainment into Slag due to Rising Gas Bubbles. *ISIJ Int.* **2003**, *43*, 292–297. [CrossRef]
38. Soder, M.; Jonsson, P.; Jonsson, L. Inclusion Growth and Removal in Gas-Stirred Ladles. *Steel Res. Int.* **2004**, *75*, 128–138. [CrossRef]

© 2019 by the authors. Licensee MDPI, Basel, Switzerland. This article is an open access article distributed under the terms and conditions of the Creative Commons Attribution (CC BY) license (http://creativecommons.org/licenses/by/4.0/).

Article

Influence of Al on Evolution of the Inclusions in Ti-Bearing Steel with Ca Treatment

Yandong Li [1], Tongsheng Zhang [2,*] and Huamei Duan [3]

1. College of Materials Science and Engineering, Yangtze Normal University, Chongqing 408000, China; andyydlee@gmail.com
2. School of Metallurgy and Environment, Central South University, Changsha 410083, China
3. College of Materials Science and Engineering, Chongqing University, Chongqing 400044, China; duanhuamei@cqu.edu.cn
* Correspondence: tongsheng.zhang@csu.edu.cn; Tel.: +86-139-7498-2473

Received: 29 December 2018; Accepted: 18 January 2019; Published: 21 January 2019

Abstract: Experimental simulations of steelmaking with different amounts of aluminum were achieved in the tube furnace at 1873 K and field scanning electron microscopy and energy dispersive X-ray spectroscopy (FE-SEM and EDX) were employed to explore the characteristics of the inclusions in Ti-bearing steel during the calcium treatment process. It was found that morphologies, chemical compositions, and the size distribution of the inclusions were obviously different before and after calcium treatment. The calcium addition need be carefully considered regarding the mass fraction of aluminum with the purpose of modifying the solid inclusions to liquid phases. The thermodynamic analysis of inclusion formation in the Al–Ti–Ca–O system at 1873 K was conducted, as well as transformation behaviors of inclusions including all types of solid inclusions and liquid phases during solidification. The thermodynamic equilibrium calculations are in good agreement with experimental data, which can be used to estimate inclusion formation in Ti-bearing steel.

Keywords: deoxidation; inclusions; thermodynamics; Ti-bearing steel; Ca treatment

1. Introduction

The quality of steel products can be effectively increased after being treated with the element titanium [1–4]. Titanium oxides and titanium carbides can be generated during the deoxidation process, which can promote the mechanical properties around the region of welded steels by improving the nucleation ratio of intragranular bainite [5–9]. However, TiO_x inclusions have a high probability of being agglomerate and subsequently form clusters, which results in the serious clogging issue of submerged entry nozzle and, consequently, lower the productivity of the continuous casting process as well [9–11]. Owing to the importance of Ti-containing inclusions, scholars have conducted a number of studies on inclusion control in Ti-bearing Al-killed steel, such as on thermodynamic computation and analysis [12–22], size distribution statistics [23–26], precipitation mechanisms [10,11,16,20], reaction kinetics, and evolution trajectory [14], mostly within the alumina and titanate system. In other words, it is usually difficult to keep the Al–Ti–O system inclusions as liquid phases during the production process at the present period.

It is generally the technique of calcium treatment that has been introduced to get liquid phases of calcium aluminate at casting temperature, which relieves the nozzle clogging issue during continuous process of steels, especially of grades of Al-killed steel, and also has benefits for the mechanical performance of final steel products [27–37]. However, the research involving the calcium treatment process of Ti-bearing Al-killed steel have been limited, until now [38–42]. In addition, the types of inclusions at the solidification temperature are directly related to the properties of the inclusions in the final product.

In this work, the laboratory-scale melts, with different mass fractions of calcium and aluminum, were prepared in order to clarify the influence of aluminum element on inclusion properties during the calcium treatment process of Ti-bearing Al-killed steels. Then, FE-SEM and EDX were employed to observe and analyze morphologies, chemical compositions, and number and size of inclusions in steel cylinder sampled from high-temperature melts before and after calcium treatment. In addition, equilibrium phases of inclusions at smelting temperature were calculated, and the transformation behaviors of inclusions during the process of solidification were estimated by the commercial software FactSage. Present works will lay the experimental and thermodynamic foundation on expanding the combined treatment of titanium and calcium to industrial-scale production, and suggest an alternative way to eliminate the nozzle clogging issue during the continuous casting process of Ti-bearing Al-killed steel.

2. Experimental Methods

Three sets of experiments were conducted in the furnace according to the schematic diagram shown in Figure 1. A 350 gram plate of iron was loaded in the Al_2O_3 crucible surrounded by the graphite crucible, and then the graphite crucible was placed into the furnace which was heated to 1873 K in the protective atmosphere of high-purity argon gas with a constant flow rate. Thereafter, the furnace temperature was maintained for 0.5 h after the raw materials were completely melted at 1873 K to reach the full homogenization of solutes. At the time node of before and after deoxidant (Al and Ti) addition, the activity of dissolve oxygen was determined by the oxygen probe with a resolution of $\pm 10^{-6}$. The cylinder steel sample was extracted by the quartz tube with the inner diameter of 3 millimeters, and then immediately quenched into water at ambient temperature. At last, Ca treatment was carried out by adding Ca–Fe alloy. Another steel sample was obtained in the above way just 10 min after Ca treatment. The experimental details are shown in Figure 2, and the compositions of the raw materials in the work involving ARMCO iron, Ca–Fe alloy, and Al wire are listed in Table 1.

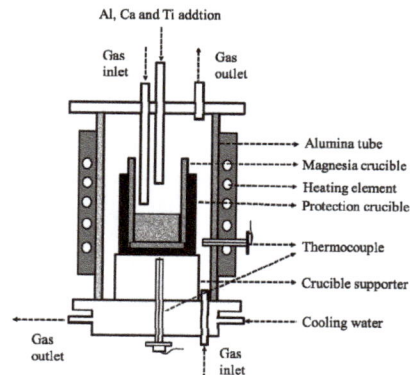

Figure 1. The schematic diagram of the experimental set-up.

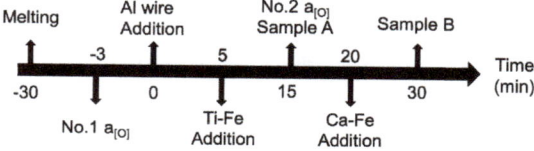

Figure 2. The melting sequence of experiment process and sampling.

Table 1. The compositions of experimental materials (mass %).

Type	Fe	C	Ti	Mn	Si	Ca	Al	P	S	Others
Ca–Fe alloy	69.82	-	-	-	-	30.10	-	-	-	0.08
Ti–Fe alloy	26.513	0.130	69.594	0.241	0.030	-	3.192	0.025	0.011	0.284
Al wire	-	-	-	-	-	-	99.99	-	-	0.01
Pure iron	99.944	0.002	-	0.03	0.01	-	0.001	0.007	0.007	0.043

The obtained steel samples were machined to two cylinders. One was adopted to analyze the chemical compositions by inductively coupled plasma-optical emission spectroscopy (ICP-OES) (Thermo Fisher Scientific ICAP6300, Waltham, MA, USA) with the resolution of $\pm 5 \times 10^{-7}$. The total oxygen levels (T.[O]) of the steel samples were measured by inert gas fusion-infrared absorptiometry with the resolution of $\pm 10^{-6}$. The chemical compositions of all steel samples are given in Table 2. Another was polished for monitoring the inclusion properties, such as morphologies, size, and chemical compositions by scanning electron microscopy and energy dispersive spectroscopy (Zeiss Ultra-Plus, ZEISS, Jena, Germany). The size distribution of precipitates was surveyed in the observed regions of 19.86 mm^2, magnified 500× in the cross-section.

Table 2. The compositions of the obtained sample (mass %).

No.	[Al]	[Ca]	[Ti]	$a_{[O]}$	T.[O]
1-A	0.0042	-	0.0105	0.0005	0.0053
1-B	0.0036	0.0005	0.0094	-	0.0049
2-A	0.0053	-	0.0118	0.0006	0.0045
2-B	0.0051	0.0038	0.0115	-	0.0038
3-A	0.0430	-	0.0112	0.0002	0.0038
3-B	0.0413	0.0025	0.0106	-	0.0032

3. Results and Discussion

3.1. Chemical Composition, Size Distribution, and Morphologies of Inclusions

Titanium in the steel production process shows many valence states, such as Ti^{2+}, Ti^{3+}, and Ti^{4+} [43–45]. In addition, the types of titanium oxides are determined by combining the partial pressure of oxygen with mass fraction of titanium in steel [46]. Although a number of Ti-containing oxides (TiO, TiO_2, Ti_3O_5, Ti_2O_3, etc.) can exist as products of deoxidation reactions by titanium, when mass % Ti was between 0.0004 to 0.36, Ti_3O_5 was the only stable equilibrium oxide in steel, as demonstrated by the electron backscatter diffraction technique [43–46]. Based on the compositions of steel in the current research, the isothermal section of the Al_2O_3–Ti_3O_5–CaO ternary phase diagram at 1873 K with $p(O_2) = 10^{-14}$ atm (computed by FactSage 7.0, THERMFACT LTD, Quebec, Canada), as shown in Figure 3, was introduced to analyze the composition of inclusions. It can be easily seen from Figure 3 that solid phases (CaO, Ti_3O_5, Al_2O_3, perovskite, calcium aluminates, and titanium aluminates) coexist with liquid phases in the two- or three-phase zones. It is noticeable that two liquid phases emerge at 1873 K, which are locate in the regions with a small amount of CaO and Ti_3O_5.

The chemical elements, as well as mole ratios in each observed precipitate, were determined by SEM-EDS (Zeiss Ultra-Plus, ZEISS, Jena, Germany), and then the data were converted to mass fraction of CaO, Ti_3O_5, and Al_2O_3. As presented in Figure 4, each plot in the phase diagram represents an individual inclusion to assess inclusion behavior and transformation, and the thick red lines are liquiduses at 1873 K. As shown in Figure 4a, the inclusions in No.1 melt after Ti addition are mainly spherical titanium aluminates. There is no significant change in the morphologies of the inclusions after calcium treatment. Only a small amount of compositional changes occur in these inclusions which are still in liquid phases. By contrast, a mass of irregular inclusions of Al–Ti–Ca–O system are generated in No. 2 melt as result of the increased mass fraction of 40 ppm [Ca] by calcium treatment. As the mass fraction of CaO in these inclusions is significantly increased, the locations of inclusion compositions in the isothermal section diagram are beyond the liquid region. From the view of the phase types, they mostly locate in perovskite and $(CaO)_3 \cdot (TiO_x)_2$. When the mass fraction of [Al] was increased from 40 ppm to 400 ppm, almost no liquid inclusions are found in No. 3 melt. Compared with No. 1 and No. 2 melts, the locations of inclusion compositions in No. 3 melt are outside of the liquidus in the isothermal section before calcium treatment, and bring on the corresponding irregular appearance. However, inclusions are evolved to calcium aluminates of liquid state after calcium treatment.

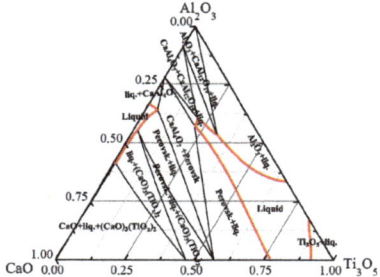

Figure 3. Isothermal section diagram of Ti_3O_5–Al_2O_3–CaO system at 1873 K with $p(O_2) = 10^{-14}$ atm.

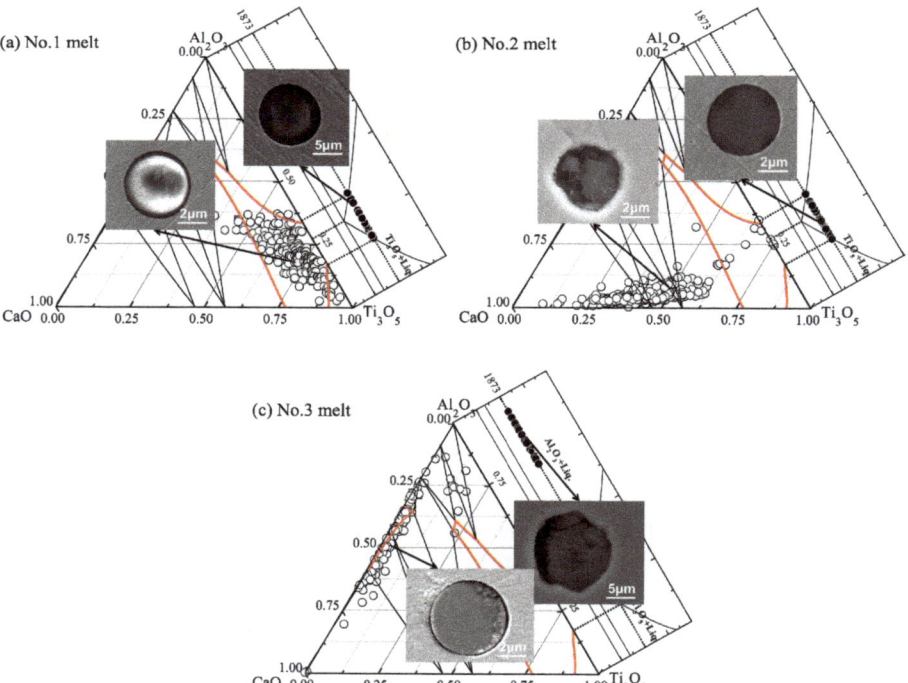

Figure 4. Morphologies and locations of inclusions in the isothermal sections. Thick red lines are the liquidus at 1873 K (1600 °C). The black plots represent the sample before calcium, and the white ones represent the sample after calcium. (**a**) No. 1 melt, (**b**) No. 2 melt, (**c**) No. 3 melt.

Figure 5 gives the size distributions of inclusions in the melts. The results indicate that more than half of the inclusions in sample 1-A are larger than 5 µm, and about 20% of the inclusions are smaller than 2 µm. There is no significant change in the size of inclusions after calcium treatment. There is just a little increase in the proportions of the larger inclusions (>10 µm) in sample 1-B. This may be due to collision and aggregation of the inclusions. The situation of No. 2 melt before calcium treatment is nearly identical to the former one. Nevertheless, the number of inclusions that are smaller than 5 µm in sample 2-B is more than 65%. This illustrates that the solid inclusions of Al_2O_3–CaO–TiO_x system trend to be fine and well-dispersed in the melts of Ti-bearing steel as treated by calcium. Apparently, the inclusions in sample 3-A also have a relatively small size. After calcium treatment, the proportion of the smaller inclusion in sample 3-B is obviously decreased and the numbers of inclusions are present as liquid state, combined with the results in Figure 4.

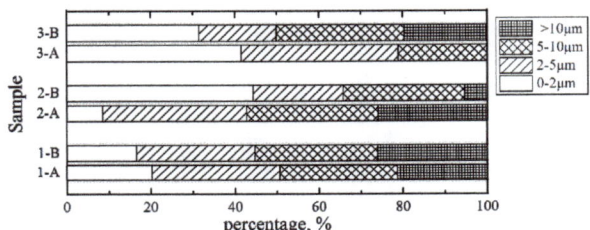

Figure 5. Size distributions of inclusions in all steel samples.

3.2. Thermodynamic Analysis of Al–Ti–Ca–O System

A series of generating reactions of inclusions, as listed in Table 2, were considered to understand the transformation process of Al–Ti–Ca–O system. The intermediate products, titanium oxides, CaO and calcium titanates in Table 3, could be firstly generated according to the very negative Gibbs energies of reactions between titanium/calcium and oxygen at 1873 K. As $12CaO \cdot 7Al_2O_3$ is the only compound located inner of the liquidus in the phase diagrams of Al_2O_3–TiO_x and Al_2O_3–CaO–TiO_x systems at 1873 K, it is introduced as liquid calcium aluminate to calculate the equilibrium state with solid inclusions.

Table 3. Standard Gibbs energies of inclusions formation.

No.	Reactions	$\Delta G^\theta = A + B \times T$ /J·mol^{-1}		References
		A	B	
1	$Al_2O_3 = 2[Al] + 3[O]$	867,500	−222.5	[27]
2	$CaO = [Ca] + [O]$	138,227	63.0	[47]
3	$Al_2TiO_5 = [Ti] + 2[Al] + 5[O]$	1,435,000	−40.5	[23]
4	$CaO \cdot 6Al_2O_3 = CaO_{(s)} + 6Al_2O_{3(s)}$	16,380	37.58	[48]
5	$CaO \cdot 2Al_2O_3 = CaO_{(s)} + 2Al_2O_{3(s)}$	15,650	25.82	[48]
6	$CaO \cdot Al_2O_3 = CaO_{(s)} + Al_2O_{3(s)}$	17,910	17.38	[48]
7	$12CaO \cdot 7Al_2O_3 = 12CaO_{(s)} + 7Al_2O_{3(s)}$	−618,000	612.1	[49]
8	$Ti_2O_3 = 2[Ti] + 3[O]$	822,000	−247.7	[50]
9	$Ti_3O_5 = 3[Ti] + 5[O]$	1,307,000	−381.8	[50]
10	$TiO_2 = [Ti] + 2[O]$	−675,600	234	[51]
11	$3CaO \cdot Ti_2O_3 = 3CaO_{(s)} + Ti_2O_{3(s)}$	192,745 (1873K)		[52]
12	$3CaO \cdot 2TiO_2 = 3CaO_{(s)} + 2TiO_{2(s)}$	148,365	24.14	[52]
13	$CaO \cdot TiO_2 = CaO_{(s)} + TiO_{2(s)}$	74,392	10.13	[53]

The phase equilibrium calculations are based on the minimum ΔG theory, since the elements (such as Ca, Ti, Al, etc.) involved in the present research are of low concentration in molten steel. Therefore, the molten steel can be assumed to be an ideal solution and to follow Henry's law. The component activity coefficient was calculated by using 1% (mass) extremely dilute solution as the standard state. The activity coefficient between the contents of all elements and the interaction coefficient (shown in Table 4) was expressed by Wagner's formula, as shown in Formula (1).

$$\log f_i = \sum_{j=2}^{n} e_i^j [\%j] + \sum_{j=2}^{n} r_i^j [\%j] + \sum_{j=2}^{n}\sum_{k=2}^{n} r_i^{j,k} [\%j][\%k] \quad (1)$$

Table 4. The interaction coefficients of Fe–Al–Ti–Ca–O system molten steel at 1873 K [54–69].

i	j	k	e_i^j	r_i^j	$r_i^{(j,K)}$	i	j	k	e_i^j	r_i^j	$r_i^{(j,K)}$
O	Al	Ca			0	Al	Al	O	0.043	−0.001	−0.028
		Ti	−3.9	−0.01	-		Ca	O	−0.047	0	0
		O			47.45		Ti	-	0.004	-	-
	Ca	Al	−310	−17,984	0		Al	O	−1.98	39.82	−0.028
		O			519,903		Ca				0
	Ti	O	−0.34	0.031	0.026		Al	-	−0.072	0.0007	-
	O	Al	−0.20	0	47.45	Ca	Ca	O	−0.002	-	−90,227
		Ca			520,000		Ti	-	−0.13	-	-
Ti		Al	-	0.0037	-		O	Ca	−580	650,129	−90,056
	Ti	O	0.042	−0.001	0.20						
	O	Ti	−3.4	−0.0355	0.20						

According to the activity of all elements and the free energy change for stable oxides' region transformation, the phase diagrams can be calculated, and the details of the calculation methods are mentioned in other work and my former work [70,71]. The phase diagrams involving Al–Ti–Ca–O system at 1873 K were worked out to estimate potential oxides in molten steel, and are shown in Figure 6. It can be seen from Figure 6a that CaO and CT ($3CaO \cdot 2TiO_2$, $3CaO \cdot Ti_2O_3$, $CaO \cdot TiO_2$) are the main inclusions in the molten steel, as the mass fraction of calcium and titanium increase from $10^{-5}\%$ to $10^{-2}\%$, and $10^{-4}\%$ to 1%, respectively. When 0.005% aluminum is added into the Ca–Ti–O system as shown in Figure 6b, LCA (liquid calcium aluminate) precipitates, which squeezes the region of CT. It is obvious that aluminum content in steel has an important influence on stable phases of inclusions. As the mass fraction of aluminum increases to 0.05%, SCA (solid calcium aluminate) is present, which results in the further reduction of CT region, as shown in Figure 6c. It is noticeable that the stability region of liquid calcium aluminate is located in a common calcium content range. As the compositions of the steel samples in the current work (based on the content of the calcium and titanium) and experimental results introduced from Seo and Zheng [38,39] were marked in the phase diagrams, a good consistency in the above data is conspicuous, as shown in Figure 6b,c. Consequently, the calculated phase diagrams of Al–Ca–Ti–O system are reliable for effectively estimating the evolution behavior of inclusions at steelmaking temperature.

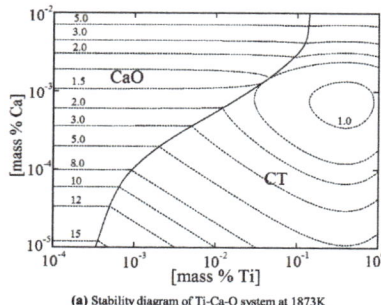
(a) Stability diagram of Ti-Ca-O system at 1873K

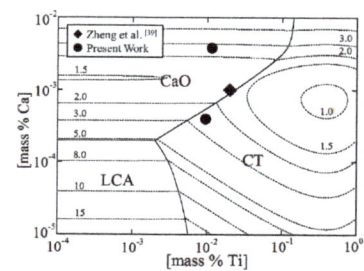
(b) Stability diagram of Al-Ti-Ca-O system at 1873K for [Al]=0.005 pct with experimental data

Figure 6. Cont.

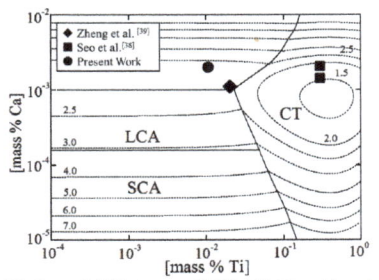

(c) Stability diagram of Al-Ti-Ca-O system at 1873K for [Al]=0.05 pct with experimental data

Figure 6. Calculated diagrams of stable oxides in the Al–Ca–Ti–O system at 1873 K (**a**) [Al] = 0, (**b**) [Al] = 0.005%, (**c**) [Al] = 0.05%.

Some transformation behaviors of precipitates are overlooked in this work, due to the extreme cooling speed during the sampling process. For this reason, the phase transformation of inclusions during solidification of melts was computed by FactSage 7.0 as the FSstel, FactPS, and FToxid databases were employed, and the relevant results are present in Figure 7. The mass fraction of aluminum and calcium varies in the Fe–Al–Ca–0.01Ti–0.005O systems when temperature decreases from 1873 K to 1473 K. It can be seen from Figure 7a that liquid inclusions are present in the steel at a wide temperature range, from 1873 K to about 1623 K, when the mass fractions of aluminum and calcium are both small, around 0.005%, and the transformation process follows liquid inclusions → Al_2O_3 → $2CaO·Ti_2O_3$ → Ti_2O_3 during solidification. Nevertheless, the liquid inclusions only exist around 1873 K as the mass fraction of calcium increases to 0.003% and the formation of solid calcium titanates is favorable, as shown in Figure 7b. As the mass fraction of aluminum increases to 0.03%, while that of calcium is 0.005%, only alumina and calcium aluminates precipitate at 1873 K and no other inclusions are formed during the cooling process, as shown in Figure 7c. However, the liquid phase appears again at a higher temperature as the mass fraction of calcium increases 0.003% in Figure 7d. The liquid inclusions are modified into solid calcium aluminates during the solidification process. When the mass fraction of calcium increases further to 0.007%, the main stable phases become CaO and solid calcium titanates in steel, replacing the calcium aluminates as shown in Figure 7e.

From the found inclusions of Al–Ti–Ca–O system and the above thermodynamic analysis, it is suggested that the calcium treatment technique associated with the right aluminum addition is available to get the liquid phase inclusions at the casting temperature. The calcium addition needs to be reconsidered as the amount of aluminum varies in the steelmaking process.

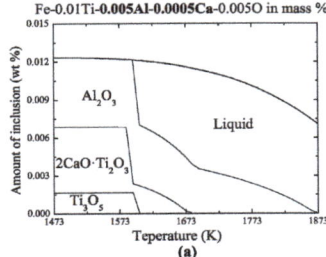
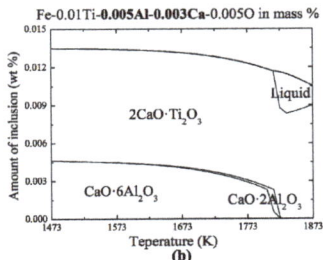

Figure 7. *Cont.*

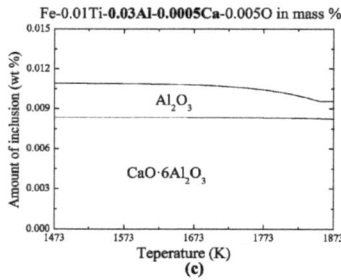
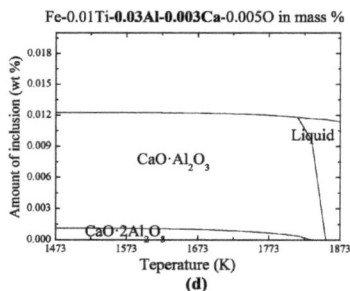

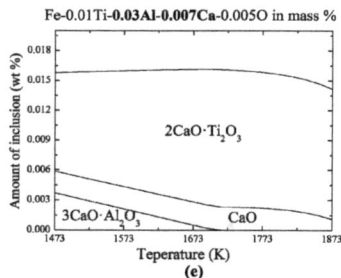

Figure 7. Transformation of inclusions during solidification as different compositions of steel. (**a**) [Al] = 0.005%, [Ca] = 0.0005%; (**b**) [Al] = 0.005%, [Ca] = 0.003%; (**c**) [Al] = 0.03%, [Ca] = 0.0005%; (**d**) [Al] = 0.03%, [Ca] = 0.003%; (**e**) [Al] = 0.03%, [Ca] = 0.007%.

4. Conclusions

The characteristics and transformation behaviors of the inclusions in Ti-containing steel after calcium addition with different aluminum amount have been discussed by physical simulations and thermodynamic analysis at 1873 K (1600 °C), as well as during solidification. The main results are summarized as follows.

The morphologies, chemical compositions, and size distribution of the inclusions are dramatically different before and after calcium treatment, and the calcium addition should be reconsidered according to the mass fraction of aluminum in order to get liquid phase inclusions. The generation of liquid inclusions is more favorable as less calcium addition is needed at the lower amount of aluminum, and as more calcium is appropriate for a higher amount of aluminum. That is, 0.0005% calcium for 0.0036% aluminum, and 0.0025% calcium for 0.0413% aluminum in this study. The inappropriate calcium treatment level can induce the generating trend of solid inclusions in melts. The inclusion-oriented diagrams of Al–Ti–Ca–O system in melts at 1873 K, and the transformation behaviors of inclusions during solidification of steel, were systemically computed involving all types of solid inclusions and liquid phases. The thermodynamic equilibrium calculations are in good agreement with experimental data, and the liquid inclusions can exist during the whole cooling process, as formed at steelmaking temperature.

Author Contributions: Conceptualization, Methodology, T.Z.; Writing—review & editing, Y.L., T.Z. and H.D., Funding Acquisition, T.Z. and Y.L.

Funding: Natural Science Foundation of Chongqing (cstc2018jcyjAX0792) and the Introduce Talents Research Start-up Fund in Central South University of China.

Conflicts of Interest: The authors declare no conflict of interest.

References

1. Liu, H.; Wang, H.; Li, L.; Zheng, J.; Li, Y.; Zeng, X. Investigation of Ti inclusions in wire cord steel. *Ironmak. Steelmak.* **2011**, *38*, 53–58. [CrossRef]
2. Liu, T.; Chen, L.; Bi, H.; Che, X. Effect of Mo on high-temperature fatigue behavior of 15CrNbTi ferritic stainless steel. *Acta Metall. Sin.* **2014**, *27*, 452–456. [CrossRef]
3. Zhang, J.; Cao, Y.; Jiang, G.; Di, H. Effect of annealing temperature on the precipitation behavior and texture evolution in a warm rolled P-containing interstitial-free high strength steel. *Acta Metall. Sin.* **2014**, *27*, 395–400. [CrossRef]
4. Lee, Y.D.; Park, S.H. Effects of titanium and niobium on the weldability of 11% Cr ferritic stainless steel. *J. Korean Inst. Met. Mater.* **1993**, *31*, 984–987.
5. Kim, H.S.; Chang, C.-H.; Lee, H.-G. Evolution of inclusions and resultant microstructural change with Mg addition in Mn/Si/Ti deoxidized steels. *Scr. Mater.* **2005**, *53*, 1253–1258. [CrossRef]
6. Wang, C.; Misra, R.; Shi, M.; Zhang, P.; Wang, Z.; Zhu, F.; Wang, G. Transformation behavior of a Ti–Zr deoxidized steel: Microstructure and toughness of simulated coarse grain heat affected zone. *Mater. Sci. Eng. A* **2014**, *594*, 218–228. [CrossRef]
7. Nakai, K.; Yudate, A.; Kobayashi, S.; Hamada, M.; Komizo, Y. Effects of Ti-based oxide inclusions on formation of intragranular ferrite in steel. *Tetsu-to-Hagané* **2004**, *90*, 141–145. [CrossRef]
8. Kiviö, M.; Holappa, L.; Iung, T. Addition of dispersoid titanium oxide inclusions in steel and their influence on grain refinement. *Metall. Mater. Trans. B* **2010**, *41*, 1194–1204. [CrossRef]
9. Kiviö, M.; Holappa, L. Addition of titanium oxide inclusions into liquid steel to control nonmetallic inclusions. *Metall. Mater. Trans. B* **2012**, *43*, 233–240. [CrossRef]
10. Wang, C.; Verma, N.; Kwon, Y.; Tiekink, W.; Kikuchi, N.; Sridhar, S. A study on the transient inclusion evolution during reoxidation of a Fe-Al-Ti-O melt. *ISIJ Int.* **2011**, *51*, 375–381. [CrossRef]
11. Wang, C.; Nuhfer, N.T.; Sridhar, S. Transient behavior of inclusion chemistry, shape, and structure in Fe-Al-Ti-O melts: Effect of titanium/aluminum ratio. *Metall. Mater. Trans. B* **2009**, *40*, 1022–1034. [CrossRef]
12. Wang, C.; Nuhfer, N.T.; Sridhar, S. Transient behavior of inclusion chemistry, shape, and structure in Fe-Al-Ti-O melts: Effect of titanium source and laboratory deoxidation simulation. *Metall. Mater. Trans. B* **2009**, *40*, 1005–1021. [CrossRef]
13. Zhang, T.; Liu, C.; Jiang, M. Effect of Mg on behavior and particle size of inclusions in Al-Ti deoxidized molten steel. *Metall. Mater. Trans. B* **2016**, *47*, 2253–2262. [CrossRef]
14. Wang, D.; Jiang, M.; Matsuura, H.; Tsukihashi, F. Dynamic evolution of inclusions in Ti-bearing Al-deoxidized molten irons at 1873 K. *Steel Res. Int.* **2014**, *85*, 16–25. [CrossRef]
15. Basu, S.; Choudhary, S.K.; Narendra, U. Nozzle Clogging Behaviour of Ti-bearing Al-killed Ultra Low Carbon Steel. *ISIJ Int.* **2004**, *44*, 1653–1660. [CrossRef]
16. Kawashima, Y.; Nagata, Y.; Shinme, K. Influence of Ti concentration on nozzle clogging on Al-Ti deoxidation: behavior of inclusion on Al-Ti deoxidation-2. *CAMP-ISIJ* **1991**, *4*, 1237–1242.
17. Jung, I.H.; Eriksson, G.; Wu, P.; Pelton, A. Thermodynamic modeling of the Al_2O_3-Ti_2O_3-TiO_2 system and its applications to the FeAl-Ti-O inclusion diagram. *ISIJ Int.* **2009**, *49*, 1290–1297. [CrossRef]
18. Ruby-Meyer, F.; Lehmann, J.; Gaye, H. Thermodynamic analysis of inclusions in Ti-deoxidised steels. *Scand. J. Metall.* **2000**, *29*, 206–212. [CrossRef]
19. Park, D.-C.; Jung, I.-H.; Rhee, P.C.H.; Lee, H.-G. Reoxidation of Al-Ti containing steels by CaO-Al_2O_3-MgO-SiO_2 slag. *ISIJ Int.* **2004**, *44*, 1669–1678. [CrossRef]
20. Doo, W.-C.; Kim, D.-Y.; Kang, S.-C.; Yi, K.-W. Measurement of the 2-dimensional fractal dimensions of alumina clusters formed in an ultra low carbon steel melt during RH process. *ISIJ Int.* **2007**, *13*, 249–258. [CrossRef]
21. Zinngrebe, E.; Van Hoek, C.; Visser, H.; Westendorp, A.; Jung, I.-H. Inclusion population evolution in Ti-alloyed Al-killed steel during secondary steelmaking process. *ISIJ Int.* **2012**, *52*, 52–61. [CrossRef]
22. Van Ende, M.-A.; Guo, M.; Dekkers, R.; Burty, M.; Van Dyck, J.; Jones, P.; Blanpain, B.; Wollants, P. Formation and evolution of Al-Ti oxide inclusions during secondary steel refining. *ISIJ Int.* **2009**, *49*, 1133–1140. [CrossRef]
23. Matsuura, H.; Wang, C.; Wen, G.; Sridhar, S. The transient stages of inclusion evolution during Al and/or Ti additions to molten iron. *ISIJ Int.* **2007**, *47*, 1265–1274. [CrossRef]

24. Seo, M.-D.; Cho, J.-W.; Kim, K.-C.; Kim, S.-H. Evolution of nonmetallic inclusions in ultra low carbon steel after aluminum deoxidization. *ISIJ Int.* **2014**, *54*, 475–481. [CrossRef]
25. Wu, Z.; Zheng, W.; Li, G.; Matsuura, H.; Tsukihashi, F. Effect of inclusions' behavior on the microstructure in Al-Ti deoxidized and magnesium-treated steel with different aluminum contents. *Metall. Mater. Trans. B* **2015**, *46*, 1226–1241. [CrossRef]
26. Wang, C.; Nuhfer, N.T.; Sridhar, S. Transient behavior of inclusion chemistry, shape, and structure in Fe-Al-Ti-O melts: Effect of gradual increase in Ti. *Metall. Mater. Trans. B* **2010**, *41*, 1084–1094. [CrossRef]
27. Itoh, H.; Hino, M. Assessment of Al deoxidation equilibrium in liquid iron. *Tetsu-to-Hagané* **1997**, *83*, 773–778. [CrossRef]
28. Zhang, L.; Thomas, B.G. State of the art in evaluation and control of steel cleanliness. *ISIJ Int.* **2003**, *43*, 271–291. [CrossRef]
29. Park, J.H.; Todoroki, H. Control of MgO Al_2O_3 Spinel Inclusions in stainless steels. *ISIJ Int.* **2010**, *50*, 1333–1346. [CrossRef]
30. Jung, I.H. Overview of the applications of thermodynamic databases to steelmaking processes. *Calphad* **2010**, *34*, 332–362. [CrossRef]
31. Taguchi, K.; Ono-nakazato, H.; Usui, T.; Marukawa, K.; Katogi, K.; Kpsaka, H. Complex deoxidation equilibria of molten iron by aluminum and calcium. *ISIJ Int.* **2005**, *45*, 1572–1576. [CrossRef]
32. Kurayasu, H.; Takayama, T.; Hinotani, S.; Shirota, Y. Phase analysis of Ca-containing inclusions in Ca-treated steels. *Tetsu-to-Hagané* **1996**, *82*, 1017–1022. [CrossRef]
33. Herrera, M.; Castro, F.; Castro, M.; Méndez, M.; Solís, H.; Castellá, A.; Barbaro, M. Modification of Al_2O_3 inclusions in medium carbon aluminium killed steels by AlCaFe additions. *Ironmak. Steelmak.* **2006**, *33*, 45–51. [CrossRef]
34. Blah, C.; Espérance, G.L.; LeHuy, H.; Forget, C. Development of an integrated method for fully characterizing multiphase inclusions and its application to calcium-treated steels. *Mater. Charact.* **1997**, *38*, 25–37.
35. Holappa, L.; Hämäläinen, M.; Liukkonen, M.; Lind, M. Thermodynamic examination of inclusion modification and precipitation from calcium treatment to solidified steel. *Ironmak. Steelmak.* **2003**, *30*, 111–115. [CrossRef]
36. Lind, M.; Holappa, L. Transformation of alumina inclusions by calcium treatment. *Metall. Mater. Tran. B* **2010**, *41*, 359–366. [CrossRef]
37. Zhang, T.; Min, Y.; Liu, C.; Jiang, M. Effect of Mg addition on the evolution of inclusions in Al–Ca deoxidized melts. *ISIJ Int.* **2015**, *55*, 1541–1548. [CrossRef]
38. Seo, C.-W.; Kim, S.-H.; Jo, S.-K.; Suk, M.-O.; Byun, S.-M. Modification and minimization of spinel ($Al_2O_3 \cdot xMgO$) inclusions formed in Ti-added steel melts. *Metall. Mater. Trans. B* **2010**, *41*, 790–797. [CrossRef]
39. Zheng, W.; Wu, Z.; Li, G.; Zhang, Z.; Zhu, C. Effect of Al content on the characteristics of inclusions in Al–Ti complex deoxidized steel with calcium treatment. *ISIJ Int.* **2014**, *54*, 1355–1364. [CrossRef]
40. Mizoguchi, T.; Ueshima, Y. Determination of the Ti_2O_3-CaO-Al_2O_3 phase diagram at steelmaking Temperature. *Tetsu-to-Hagané* **2005**, *91*, 376–382. [CrossRef]
41. Kim, K.-H.; Do, K.-H.; Choi, W.-J.; Lee, S.-B.; Kim, D.-S.; Pak, J.-J. Inclusion modification by Al deoxidation and Ca treatment in Ti containing 18%Cr stainless steel melts. *Korean J. Met. Mater.* **2013**, *51*, 113–118.
42. Zheng, H.; Chen, W. Formation of CaO·TiO_2-MgO·Al_2O_3 dual phase inclusion in Ti stabilized stainless steel. *J. Univ. Sci. Technol. B* **2006**, *13*, 16–20. [CrossRef]
43. Pak, J.-J.; Jo, J.-O.; Kim, S.-I.; Kim, W.-Y.; Chung, T.-I.; Seo, S.-M.; Park, J.-H.; Kim, D.-S. Thermodynamics of titanium and oxygen dissolved in liquid iron equilibrated with titanium oxides. *ISIJ Int.* **2007**, *47*, 16–24. [CrossRef]
44. Cha, W.-Y.; Miki, T.; Sasaki, Y.; Hino, M. Identification of titanium oxide phases equilibrated with liquid Fe-Ti alloy based on EBSD analysis. *ISIJ Int.* **2006**, *46*, 987–995. [CrossRef]
45. Cha, W.-Y.; Nagasaka, T.; Miki, T.; Sasaki, Y.; Hino, M. Equilibrium between titanium and oxygen in liquid Fe-Ti alloy coexisted with titanium oxides at 1873 K. *ISIJ Int.* **2006**, *46*, 996–1005. [CrossRef]
46. Seok, S.-H.; Miki, T.; Hino, M. Equilibrium between Ti and O in Molten Fe-Ni, Fe-Cr and Fe-Cr-Ni Alloys Equilibrated with 'Ti_3O_5' Solid Solution. *ISIJ Int.* **2011**, *51*, 566–572. [CrossRef]
47. Itoh, H.; Hino, M.; Ban-ya, S. Deoxidation equilibrium of calcium in liquid iron. *Tetsu-to-Hagané* **1997**, *83*, 695–700. [CrossRef]

48. Nagata, K.; Tanabe, J.; Goto, K.S. Standard free energies of formation of CaO-Al$_2$O$_3$ intermediate compounds by means of EMF measurement of galvanic cells. *Tetsu-to-Hagané* **1989**, *75*, 2013–2030. [CrossRef]
49. Yang, W.; Zhang, L.; Wang, X.; Ren, Y.; Liu, X.; Shan, Q. Characteristics of inclusions in low carbon Al-killed steel during ladle furnace refining and calcium treatment. *ISIJ Int.* **2013**, *53*, 1401–1410. [CrossRef]
50. Cha, W.-Y.; Miki, T.; Sasaki, Y.; Hino, M. Temperature dependence of Ti deoxidation equilibria of liquid iron in coexistence with 'Ti$_3$O$_5$' and Ti$_2$O$_3$. *ISIJ Int.* **2008**, *48*, 729–738. [CrossRef]
51. Hong, T.; Debroy, T. Time-temperature-transformation diagrams for the growth and dissolution of inclusions in liquid steels. *Scr. Mater.* **2001**, *44*, 847–852. [CrossRef]
52. The Japan Society for the Promotion of Science. *Steelmaking Data Sourcebook*; Gordon and Breach Science Publishers: New York, NY, USA, 1988.
53. Chen, J. *Handbook of Data on Steelmaking*, 2nd ed.; Metallurgical Industry Press: Beijing, China, 2010.
54. Kishi, M.; Inoue, R.; Suito, H. Thermodynamics of oxygen and nitrogen in liquid Fe-20 mass% Cr alloy equilibrated with titania based slags. *ISIJ Int.* **1994**, *34*, 859–867. [CrossRef]
55. Ohta, M.; Morita, K. Thermodynamics of the MnO-Al$_2$O$_3$-TiO$_2$ system. *ISIJ Int.* **1999**, *39*, 1231–1238. [CrossRef]
56. Morioka, Y.; Morita, K.; Tsukihashi, F.; Sano, N. Equilibria between molten steels and inclusions during deoxidation by titanium-manganese alloy. *Tetsu-to-Hagané* **1995**, *81*, 40–45. [CrossRef]
57. Karasev, A.; Suito, H. Analysis of size distributions of primary oxide inclusions in Fe-10 mass Pct Ni-M (M = Si, Ti, Al, Zr, and Ce) alloy. *Metall. Mater. Trans. B* **1999**, *30*, 149–157. [CrossRef]
58. Ohta, M.; Morita, K. Thermodynamics of the Al$_2$O$_3$-SiO$_2$-TiOX oxide system at 1873 K. *ISIJ Int.* **2002**, *42*, 474–481. [CrossRef]
59. Ma, Z.; Janke, D. Characteristics of oxide precipitation and growth during solidification of deoxidized steel. *ISIJ Int.* **1998**, *38*, 46–52. [CrossRef]
60. Cho, S.W.; Suito, H. Assessment of Aluminum-Oxygen equilibrium in liquid iron and activities in CaO-Al$_2$O$_3$-SiO$_2$ slags. *ISIJ Int.* **1994**, *34*, 177–185. [CrossRef]
61. Taguchi, K.; Ono-nakazato, H.; Nakai, D.; Usui, T.; Marukawa, K. Deoxidation and desulfurization equilibria of liquid iron by calcium. *ISIJ Int.* **2003**, *43*, 1705–1709. [CrossRef]
62. Itoh, H.; Hino, M. Thermodynamics on the formation of spinel nonmetallic inclusion in liquid steel. *Metall. Mater. Trans. B* **1997**, *28*, 953–956. [CrossRef]
63. Jo, S.K.; Song, B. Thermodynamics on the formation of spinel (MgO·Al$_2$O$_3$) inclusion in liquid iron containing Chromium. *Metall. Mater. Trans. B* **2002**, *33*, 703–709. [CrossRef]
64. Ishii, F.; Ban-ya, S. Thermodynamics of the deoxidation equilibrium of aluminum in liquid nickel and Nickel-Iron alloys. *ISIJ Int.* **1996**, *36*, 25–31. [CrossRef]
65. Higuchi, Y.; Numata, M. Effect of method of Ca treatment on composition and shape of non-metallic inclusions. *Tetsu-to-Hagané* **1996**, *82*, 671–676. [CrossRef]
66. Ohta, H.; Suito, H. Activities in CaO-MgO-Al$_2$O$_3$ slags and deoxidation equilibria of Al, Mg, and Ca. *ISIJ Int.* **1996**, *36*, 983–990. [CrossRef]
67. Itoh, H.; Hino, M. Thermodynamics on the formation of nonmetallic inclusion of spinel (MgO·Al$_2$O$_3$) in liquid steel. *Tetsu-to-Hagané* **1998**, *84*, 85–90. [CrossRef]
68. Satoh, N.; Taniguchi, T. Prediction of nonmetallic inclusion formation in Fe-40mass% Ni-5mass% Cr alloy production process. *Tetsu-to-Hagané* **2009**, *95*, 827–836. [CrossRef]
69. Cho, S.W.; Suito, H. Assessment of calcium-oxygen equilibrium in liquid iron. *ISIJ Int.* **1994**, *34*, 265–269. [CrossRef]
70. Zhang, T.; Wang, D.; Jiang, M. Effect of Magnesium on the Evolution of Oxide and Sulphide in Liquid Iron at 1873K. *J. Iron Steel Res. Int.* **2014**, *21*, 1073–1080. [CrossRef]
71. Zhang, L.; Ren, Y.; Duan, H.; Yang, W.; Sun, L. Stability Diagram of Mg-Al-O System Inclusions in Molten Steel Effect of Magnesium on the Evolution of Oxide and Sulphide in Liquid Iron at 1873K. *Metall. Mater. Trans. B* **2015**, *46*, 1809–1825. [CrossRef]

© 2019 by the authors. Licensee MDPI, Basel, Switzerland. This article is an open access article distributed under the terms and conditions of the Creative Commons Attribution (CC BY) license (http://creativecommons.org/licenses/by/4.0/).

Article

Mixing Phenomenon and Flow Field in Ladle of RH Process

Kaitian Zhang [1,2], Heng Cui [1,*], Rudong Wang [1] and Yang Liu [2]

1. Collaborative Innovation Center of Steel Technology, University of Science and Technology Beijing, Beijing 100083, China
2. Engineering Research Institute, University of Science and Technology Beijing, No.30 Xueyuan Road, Haidian District, Beijing 100083, China
* Correspondence: cuiheng@ustb.edu.cn; Tel.: +86-136-7123-9796

Received: 14 July 2019; Accepted: 13 August 2019; Published: 14 August 2019

Abstract: Particle image velocimetry (PIV) system was adopted to investigate the relationship between the mixing phenomenon and the flow field of a 210 t RH degasser by a 1:4 scale water model. The results of mixing simulation experiments indicated that the mixing time decreased with the increase of gas blowing rate. However, with the increase of Snorkel immersion depth (SID), the mixing time presented a decreasing rend firstly and then increased. The measurement of flow fields of RH ladle by PIV system can explain the phenomenon above. According to the characteristics of the flow field in RH ladle, the flow field can be divided into the mixing layer, the transition layer, and the inactive layer. On the one hand, the stirring power in RH ladle and vacuum chamber both increases with more gas blowing rate, leading to the decrease of mixing time. On the other hand, with SID increases from 400 mm to 480 mm, the gas blowing depth increase results in the mixing power increases, and the mixing time decreases at the beginning. Because of too much-molten steel in the vacuum chamber and the expanding of the inactive layer in RH ladle, however, the utilization rate of the gas driving force begins to decrease. Therefore, the mixing time started to increases with the increase of SID.

Keywords: RH degasser; physical simulation; mixing time; PIV; flow field

1. Introduction

As the the world economy develops and evolves, the pursuits of the steel products quality are gradually increasing, leading to the improvement requirements on the refining. Besides, the development of continuous casting also requires the matching of high refining efficiency. It has been proven [1–5] that Ruhrstahl–Heraeus (RH) vacuum refining, an important technology to improve product quality, reduce cost, and expand variety, has become the most widely used refining equipment in the world.

Mixing time is an important index to measure the refining efficiency of RH, and it reflects the effect of mixing of the molten steel in the RH degasser [6–10]. To understand and master the relationship between the mixing time and the process parameters is significant to the design and optimization of the RH equipment and refining process.

The chemical compositions of samples taken from different positions in the RH ladle are different. This phenomenon illustrates that the molten steel in the RH ladle is not in a thoroughly mixed state [9]. The compositions of products are quickly out of the range when the steel grades are of narrow composition range. The mixing time can reflect the macrostate of molten steel mixed in the ladle, but it cannot explain the actual situation of molten steel flow, such as the presence or absence of inactive zones in the ladle and where they are, etc. [11,12]. The flow behavior of molten steel plays a decisive role in the refining effect of RH [13]. Therefore, it is needed to study the velocity distribution of RH ladle flow field further. However, RH vacuum refining is not a transparent process with complex reaction and transmission. The most direct and effective research method for RH ladle flow field is the simulation. Through simulation research, rules can be found out from the complicated process to serve the actual production process. There are many studies on RH vacuum refining process by physical [14–16] and numerical [17–20] model. For example, Li et al. [14] studied RH flow and mixing phenomenon and found that mixing time was usually 3 to 4 times of circulation time, and mixing time depends on its positions. The influence of snorkels shape and number on RH mixing time was studied by Ling et al. [15], and the relationship between optimal mixing conditions and numbers of snorkels was obtained. Moreover, He et al. [17] studied the mixing behavior and slag layer behavior under RH bottom blowing process by numerical simulation.

With the development of measurement technology, many advanced devices are used in research to make physical simulations quantifiable. Particle image velocimetry (PIV) is a transient, multi-point, and non-contact method for measuring the velocity of the flow field. It can record the velocity distribution of a large number of spatial points in a moment and provide abundant information on space structure and flow characteristics of the flow field. In this work, a 1:4 scale water RH model was established to simulate the process of refining. The velocity distribution of the flow field in the model was obtained by PIV system to explore the relationship between the flow field structure and the mixing phenomenon.

2. Experiments

2.1. Similarity Principle

According to similarity theory, the geometric dimension of the model and original RH must be kept similar. Moreover, the flow of molten steel in RH degasser is ruled by gravity, viscosity, and inertia, therefore the dimensionless numbers, Reynolds (Re) and fixed Froude (Fr), also should be kept equal.

2.1.1. Geometric Similarity

To understand the characteristic of mixing in the 210 t RH degasser, a 1:4 scale water model was established. Table 1 gives the main parameters of industrial and model RH.

Table 1. Characteristic parameters of industrial and model RH.

Main Parameters	Ladle/mm				Vacuum Chamber/mm		Snorkels/mm		
	Upper Internal Diameter	Lower Internal Diameter	Height	Liquid Level	Internal Diameter	External Diameter	Internal Diameter	Length
Prototype	3884	3222	4060	3300	2138	1294	650	1075
Model	971	805.5	1015	825	534.5	323.5	162.5	268.75

2.1.2. Dynamic Similarity

Because of the Re of model and prototype RH in the second self-modeling region, so the disorder degree and the velocity distribution of fluid no longer change. Therefore, the researchers only need to keep the modified Fr of model and prototype equal in this study [10]. The relationship between model and the actual gas flow rate is shown as follows, which is derived from the modified Fr showing as following:

$$Q_m^o = Q_p^o \sqrt{\left(\frac{1}{\lambda}\right)^5 \frac{\rho_w}{\rho_s} \frac{\rho_{Ar}^o}{\rho_a^o} \frac{P_m}{P_p} \frac{T_p}{T_m}} = K Q_p^o \qquad (1)$$

where, λ are similarity scale, Q_m^o and Q_p^o are the flow rate of the air for model and Argon for prototype in the standard state, respectively, L·min^{-1}; ρ_w, ρ_s, $\rho_{Ar'}^o$ and ρ_a^o are the density of water, molten steel, standard Argon, and standard air respectively, kg·m^{-3}; P_m and P_p is pressure of the air for model and Argon for prototype at outlet, respectively, Pa; T_m and T_p is temperature of the air for model and Argon for prototype at outlet, respectively, K. As the pressure at the air outlet changes with the immersion depth, different immersion depths correspond to different gas conversion coefficient K, and it is obtained by substituting relevant parameters into the above equation showing in Table 2. Furthermore, all results by simulation experiments in this work were converted to prototype data by similarity coefficient for display.

Table 2. The conversion of gas flow rate between prototype and mold with different Snorkel immersion depths (SIDs).

SID (mm)		Outlet Pressure (Pa)		Gas Flow Rate (L/min)		K
Prototype	Model	Prototype	Model	Prototype	Model	
400	100	92,407	101,006.5	1000	36.67	0.03667
480	120	97,895	101,202.5	1200	42.79	0.03566
560	140	103,383	101,398.5	1600	55.57	0.03473
640	160	108,871	101,594.5	2000	67.76	0.03388
720	180	114,359	101,790.5	2400	79.42	0.03309

2.2. Experimental Apparatus and Method

2.2.1. Experiment of Mixing

The experimental devices of mixing are showing in Figure 1. A cube tank cover placed on the outside of the RH model. During the experiment, water fills up between the tank and RH model to prevent refraction of light, which affects PIV observation, from the cylinder wall of the RH model. A vacuum pump is used to keep the low air pressure in the vacuum chamber. Besides, Argon is blown out at the up-leg from a gas supply system to adjust the influence of gas flow rate on the flow field in RH. The mixing time transformed from electrical conductivity measured by a computer.

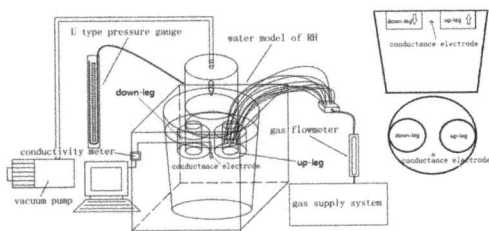

Figure 1. Schematic of the experimental apparatus.

When the devices stably, 200 mL of saturated sodium chloride solution was poured instantaneously into the vacuum chamber as a tracer. Then the conductivity instrument was applied to measure the changes in water solution conductivity. It is the mixing time when the conductivity change does not exceed the stable value plus or minus 5%. The sampling time was 120 s, and the sampling frequency was 25 Hz. Each test repeated three times, and the average mixing time was the final result of the experiment, to obtain accurate results. Moreover, the position of the conductance electrode usually is the sampling location in the actual production process. Therefore, this position selected as the measurement point.

2.2.2. PIV System

The experimental devices of PIV are showing in Figure 2. Amount of tiny hollow glass particles of 50 microns in diameter poured into the ladle as tracer particles. The tiny particles can follow the fluid and reflect the velocity of flow in the RH model. Then the dual-cavity laser shined on the ladle and, meanwhile, the CCD cameras take 100 photos in 50 s continuously. The PIV image data processing system calculated the average velocity of the 100 photos the, and it is the flow velocity distribution of the fluid in the ladle. Besides, the double frame mode was adopted to record the flow field from two angles, thus forming a cross record angle and achieving better accuracy.

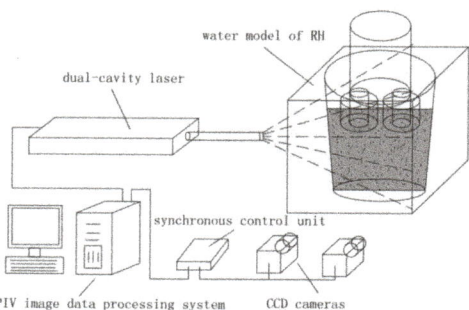

Figure 2. Particle image velocimetry (PIV) velocity measurement system of RH.

3. Results

3.1. Effect of Gas Flow Rate and Snorkel Immersion Depth on Mixing Time

Figure 3 shows the variation of mixing time against gas flow rate at five different SIDs. When the SID is 400 mm, the mixing time significantly decreases with the increase of gas flow rate. Moreover, the decrease in mixing time becomes smaller and smaller. For the case of 480 mm and 560 mm SID, the mixing time decreases fast when the gas flow rate is less than 1200 L/min; meanwhile, the mixing time decreases slowly when the gas flow rate is higher than 1200 L/min. For the case of 640 mm and 720 mm SID, the mixing time decreases slowly with the increase of gas flow rate. Under different gas flow rates, the mixing time decreases first and then increases with the increase of SID. The mixing time of each gas flow rate in the 720 mm SID is the maximum and that for the case of 480 mm SID is the minimum.

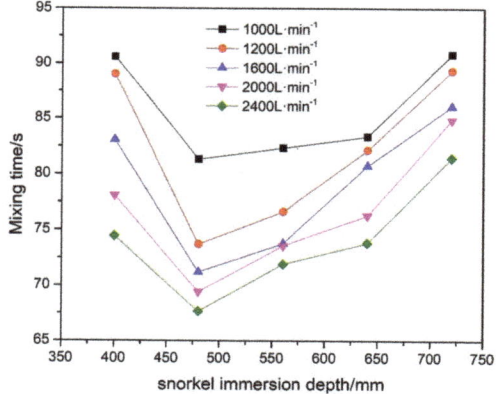

Figure 3. The effect of gas flow rate and SID on mixing time.

3.2. Characteristics of RH Ladle Flow Field

In this work, the flow field of RH ladle with different SIDs and gas flow rate were carried out by the PIV system. Take the 560 mm SID and the gas flow rate of 1600 L/min as an example for analysis. Figure 4 is the result of the PIV system of different sections of the RH ladle flow field.

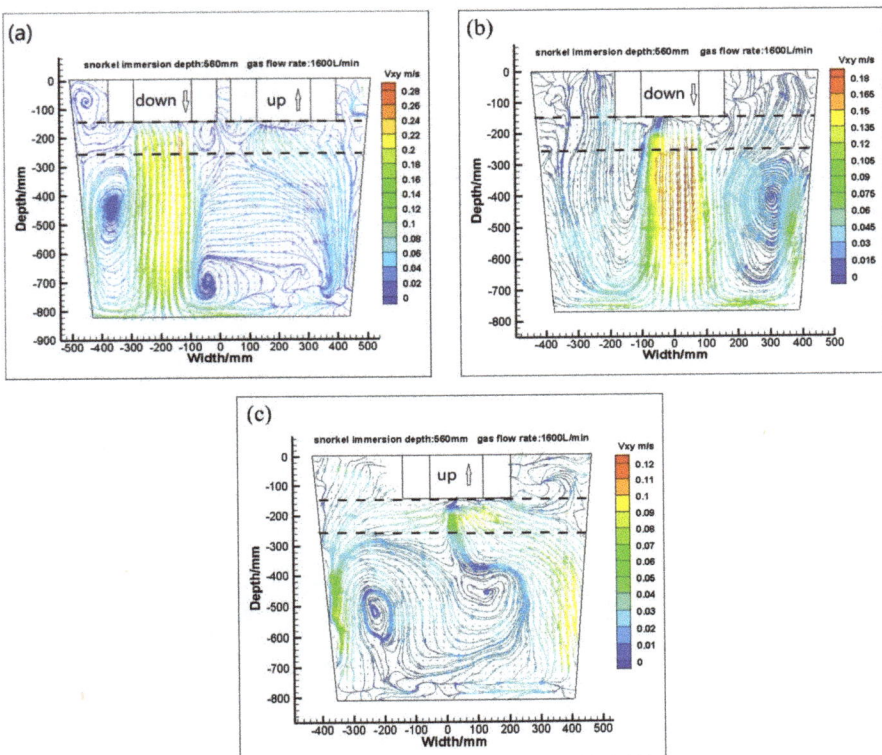

Figure 4. RH flow field by PIV system: (**a**) section of up-leg and down-leg; (**b**) section of down-leg; (**c**) section of up-leg.

The flow field of the section of up-leg and down-leg is showing in Figure 4a. In this section, the flow pattern mainly composed of the circulation flow from the down-leg to the up-leg and the return flow between the downward flow and the ladle wall. The velocity of the circulation flow and return flow is greater than 0.05 m/s. Among them, the velocity of downward flow is greater than 0.1 m/s and the maximum velocity located at the outlet of the down-leg for 0.26 m/s. There are two other return flows, which are located between the downward flow and the upward flow and between the down-leg and the wall. The velocity of the two return flows is less than 0.05 m/s. The flow velocity of the areas between the two snorkels and between the snorkel and the wall of the ladle is less than 0.01 m/s, which identified as the inactive regions.

Figure 4b shows the flow field of the section of the down-leg. The molten steel forms return flows between the downward flow and the wall. Besides, the velocity of downward flow is greater than 0.09 m/s, and the velocity of return flows between 0.01 m/s, and 0.09 m/s. The areas between the downward flow and the wall are the inactive regions, and their velocity is less than 0.01 m/s.

Figure 4c is the flow field of the section of the up-leg. The molten steel climb along the wall. Among them, a part of the flow into the up-leg and the velocity is greater than 0.03 m/s; the other part flows down and forms a return flow with the velocity of between 0.01~0.03 m/s. Furthermore, the velocity of the right-side return flow is slightly larger than the left side.

Synthesizing the three flow fields mentioned above, it can be found that the flow of molten steel in RH ladle is mainly the circulation flow from the down-leg to the up-leg and the return flow around the downward flow. The inactive region is mainly within the range of snorkel immersion depth. According to the characteristics of the flow field in RH ladle, the flow field can be divided into three different layers, as shown by the dotted line in Figure 4. From the bottom of the ladle to the upper boundary of the recirculation zone is the mixing layer. This layer is the main area of the mixing of the molten steel in the ladle because the circulation flow from the down-leg to the up-leg and the return flow around the downward flow greatly promote the mixing effect. Above the mixing layer and below the snorkel bottom is the transition layer for the exchange of molten steel between the vacuum chamber and the ladle in this layer, while the flow of the molten steel flowing out (in) the snorkel is almost no material and energy exchange because the flow is fast and consistent. The inactive layer is from the snorkel bottom to the surface of molten steel with a low velocity of disorderly flow. Because the snorkels divided this layer into many scattered small areas, the exchange of material and energy within the layer is blocked.

3.3. RH Ladle Flow Field under Different Gas Flow Rate

Figure 5 is the flow field of RH ladle at 560 mm SID and different gas flow rates by the PIV system. With the increase of gas flow rate, the structure of RH ladle flow field is consistent, and the maximum velocity of the downward flow is 0.248 m/s, 0.251 m/s, 0.254 m/s, and 0.257 m/s, respectively.

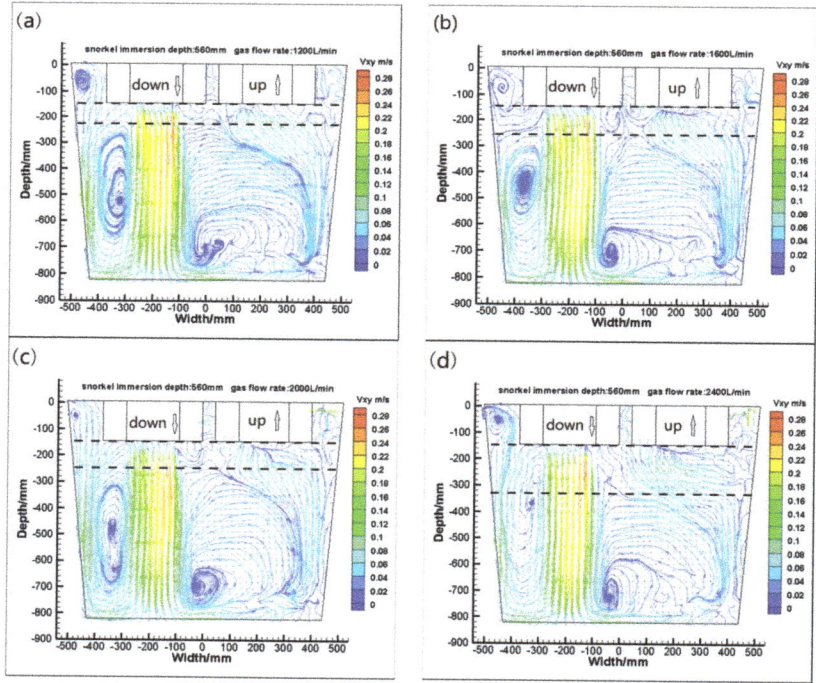

Figure 5. RH ladle flow field under different gas flow rates by PIV system: (**a**) 1200 L/min; (**b**) 1600 L/min; (**c**) 2000 L/min; (**d**) 2400 L/min.

Figure 6 is the average of vertical velocity in the transition layer with different gas flow rate conditions. The vertical velocity on the left side of the downward flow is higher than zero, which means the molten steel flows upwards. The velocity of the downward flow is less than zero, which means the molten steel flows downward. Besides, the flow velocities with different gas flow rates are relatively consistent on the edge of the downward, but there is a separation phenomenon in the middle part. Maximum velocity of each operating condition does not locate in the center of the downward flow but the center-right. The vertical velocity in the middle of two snorkels is less than zero. Due to the pumping action of the up-leg to the molten steel, the flow is faster near the up-leg. Below the up-leg, the velocity curves in each operating condition relatively scatter, which indicates that the primary distinction of the molten steel flow on the transition layer results from the pumping action of the up-leg.

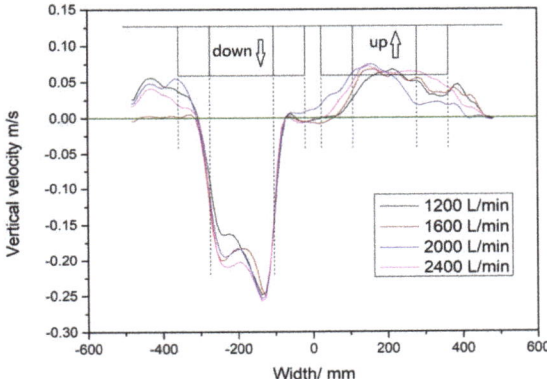

Figure 6. Vertical velocity distribution of flow field in transition layer under different gas flow rates.

Figure 7 is the average of vertical velocity in the mixing layer with different gas flow rates. With different gas flow rates, the velocity curves of the downward flow more comparatively disperse than that of in the transition layer. Besides, the maximum flow velocity of the mixing layer is less than the transition layer, and the maximum velocity is on the center of the downward flow. These phenomena above indicate that there is an oscillation in the process of the downward flow.

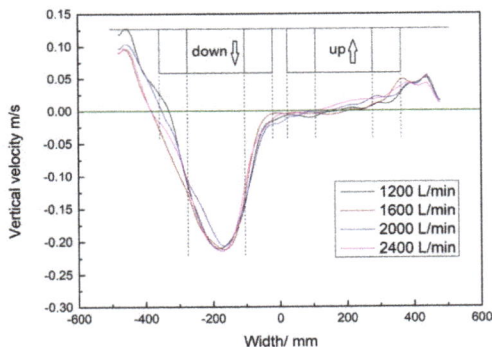

Figure 7. Vertical velocity distribution of flow field in mixing layer under different gas flow rates.

3.4. RH Ladle Flow Field under Different SIDs

Figure 8 shows the flow fields of the RH ladle with the gas flow rate of 1600 L/min and the SID changing from 400 to 720 mm. With the increase of the SID, the patterns of RH ladle flow field are almost consistent, and the maximum velocity of the downward flow are 0.22 m/s, 0.239 m/s, 0.251 m/s, 0.273 m/s and 0.294 m/s respectively. Besides, the inactive layer and the transition layer become thicker with the increase of SID.

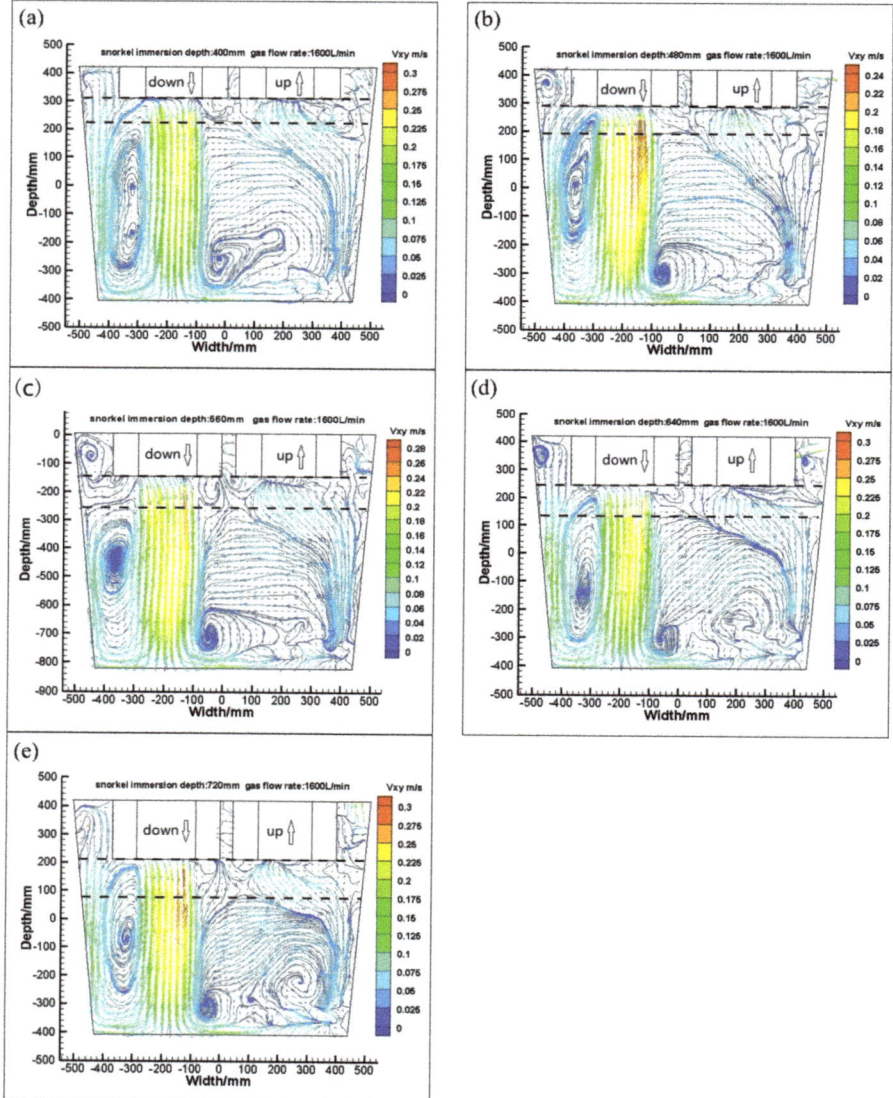

Figure 8. RH ladle flow field with different SIDs: (**a**) 400 mm; (**b**) 480 mm; (**c**) 560 mm; (**d**) 640 mm; (**e**) 720 mm.

Figure 9 is the average of vertical velocity in the transition layer with different SIDs. It shows that the velocity curves on the downward flow left are quite dispersive with different SIDs. The velocity is close to zero when the SIDs is 560 mm, and it increases with the SIDs increasing or decreasing. On the edge of the downward flow, the velocities of different SIDs are relatively coincident. Same as the left part, there is a separation phenomenon in the middle part. The maximum velocity is not located in the center of the downward flow either but on the center-right. The maximum velocity is increasing with the SIDs increasing.

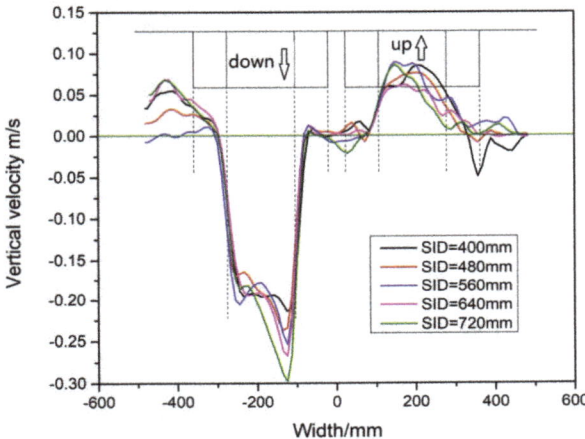

Figure 9. Vertical velocity distribution of flow field in transition layer with different SIDs.

Figure 10 is the average of vertical velocity in the mixing layer with different SIDs. It can be found that the curves are more dispersed than that of the transition layer, while the maximum velocity in the mixing layer is slower than in the transition layer. In addition, the maximum velocity is increasing with the SIDs increasing and the position is closed to the center of the downward flow. On the right of up-leg, near to snorkel wall, the vertical velocity is greater than zero, which means the molten steel under up-leg flows upward along the snorkel wall in the mixing layer.

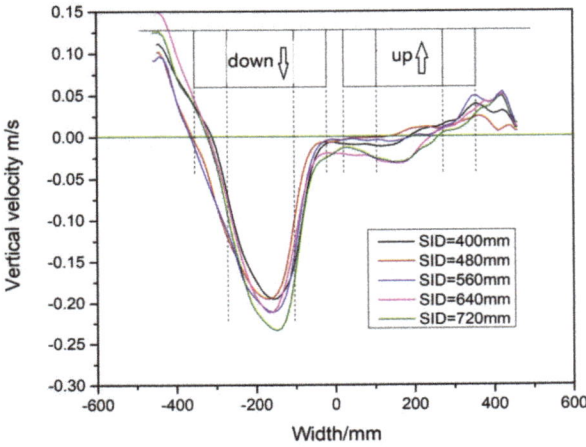

Figure 10. Vertical velocity distribution of flow field in mixing layer with different SIDs.

4. Discussion

An essential function of refining technology is to stir the molten steel well, to make the temperature and composition of molten steel uniform, and to promote the refining reaction to proceed smoothly. A critical index to judge the stirring effect is the mixing time, which is directly related to the homogenization of molten steel composition and temperature. Previous studies [6–10] have shown that mixing time is inversely proportional to the power of stirring power. In fluid mechanics, stirring power, also known as turbulent kinetic energy dissipation rate, can better analyze RH refining process in which turbulent diffusion plays a significant role in mass transfer. Reference [9] shows that the stirring power in RH device consists of two parts. One part is stirring of molten steel in ladle caused by liquid flow in down-leg, the other part is stirring caused by liquid flow in up-leg in the vacuum chamber. They can be represented by equations respectively as follows:

$$\varepsilon_M = \frac{0.5Qv^2}{W} = \frac{0.375Q^3}{d^3 W} \tag{2}$$

$$\varepsilon_v = \frac{6.18GT_1}{W}\left[\ln\left(1+\frac{\rho g h}{p}\right)+\left(1-\frac{T_0}{T_1}\right)\right] \tag{3}$$

where, ε_M and ε_v are the stirring power in RH ladle and vacuum chamber respectively; W/t; Q is the circulation flow, L/min; v is the flow velocity of down-leg, m/s; W is the mass of circulating fluid, t; d is the inner diameter of the down-leg, m; G is the gas blowing flow rate, L/min; T_0 and T_1 are molten steel temperature and room temperature, K; ρ is the density of liquid steel, kg/m^3; g is the acceleration of gravity, m/s^2; h is the gas blowing depth, m; p is the vacuum chamber pressure, Pa. Therefore, in this work, the main factors affecting RH ladle stirring power are circulation flow Q, and the main factors affecting vacuum chamber stirring power are gas blowing volume G and gas blowing depth h.

When the gas blowing rate is increased from 1200 L/min to 2400 L/min, on the one hand, the driving force of gas on the molten steel in the up-leg enhances, and the maximum velocity in RH ladle increases from 0.248 m/s to 0.257 m/s, indicating that the circulation flow in the whole RH ladle increases. As a result, the stirring power in RH ladle increases. On the other hand, when the blowing rate G increases, and the blowing depth h is unchanged, the stirring power in the vacuum chamber increases directly. Therefore, ε_M and ε_M both increase with the increase of blowing rate, leading to a decrease of mixing time in Figure 2.

When the SID increases from 400 mm to 720 mm, on the one hand, the gas blowing depth h increase, leading to the utilization rate of the driving force, generated by the gas, increases in the process of driving the liquid steel circulation. Therefore, the flow velocity of both the transition layer and the mixing layer increases, which means that the circulation flow Q increases, the mixing power increases, and the mixing time decreases at the beginning. On the other hand, as the SID continues to increase, the mass of liquid steel in the vacuum chamber is too much. The gas flow rate is constant, the driving force generated by the gas begins to decrease in the utilization rate of the driving force, result in a decrease of ε_v. Meanwhile, as showing in Figure 8, the inactive layer of the flow field expands with the increases of SID, leading to a reduction in mixing effect here. Therefore, when the SID increases to 480 mm, the mixing time is the minimum. After that, the mixing time increases with the increase of SID. These phenomena indicate that in addition to the stirring power, the flow field structure also has a significant impact on the mixing process in the RH ladle.

5. Conclusions

(1) At different SIDs, the mixing time decreases with the increase of gas flow rate. Furthermore, the decrease in mixing time becomes smaller and smaller. Under different gas flow rates, the mixing time decreases first and then increases with the increase of SID.

(2) During RH refining process, the flow of molten steel in the ladle is mainly the circulation flow from the down-leg to the up-leg and the return flow around the downward flow. The inactive

region is mainly within the range of snorkel immersion depth. According to the characteristics of the flow field in RH ladle, the flow field can be divided into the mixing layer, the transition layer, and the inactive layer.

(3) The stirring power and flow field structure have a great impact on the mixing process in the RH ladle. With the increase of gas blowing rate, the circulation flow in the whole RH ladle increases. The stirring power in RH ladle and vacuum chamber both increases, leading to the decrease of mixing time. With SID increases from 400 mm to 480 mm, the gas blowing depth increase. So, the circulation flow increases, the mixing power increases, and the mixing time decrease at the beginning. However, as the SID continues to increase, the utilization rate of the gas driving force begins to decrease because of too-much molten steel. Meanwhile, the inactive layer of the flow field expands with the increases of SID. Therefore, the mixing time increases with the increase of SID.

Author Contributions: Investigation, Y.L.; Project administration, H.C.; Writing – original draft, K.Z. and H.C.; Writing – review & editing, K.Z., H.C. and R.W.

Funding: This research was funded by the National Natural Science Foundation of China (No. U1860106).

Conflicts of Interest: The authors declare no competing interests.

References

1. Ajmani, S.K.; Dash, S.K.; Chandra, S.; Bhanu, C. Mixing evaluation in the RH process using mathematical modelling. *ISIJ Int.* **2004**, *44*, 82–89.
2. Yamaguchi, K.; Kishimoto, Y.; Sakuraya, T.; Fujii, T.; Aratani, M.; Nishikawa, H. Effect of refining conditions for ultra low carbon steel on decarburization reaction in RH degasser. *ISIJ Int.* **1992**, *32*, 126–135.
3. Geng, D.Q.; Lei, H.; He, J.C. Effect of traveling magnetic field on flow, mixing, decarburization and inclusion removal during RH refining process. *ISIJ Int.* **2012**, *52*, 1036–1044. [CrossRef]
4. Kato, Y.; Nakato, H.; Fujii, T.; Ohmiya, S.; Takatori, S. Fluid flow in ladle and its effects on decarburization rate in RH degasser. *ISIJ Int.* **1993**, *33*, 1088–1094.
5. Ai, X.G.; Bao, Y.P.; Jiang, W.; Liu, J.H.; Li, P.H.; Li, T.Q. Periodic flow characteristics during RH vacuum circulation refining. *Int. J. Miner. Metall. Mater.* **2010**, *17*, 17–21. [CrossRef]
6. Mukherjee, D.; Shukla, A.K.; Senk, D.G. Cold Model-Based Investigations to Study the Effects of Operational and Nonoperational Parameters on the Ruhrstahl–Heraeus Degassing Process. *MMTB* **2017**, *48*, 763–771. [CrossRef]
7. Lin, L.; Bao, Y.P.; Yue, F.; Zhang, L.Q.; Qu, H.L. Physical model of fluid characteristics in RH-TOP vacuum refining process. *Int. J. Miner. Metall. Mater.* **2012**, *19*, 483–489. [CrossRef]
8. Nakanishi, K.; Szekely, J.; Chang, C. Experiments and theoretical investigation of mixing phenomena in the RH-vacuum process. *Ironmak. Steelmak.* **1975**, *2*, 115–124.
9. Zhang, L.F.; LI, F. Investigation on the fluid flow and Mixing phenomena in a Ruhrstahl-Hereaus (RH) steel degasser using physical modeling. *JOM* **2014**, *66*, 1227–1240. [CrossRef]
10. Niu, D.L.; Ma, D.G.; Yang, J.; Xu, M.R.; Wang, Z. Characteristics of the dead zones in RH desgasser with simulation methods. *Mater. Sci. Forum* **2015**, *817*, 755–763. [CrossRef]
11. Chen, G.J.; He, S.P. Mixing behavior in the RH degasser with bottom gas injection. *Vacuum* **2016**, *130*, 48–55. [CrossRef]
12. Kishan, P.A.; Dash, S.K. Mixing time in RH ladle with upleg size and immersion depth: A new correlation. *ISIJ Int.* **2007**, *47*, 1549–1551. [CrossRef]
13. Zhu, B.H.; Chattopadhyay, K.; Hu, X.P.; Zhang, B.; Liu, Q.C. Optimization of sampling location in the ladle during RH vacuum refining process. *Vacuum* **2018**, *152*, 30–39. [CrossRef]
14. Li, F.; Zhang, L.F.; Liu, Y.; Li, Y.L. Study of Mixing Phenomena during RH Refining Using Water Modeling. *TMS (The Minerals, Metals & Materials Society)* **2014**, *57*, 459–466.
15. Ling, H.T.; Guo, C.B.; Conejo, A.N.; Li, F.; Zhang, L.F. Effect of snorkel shape and number of nozzles on mixing phenomena in the RH process by physical modeling. *Metall. Res. Technol.* **2017**, *114*, 111–123. [CrossRef]
16. Pieprzyca, J.; Merder, T.; Saternus, M.; Michalek, K. Physical modelling of the steel flow in RH apparatus. *Arch. Metall. Mater.* **2015**, *60*, 1861–1864. [CrossRef]

17. He, S.; Chen, G.; Guo, C. Investigation of mixing and slag layer behaviours in the RH degasser with bottom gas injection by using the VOF–DPM coupled model. *Ironmak. Steelmak.* **2017**, *44*, 1–6. [CrossRef]
18. Geng, D.Q.; Lei, H.; He, J.C. Numerical simulation of the multiphase flow in the Rheinsahl-Heraeus (RH) system. *MMTB* **2009**, *41*, 234–247. [CrossRef]
19. Tsujino, R.; Nakashima, J.; Hirai, M.; Sawada, I. Numerical analysis of molten steel flow in ladle of RH process. *ISIJ Int.* **1989**, *29*, 589–595.
20. Liu, C.; Duan, H.J.; Zhang, L.F. Modeling of the melting of aluminum particles during the RH refining process. *Metals* **2019**, *9*, 442. [CrossRef]

© 2019 by the authors. Licensee MDPI, Basel, Switzerland. This article is an open access article distributed under the terms and conditions of the Creative Commons Attribution (CC BY) license (http://creativecommons.org/licenses/by/4.0/).

Article

Geometric Modification of the Tundish Impact Point

Branislav Buľko [1],*, Ivan Priesol [2], Peter Demeter [1], Peter Gašparovič [3], Dana Baricová [1] and Martina Hrubovčáková [1]

[1] Faculty of Metallurgy, Technical University of Košice, 04200 Košice, Slovakia; peter.demeter@tuke.sk (P.D.); dana.baricova@tuke.sk (D.B.); martina.hrubovcakova@tuke.sk (M.H.)
[2] IPC REFRACTORIES s.r.o., Magnezitárska 11, 04013 Košice, Slovakia; ipc@ipc.sk
[3] Faculty of Aeronautics, Technical University of Košice, 04200 Košice, Slovakia; peter.gasparovic@tuke.sk
* Correspondence: branislav.bulko@tuke.sk; Tel.: +421-55-602-3169

Received: 28 September 2018; Accepted: 8 November 2018; Published: 14 November 2018

Abstract: In connection with the increasing requirements for cleanliness in conticast steel, it is necessary to develop original solutions. The tundish, as the last refractory-lined reactor, gives enough space to remove inclusions by optimizing the flow of steel. The basic component of the tundish is the impact pad, the shape of which creates a suitable flow of steel, thus making it part of the tundish metallurgy. The optimal steel flow in the tundish must avoid creating dead zone areas, or the slag "eye" phenomenon in the slag layer around the ladle shroud, and is intended to create conditions for the release of inclusions by promoting reactions at the steel-slag phase interface. The flow also has to prevent excessive erosion of the tundish refractory lining. This paper compares the standard impact pad with the "Spheric" spherical impact pad using computional fluid dynamiscs (CFD) tools and physical modelling. The evaluation criteria are residence time and flow in the tundish at three different casting speeds.

Keywords: continuous casting; tundish; residence time; computational fluid dynamiscs (CFD)

1. Introduction

Current trends show that more than 96% of the steel produced in the world is processed by continuous casting [1]. In view of this, there is a naturally increasing pressure on producers of refractory materials used in the continuous casting process. A key part of the continuous casting plant is the tundish, which can significantly affect steel cleanliness. In connection with the constantly increasing ratio of high-grade steel in the product portfolio, development in the field of tundish metallurgy is essential. A fully operational tundish is chosen in terms of covering and refining powders and the proper slag regime. The basic requirement for a properly functioning slag system is the controlled flow of steel in the tundish so that inclusions can be released from the steel into the slag and chemical reactions have good conditions to run at the steel–slag phase interface [2]. From this perspective, the most important criterion is the geometrical adjustment of the steel impact point in the tundish. In practice, this is solved by the use of an impact pad, which has the role of reducing the erosion of the bottom of the tundish refractory lining [3–5]. Swirl flow at the point of impact is due to the high kinetic energy of the incoming steel. The low momentum of diffusivity of the input steel causes a relatively slow transfer of fluid from the input stream with high kinetic energy to the surrounding liquid steel. In the case of a suitably shaped impact pad, a so-called "piston flow" area is created. One of the main indicators of the quality of the flow adjustment in the tundish is the residence time, which is defined as the duration of stay of steel particles in the tundish [6]. The longer the residence time, the more time inclusions have to flow from the steel into the slag.

In recent years, impact pads have undergone considerable development, especially in terms of their design, changing from simple pads through ribbed pads to the most sophisticated shapes that use the latest knowledge from mathematical and physical modelling as well.

As mentioned above, the impact pad is one of the key parts of the tundish furniture affecting the flow of liquid steel. It is mostly used with suitably selected dams, weirs, and baffles, which can significantly prolong the residence time of steel in the tundish [7–9]. In order to accurately compare the properties of the spherical impact pad with those of the standard impact pad, this article contains the results of the comparison of these impact pads without the use of other flow modifiers.

The aim of this research was to point out a new, innovative solution for the impact pad using a convex hemispherical shape. In the case of a symmetrical two-strand boat-type tundish, a more advantageous character of steel flow is assumed using a spherical impact pad.

2. Experimental Materials and Methods

2.1. The "Spheric" Impact Pad

The shape of the spherical impact pad was developed in order to decrease the hydrodynamic drag force of the impinging stream of molten steel. Dimensional analysis of the drag force F provides the dependence.

$$F = \frac{1}{2} C \cdot \rho \cdot S \cdot v \qquad (1)$$

where: C—coefficient of drag, ρ—specific mass of fluid, S—size of the reference area (planform area of the pad), and v—references the velocity of impinging stream.

The coefficient C expresses the influence of the shape of the pad on the drag force. The coefficient C is a dimensionless parameter that can be assumed constant for small changes in velocity. The experimental values of the coefficient of drag of objects in a free stream are 1.17 for the square flat plate and 0.40 for the convex hemisphere [10]. The proposed spherical pad has a square planform and the upper surface shape of a large-radius hemisphere.

The shape of the spherical impact pad should, in comparison with the standard impact pad, cause less erosion of the pad surface. Smaller deflection of the stream should reduce the creation of large intensive vortices at the surface of the fluid level. Short-path flow should be suppressed by more intensive mixing at the core of the fluid volume.

Physical modelling is done in a scaled-down model of the tundish at the scale 1:3, made of transparent plastic (PMMA), with water as the fluid medium. The description of the physical model and the experimental method is given in [7,11].

The dimensions of the impact plate were calculated with respect to the scale of the tundish at the ratio 1:3, with the height of the impact plate (Figure 1) set at 9.96 mm due to its position on the bottom of the real tundish (Figure 2). The flow of steel in the tundish equipped with the "Spheric" impact pad is optimized not only for the residence time but also for the nature of the flow, so that this flow promotes the removal of inclusions into the slag and the best conditions for the slag-metal phase interface. This method of modifying the flow is given in [12,13].

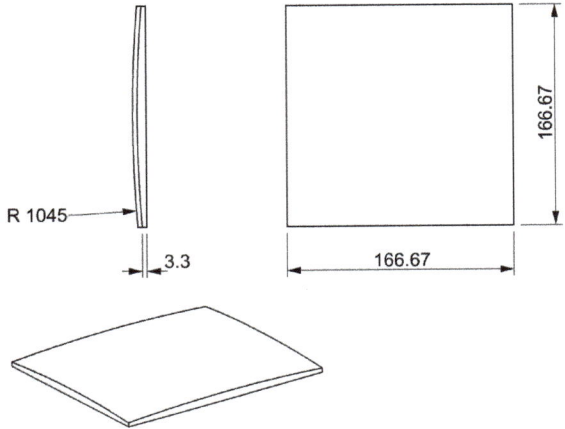

Figure 1. The "Spheric" impact pad.

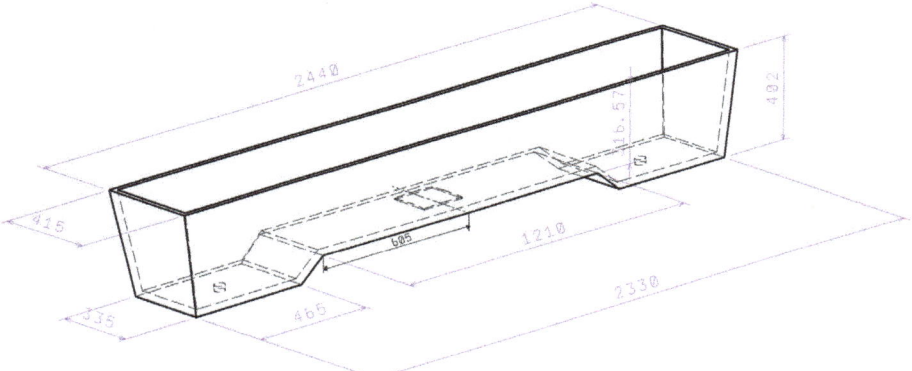

Figure 2. Position of the "Spheric" impact pad in the tundish.

2.2. *The Standard Impact Pad*

The dimensions and position of the standard impact pad in the tundish are shown in Figure 3.

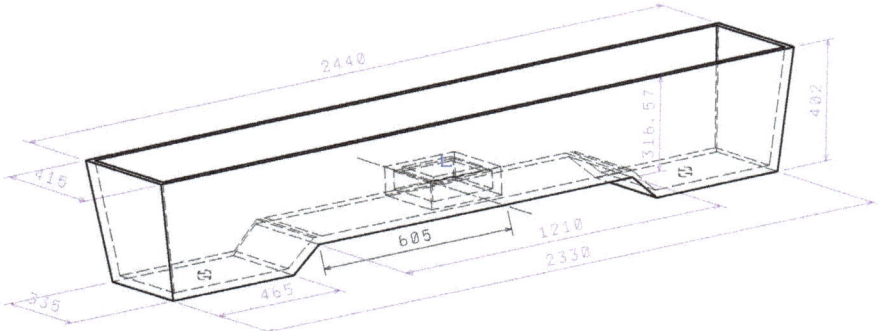

Figure 3. Position of the standard impact pad.

The measurements were performed in steady-state conditions, so the steel level in the tundish was constant and the amount of steel flowing into the tundish was equal to the amount of steel flowing out from the tundish into the molds.

The C-curve method was used to define the characteristics of the steel flow in the tundish under constant (steady) casting conditions [14–16]. Upon reaching the desired level in the mold and stabilizing the casting flow rate, a measured amount of aqueous KCl solution was injected into the ladle shroud. In the ladle shroud and in the submerged entry nozzles, conductivity probes were mounted to measure the change in conductivity of the water due to the added salt, thus obtaining the C-curve [17]. From this curve, we could determine the minimum residence time τ_{min}, which is the minimum time that the impulse of the tracer injected into the ladle shroud ($\tau_0 = 0$ s) appears on the probe located in the submerged entry nozzle. Minimum residence time has a significant effect on the duration of flow of inclusions from the steel into the slag [18]. The maximum residence time τ_{max} is the time between t_0 and the time of maximum measured concentration of the tracer in the output of the tundish. The maximum residence time refers to the time taken to reach the maximum concentration of the trace element at the output of the tundish [19,20].

Measurements were performed for the tested configurations at flows corresponding to casting speeds of 0.8 m·min^{-1}, 1.2 m·min^{-1}, and 1.6 m·min^{-1} on a real continuous-casting machine. The length of the ladle shroud on the model corresponded to a real length of 1700 mm. In both configurations the same ladle shroud was used. The distance of the nose of the ladle shroud from the tundish bottom was 203 mm. Thus, when using the standard impact pad it was 183 mm and 194 mm when the spherical impact pad was used.

Each configuration was simulated three times for more accurate statistical evaluation and comparison of the results, and these measurements were then used in each configuration for calculating the mean values reported in the results and graphs. More detailed specifications of each simulated configuration and specific results are presented below.

3. Results and Discussion

The initial idea of the "Spheric" impact pad was verified using CFD simulation tools [21,22], (Figures 4 and 5).

Numerical simulation of the flow was computed in the ANSYS Fluent v19.2, the software produced by ANSYS, Canonsburg, PA, USA. ANSYS Fluent computes discrete values of time-dependent Navier–Stokes equations, that is conservation equations of momentum in x-, y-, and z-directions, and the conservation equation of mass. Time-dependent details of turbulent eddies are removed from the Navier–Stokes equations with Reynolds averaging, and the effect of turbulent motion on transport of momentum in the averaged flow is assumed by means of the Boussinesq hypothesis which defines turbulent viscosity. Turbulent viscosity increases the basic, molecular viscosity of the fluid. Computation of turbulent viscosity requires additional equations which are based on the k-omega SST model (Shear Stress Transport, the Menter's variant of the k-omega model). Change in concentration of the solution in the water is modelled using species transport equations. The solver of the Ansys Fluent is pressure based, so the velocity is obtained from the momentum equation and the pressure is obtained from the pressure equation, which is derived from the continuity equation and the momentum equation. The equations are discretised using the control volume method (CVM). The volume of the domain is divided into discrete control volumes using a computational grid, and the equations are integrated on these volumes and linearized to create algebraic equations for unknown values of velocity, pressure and species fraction. The resulting system of equations has a sparse matrix of coefficients and is solved iteratively using the Gauss–Seidel method. Updating of unknown values is done with the coupled algorithm in the case of the velocity and pressure and sequentially in the case of the species fraction. The values are stored in cell centers. Values at cell faces, needed in convective terms of the equations, are interpolated from the cell-center values by means of upwind discretization schemes. Spatial discretization of the gradient is achieved using the least-square cell-based method. The pressure, momentum, turbulent kinetic energy and specific dissipation use second-order discretization. The species fraction also uses second-order discretization, as does the temporal discretization. The water is modelled as an incompressible fluid with constant density of 998.2 kg m^{-3} and constant viscosity of 1.002×10^{-3} Pa s. The boundary condition at the inlet is uniform velocity of 0.623 m s^{-1}. Both outlets have a predefined velocity of flow equal to half the value of the inlet velocity. The intensity of turbulence at the inlet is 0.1%. The walls are defined with zero slip condition, that is the velocity of the fluid immediately attaching to the wall is equal to zero. The time step is 0.1 s, and the flow is initialized by the flow field developing naturally after 200 s from the steady-state solution. Each time step was computed in 10 iterations. The mesh was created in the ICEM CFD software. The mesh is structured so that is it is composed of only hexahedron discrete volumes arranged in blocks with regular orthogonal structure. The orthogonal geometry of the blocks is projected onto the surface of the walls and the inner geometry of the blocks' volumes is interpolated from the boundaries. The mesh adjacent to the walls in the region of the boundary layer has perpendicular geometric spacing with the multiple of 1.15 between the heights of consecutive layers of volumes, and the first layer of volume has the height of 0.035 mm in the region of impact flow, and the height of 0.45 mm elsewhere. The worst value of y+ is 1.6 in the region of impact flow, and 0.02–1.0 elsewhere. The mesh contained 2.5 million cells and 2.5 million nodes. The computation ran in the High Performance Computing Centre at the Technical University of Kosice [23–25].

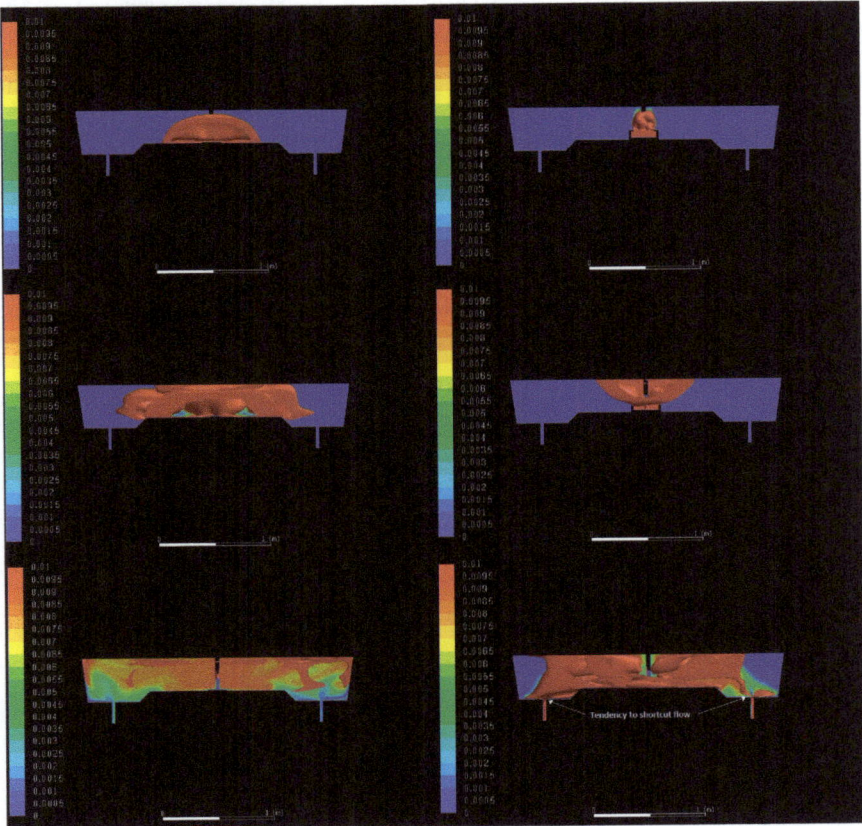

Figure 4. Comparison of fluid flow for "Spheric" impact pad and for standard impact pad at simulated casting speed 0.8 m·min−1—CFD simulation.

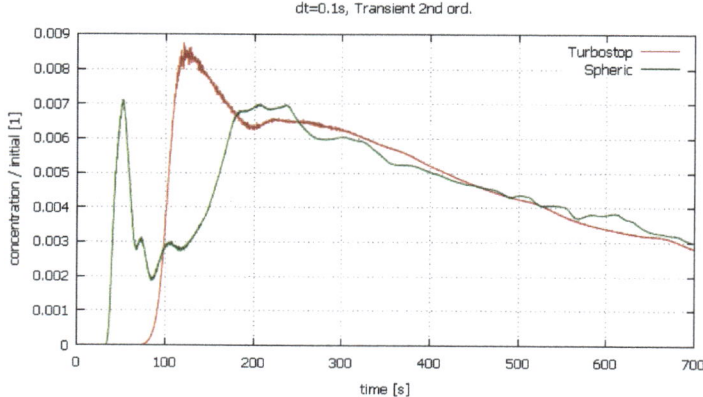

Figure 5. Comparison of C-curves for the "Spheric" impact pad and for standard impact pad @ 0.8 m·min^{-1}—CFD simulation.

Based on the results of CFD simulations, using the "Spheric" impact pad is expected to shorten the residence time compared to the standard impact pad, but on the other hand it is also expected to reduce the swirl of steel around the ladle shroud and reduce the so-called slag "eye" phenomenon. It is assumed that when using a "Spheric" impact pad the mixing area will predominate and the area of dead zones will be intensively reduced. It has also been found that the standard impact pad has a tendency to create a shortcut flow at lower casting speeds.

The proposed impact pad was tested using a model of the real symmetrical, two-strand boat-type tundish at scale 1:3 for the three default casting speeds. The C-curve, residence time, and visual evaluation of the flow in the tundish were selected as comparison criteria. The tracer is an aqueous salt solution of KCl, the concentration of which is monitored by means of the conductivity measurement system, while the flow is evaluated visually using KMnO$_4$ as tracer. Figures 6–8 show the results of the simulations comparing the standard and Spheric impact pads. For visual flow comparison, Figures 6–8 show the flow of tracer at time intervals of 5, 20 and 80 s after tracer injection. Below these pictures are the corresponding C-curves with highlighted minimum and maximum residence times for each configuration and casting speed.

Table 1 gives a comparison of minimum and maximum residence times for each configuration. The numbers in brackets indicate the percentage difference related to the minimum residence time of the alternative with standard impact pad under similar conditions. Graphical comparison of residence times of all tested configurations is presented in Figure 9.

Table 1. Comparison of residence times for all tested configurations.

Configuration	Casting Speed	Minimal Residence Time (s)	Maximal Residence Time (s)
Standard Impact Pad	0.8 m·min^{-1}	57	98
	1.2 m·min^{-1}	55	137
	1.6 m·min^{-1}	39.5	119
Spheric Impact Pad	0.8 m·min^{-1}	40.5 (71%)	119
	1.2 m·min^{-1}	42 (76%)	127
	1.6 m·min^{-1}	42 (106%)	104

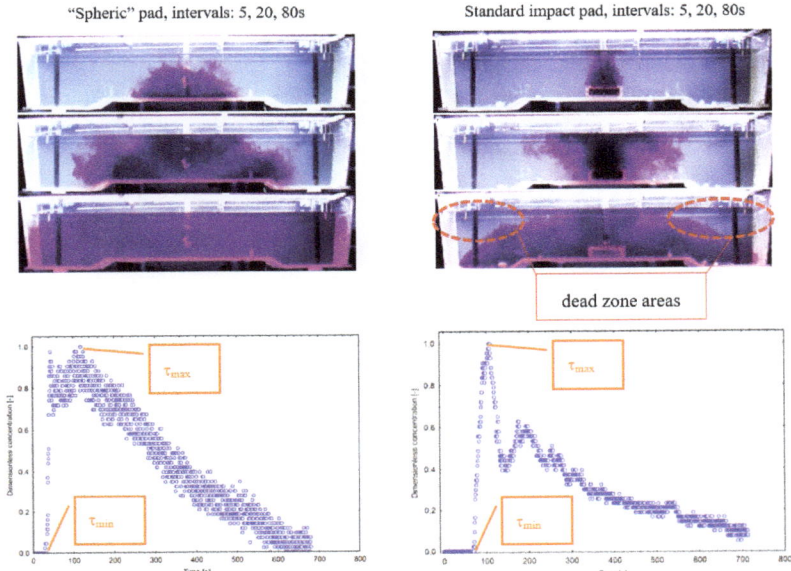

Figure 6. Visual and graphical comparison of flow for "Spheric" impact pad and for standard impact pad at simulated casting speed 0.8 m·min^{-1}—physical simulation.

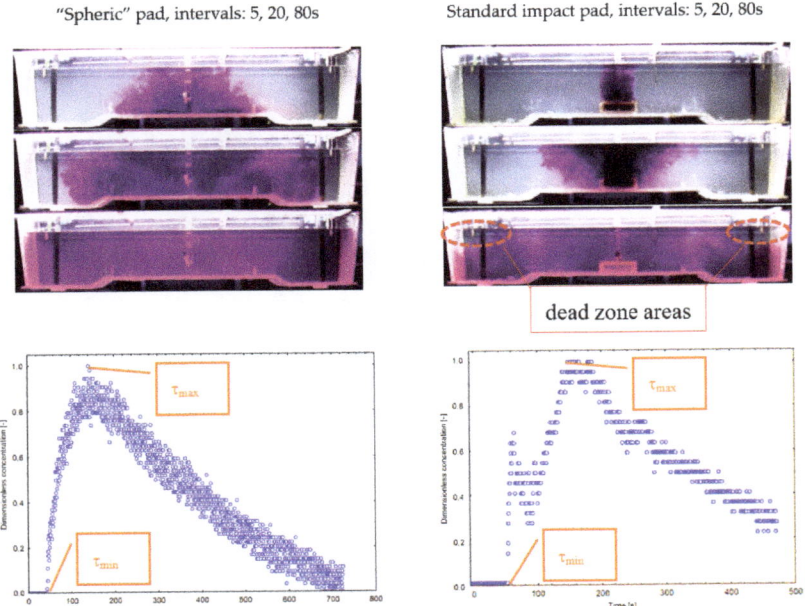

Figure 7. Visual and graphical comparison of flow for "Spheric" impact pad and for standard impact pad at simulated casting speed 1.2 m·min^{-1}—physical simulation.

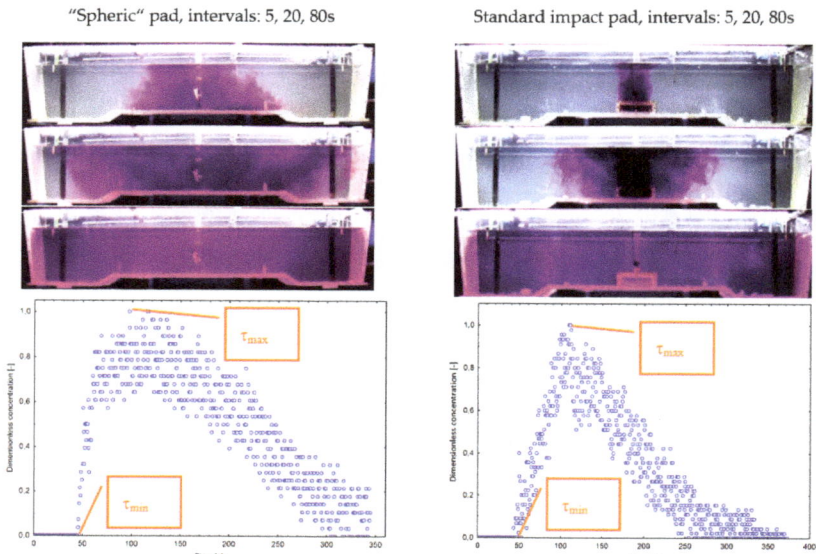

Figure 8. Visual and graphical comparison of flow for "Spheric" impact pad and for standard impact pad at simulated casting speed 1.6 m·min^{-1}—physical simulation.

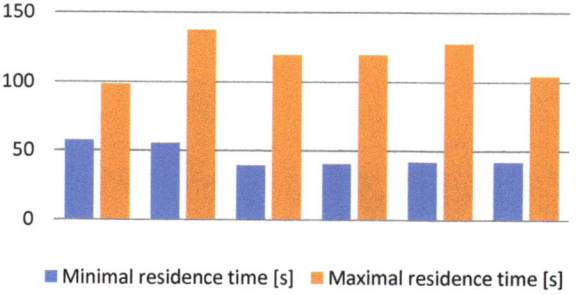

Figure 9. Graphical comparison of residence times of all tested configurations.

The difference in flow dynamics in the impact area is shown in Figures 10 and 11, where the standard and "Spheric" impact pads are compared.

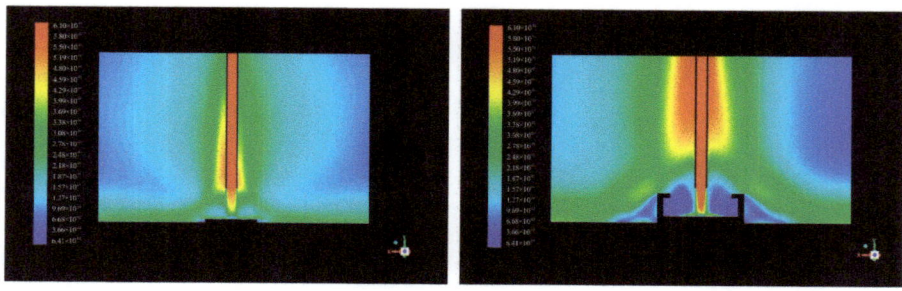

Figure 10. Comparison of velocity areas for "Spheric" and standard impact pads @ 0.8 m·min^{-1}.

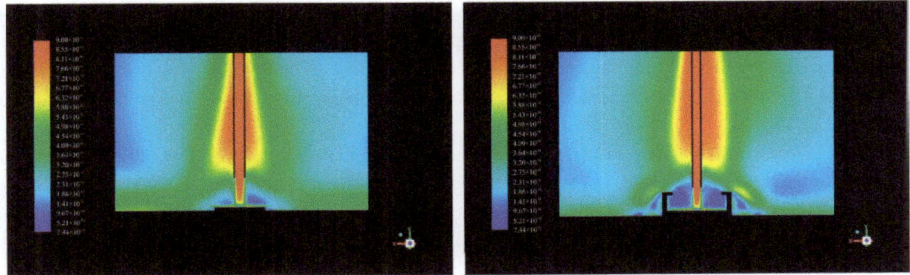

Figure 11. Comparison of velocity areas for "Spheric" and standard impact pads @ 1.2 m·min^{-1}.

The faster vertical circulation of steel up to the slag layer may cause the slag "eye" phenomenon, i.e., slag-free areas of steel surface open to air reoxidation and higher heat losses [26,27].

When using a "Spheric" impact pad, the vertical velocity of the flow around the ladle shroud is significantly lower than when using a standard impact pad. Using the "Spheric" impact pad can eliminate the so-called slag "eye" around the ladle shroud in the slag layer due to the lower vertical velocity of steel flow in this area, in contrast to the standard impact pad.

Another very important function of impact pads is to prevent splashing of molten steel while filling the empty tundish, mainly for safety reasons. A demonstration of the first seconds of filling the empty tundish is shown in Figure 12, and it is clear that the spherical impact pad safely prevents steel splashing.

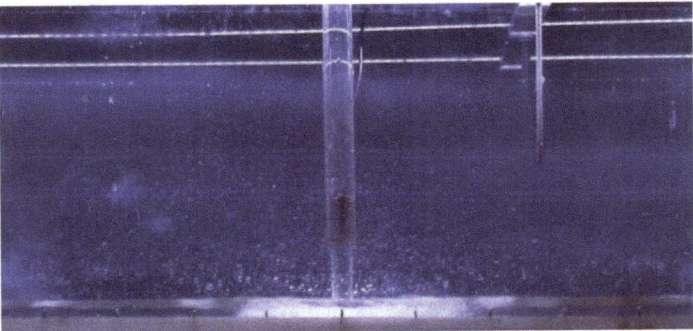

Figure 12. First seconds of filling an empty tundish, using the "Spheric" impact pad.

4. Conclusions

The design of the spherical impact pad with a convex surface was inspired by the differences between the flow past a flat plate and the flow past a sphere. CFD simulations were utilized for the initial testing and approval of this shape of impact pad. Compared to the standard impact pad at a corresponding casting speed of 0.8 m·min^{-1}, the spherical pad was found to shorten the residence time, but on the other hand the flow pattern created by this impact pad could have the benefit of reducing dead zone areas and eliminating any slag "eye" in the slag layer around the ladle shroud. The proposed impact pad has no tendency to short-circuit the flow. The "Spheric" impact pad was therefore subjected to further, more extensive testing using a 1:3 scale physical model of a tundish at flow rates simulating different casting speeds.

Compared to the standard impact pad, based on our measurements of residence time distribution (RTD) curves using the water model, the "Spheric" impact pad shortened the minimum residence times at casting speeds of 0.8 and 1.2 m·min^{-1}, which is on the level of 71% and 76% of the standard impact pad times under identical conditions. On the other hand, the "Spheric" impact pad produced a 6% longer residence time than the standard impact at casting speed 1.6 m·min^{-1}. It should be taken into account that this is only a comparison of impact pads. In both cases, it is possible to fit the tundish with flow modifiers such as dams, weirs and baffles to prolong the residence time of steel in the tundish.

From a visual comparison of flow in the tundish, we can observe that the "Spheric" impact pad produces a better flow pattern than the standard impact pad. It has no tendency to shortcut the flow at lower casting speeds. Moreover, dead zone areas are eliminated using the "Spheric" impact pad. We can predict that using this impact pad in practice will have a positive influence on steel cleanliness due to more dynamic steel flow at the steel-slag interface. Furthermore, the slag "eye" phenomenon can be reduced when the "Spheric" impact pad is used, compared to impact pads with a significant piston flow pattern.

Based on the performed measurements, it can be concluded that the "Spheric" impact pad has great potential to optimize the flow of steel in the tundish, and that in combination with the appropriate "tundish furniture" it can become a new part of modern tundish metallurgy with significant influence on the final quality and cleanliness of conticast steel.

Author Contributions: B.B. performed the physical simulations, wrote the manuscript, and arranged the funding; I.P. found and proposed the "Spheric" impact pad; P.D. evaluated the results of the physical simulations; P.G. performed the CFD simulations; D.B. designed the experiments performed on the CCM water model; M.H. revised the original manuscript.

Funding: This research was funded by [VEGA MŠ SR a SAV] grant number [1/0868/17].

Conflicts of Interest: The authors declare no conflicts of interest.

References

1. Worldsteel Association. Steel Statistical Yearbook 2016. Available online: https://www.worldsteel.org/publications/bookshop/product-details.~Steel-Statistical-Yearbook-2016~PRODUCT~SSY2016~.html (accessed on 9 November 2018).
2. Michalek, K.; Gryc, K.; Socha, L.; Tkadlečková, M.; Saternus, M.; Pieprzyca, J.; Merder, T.; Pindor, L. Physical modelling of tundish slag entrainment under various technological conditions. *Arch. Metall. Mater.* **2017**, *62*, 1467–1471. [CrossRef]
3. Warzecha, M. Numerical Modelling of Non-Metallic Inclusion Separation in a Continuous Casting Tundish. Available online: https://www.researchgate.net/publication/221913237 (accessed on 9 November 2018).
4. Braun, A.; Warzecha, M.; Pfeifer, H. Numerical and physical modeling of steel flow in a two-strand tundish for different casting conditions. *Metall. Mater. Trans. B* **2010**, *41*, 549–559. [CrossRef]
5. Chattopadhyay, K.; Isac, M.; Guthrie, R.I.L. Physical and mathematical modelling of steelmaking tundish operations: A review of the last decade (1999–2009). *ISIJ Int.* **2010**, *50*, 331–348. [CrossRef]
6. Kowitwarangkul, P.; Harnsihacacha, A. Tracer injection simulations and RTD analysis for the flow in a 3-strand steelmaking tundish. *Key Eng. Mater.* **2016**, *728*, 72–77. [CrossRef]
7. Buľko, B.; Kijac, J. Optimization of tundish equipment. *Acta Metall. Slovaca* **2010**, *16*, 76–83.
8. Buľko, B.; Molnár, M.; Demeter, P. Physical modeling of different configurations of a tundish for casting grades of steel that must satisfy stringent requirements on quality. *Metallurgist* **2014**, *57*, 976–980. [CrossRef]
9. Chatterjee, D. Designing of a novel shroud for improving the quality of steel in tundish. *Adv. Mater. Res.* **2012**, *585*, 359–363. [CrossRef]
10. Hoerner, S.F. *Fluid Dynamic Drag: Practical Information on Aerodynamic Drag and Hydrodynamic Resistance*; Hoerner Fluid Dynamics: Bakersfield, CA, USA, 1965.
11. Laboratory of Simulation of Flow Processes. Available online: https://ohaz.umet.fmmr.tuke.sk/lspp/index_en.html (accessed on 17 September 2018).
12. Priesol, I. A Method of Molten Metal Casting Utilizing an Impact Pad in the Tundish. International Patent Application No. PCT/IB2016/056207, 10 October 2016.

13. Priesol, I. Spôsob Liatia Roztaveného Kovu s Využitím Dopadovej Dosky v Medzipanve. International Patent Classification: B22D 11/10 B22D 41/00, Application No. 109-2016, 11 October 2016; B22D 11/00 B22D 41/00, Application No. 89-2016, 10 October 2016.
14. Michalek, K.; Gryc, K.; Socha, L.; Tkadlečková, M.; Saternus, M.; Pieprzyca, J.; Merder, T.; Pindor, L. Study of tundish slag entrainment using physical modeling. *Arch. Metall. Mater.* **2016**, *61*, 257–260. [CrossRef]
15. Acta Metallurgica Slovaca. Available online: http://www.ams.tuke.sk/data/ams_online/2010/number2/mag01/mag01.pdf (accessed on 17 September 2018).
16. Gryc, K.; Michalek, K.; Hudzieczek, Z.; Tkadlečková, M. Physical modelling of flow patterns in a 5-strand asymmetrical tundish with baffles. In Proceedings of the Metal 2010—19th International Conference on Metallurgy and Materials, Poruba, Czech Republic, 18–20 May 2010.
17. Sahai, Y.; Emi, T. Melt flow characterization in continuous casting tundishes. *ISIJ Int.* **1996**, *36*, 667–672. [CrossRef]
18. Väyrynen, P.; Wang, S.; Louhenkilpi, S.; Holappa, L. Modeling and removal of inclusions in continuous casting. In Proceedings of the Materials Science and Technology 2009—International Symposium on Inclusions and Clean Steel 2009, Pittsburgh, PA, USA, 25–29 October 2009.
19. Michalek, K. *Využití Fyzikálního a Numerického Modelování pro Optimalizaci Metalurgických Procesů*; Vysoká škola Báňská—Technická Univerzita: Ostrava, Czech Republic, 2001; ISBN 80-7078-861-5.
20. Falkus, J.; Lamut, J. Model testing of the bath flow through the Tundish of the continuous casting machine. *Arch. Metall. Mater.* **2005**, *50*, 709–718.
21. Ansys Fluent 12.0 User's Guide. 2009. Available online: http://users.ugent.be/~mvbelleg/flug-12-0.pdf (accessed on 22 August 2018).
22. Nastac, L.; Zhang, L.; Thomas, B.G.; Zhu, M.; Ludwig, A.; Sabau, A.S.; Pericleous, K.; Combeau, H. *CFD Modeling and Simulation in Materials Processing 2016*; Springer International Publishing: New York, NY, USA, 2016.
23. Gardin, P.; Brunet, M.; Domgin, J.F.; Pericleous, K. An experimental and numerical CFD study of turbulence in a tundish container. *Appl. Math. Model.* **2002**, *26*, 323–336. [CrossRef]
24. Mazumdar, D.; Guthrie, R.I.L. The physical and mathematical modelling of continuous casting Tundish systems. *ISIJ Int.* **1999**, *39*, 524–547. [CrossRef]
25. Liu, S.; Yang, X.; Du, L.; Li, L.; Liu, C. Hydrodynamic and mathematical simulations of flow field and temperature profile in an asymmetrical t-type single-strand continuous casting Tundish. *ISIJ Int.* **2008**, *48*, 1712–1721. [CrossRef]
26. Chatterjee, S.; Chattopadhyay, K. Formation of slag 'eye' in an inert gas shrouded Tundish. *ISIJ Int.* **2015**, *55*, 1416–1424. [CrossRef]
27. Zhang, L.; Thomas, B.G. State of the art in evaluation and control of steel cleanliness. *ISIJ Int.* **2003**, *43*, 271–291. [CrossRef]

© 2018 by the authors. Licensee MDPI, Basel, Switzerland. This article is an open access article distributed under the terms and conditions of the Creative Commons Attribution (CC BY) license (http://creativecommons.org/licenses/by/4.0/).

Article

Study of the Influence of Intermix Conditions on Steel Cleanliness

Branislav Buľko [1,*], Marek Molnár [2], Peter Demeter [1], Dana Baricová [1], Alena Pribulová [1] and Peter Futáš [1]

[1] Faculty of Metallurgy, Technical University of Košice, Institute of Metallurgy, Department of Metallurgy and Foundry, 042 00 Košice, Slovakia; peter.demeter@tuke.sk (P.D.); dana.baricova@tuke.sk (D.B.); alena.pribulova@tuke.sk (A.P.); peter.futas@tuke.sk (P.F.)
[2] U. S. Steel Košice, s.r.o., Vstupný areál U. S. Steel, 044 54 Košice, Slovakia; marek.molnar@centrum.sk
* Correspondence: branislav.bulko@tuke.sk; Tel.: +421-55-602-3169

Received: 28 September 2018; Accepted: 17 October 2018; Published: 19 October 2018

Abstract: Modern steel plants produce today a large portfolio of various steel grades, many for end-uses demanding high quality. In order to utilize the maximum productivity of the continuous-casting machine, it is sometimes necessary to cast steel grades with different chemical compositions in one sequence. It is important, therefore, to know the possibilities of a specific continuous-casting machine to make the Intermix connections as short as possible. Any interference with established procedures may, however, have a negative impact on the cleanliness of the cast steel. Using physical and numerical simulation tools, it was found that reducing the steel level in the tundish during the exchange of ladles makes it possible to shorten the transition zone. However, when the steel level is reduced, the flow of steel is impaired, which can have a negative effect on the cleanliness of the cast steel and, in extreme cases, may even lead to entrapment of slag in the mold. The cleanliness of cast steel was evaluated using one of the most advanced tools for automatic steel cleanliness evaluation, AZtecFeature (Oxford Instruments, Abingdon, UK), which enables determination of the type, size, distribution, and shape, as well as the chemical composition, of individual types of non-metal inclusions.

Keywords: continuous casting; tundish; residence time; transient casting

1. Introduction

Resolution of the Intermix problem, shortening the chemical composition transition zone in cast slabs when various steel grades are cast in succession, generally includes several options and operating parameters, which more or less affect its overall range. Apart from the chemical concept itself and the real possibilities of logistical planning of specific ways of joining different steel grades, the overall range of the Intermix is influenced mainly by the flow and mixing conditions of the liquid steel in the tundish during the casting process. For a more detailed study and definition of the Intermix range, a number of methods and procedures were published regarding the flow conditions of the steel in the tundish [1–6], as well as, to a certain extent, in the mold and the casting stream [7].

The most commonly reported operating criteria and parameters directly related to the Intermix are the chemical composition of the steel, the geometry and specific configuration of the given tundish, the volume (or operating level) of the steel, and the mass flow rate of the steel, limited by the combination of the given casting speed and the dimensions of the cast slab.

In order to verify certain operations intended to optimize the Intermix range during hard transitions in which the steel level in the tundish [8,9] is deliberately reduced, operational testing of a joining process for alternate batches was performed, in which not only the effect on the total length of the chemical composition transition zone was analyzed, but also the impact of specific casting conditions on the final cleanliness of the cast steel.

2. Materials and Methods

The main focus of this research was to verify the impact of the ultra-low tundish practice (ULT), whereby the steel level in the tundish is lowered to 15 tons during the initiation of the operating tests. Due to the specific operating conditions of the continuous-casting machine (CCM), i.e., a symmetrical, two-strand boat-type tundish with a capacity of 50 tons, the standard procedure for a hard transition is to lower the level of steel in the tundish to 20–22 tons.

Equation (1) was developed for grade transitions based on the results of a series of trials.

The equation is a fairly simple exponential function which predicts the normalized composition of the steel on the narrow face of the strand as a function of tons cast in the mold after the new ladle is opened to start the grade transition. The equation includes tundish weight as a variable and should be applicable for all types of grade transitions under normal casting conditions.

Every grade transition starts from 0% completion and the percentage grows as the number of tons cast increases. The main purpose of the transition equation is to calculate when the transition reaches a certain percentage of completion. The percentage of completion is contained in the transition equation as a normalized composition ranging from 0 to 1 instead of 0% to 100%.

$$Normalized\ composition = 1 - e^{-\frac{TM}{A.TT + B.TT^2}}, \tag{1}$$

where TM is the number of tons in the mold after ladle opening and TT is the number of tons in the tundish at ladle opening, while A and B are coefficients depending on the tundish internal design, configuration, and furniture.

The results calculated using this equation for initial tundish weight values from 10 tons to 50 tons are shown in Figure 1. The data are plotted as a normalized composition (on a 0–1 scale) versus the number of tons in the mold after ladle opening. The number of tons in the mold can be calculated from the casting speed, mold width and thickness, and the density of liquid steel, with zero tons representing the time at which the ladle is opened to start the grade transition. Note that the number of tons in the mold is for one strand, and each strand requires the same number of tons to reach the same fraction of completion of the transition. With regard to the graph of transition curves for different levels of the initial level of steel in the tundish, it is clear that, compared to the standard procedure for lowering the level of the steel to 20–22 tons, it is possible in the case of the ULT practice (15 tons) to assume additional reduction of the Intermix range, i.e., shortening of the length of the chemical composition transition zone. The alternative 10-ton ULT practice was discarded due to the risk of slag entrapment in the submerged entry nozzle, and subsequently, also in the mold.

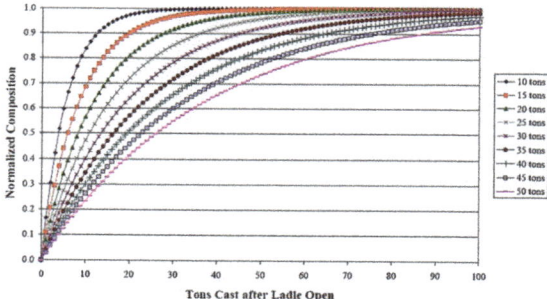

Figure 1. Basic nature of the influence of tested ultra-low tundish (ULT) practice on the Intermix (numerical simulation).

The basic nature of this tested alternative batch-joining process during the Intermix (ULT practice) is characterized in Figure 1.

The results of simple numerical simulations for the tested cases of transient casting under standard conditions (20 tons) and ULT practice (15 tons) were verified using a continuous-casting machine (CCM) physical model with a scale of 1:3 at the Faculty of Materials, Metallurgy, and Recycling, Technical University of Košice (Figure 2). The dimensions of the tundish model are given in Figure 3.

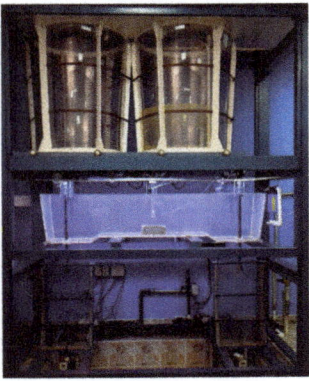

Figure 2. Water model of continuous-casting machine (scale 1:3) at the Faculty of Materials, Metallurgy, and Recycling, Technical University of Košice.

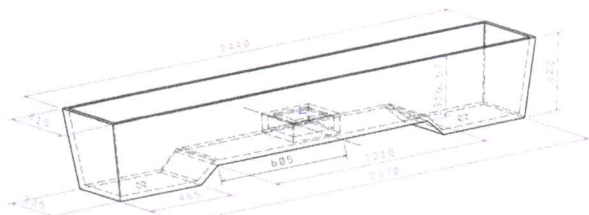

Figure 3. Dimensions of the used model of the tundish equipped with a standard impact pad.

The F-curve measurement methodology was used to verify the hypothesis that a lower initial level of steel in the tundish during transient casting can shorten the Intermix zone. Physical simulations were performed for the parameters listed in Table 1.

The measurement of F-curves on a physical model is used to describe the transition zone in the case of Intermix casting [10,11]. The tracer, 150 mL of an aqueous solution containing 10% KCl and 1% $KMnO_4$, was dissolved in the entire volume of the tundish, and when the required level of water in the tundish corresponding to the beginning of the transition was reached, the tundish was filled with pure water, resulting in a change in the concentration of the tracer at the tundish output, which very clearly defines the beginning and the end of the transition zone under specific conditions [12–16]. The water flow was scaled using Froude's dimensionless number [17].

Using this physical model, our tests confirmed that, by means of the ULT practice, it is possible to shorten the length of the transition zone (Figure 4).

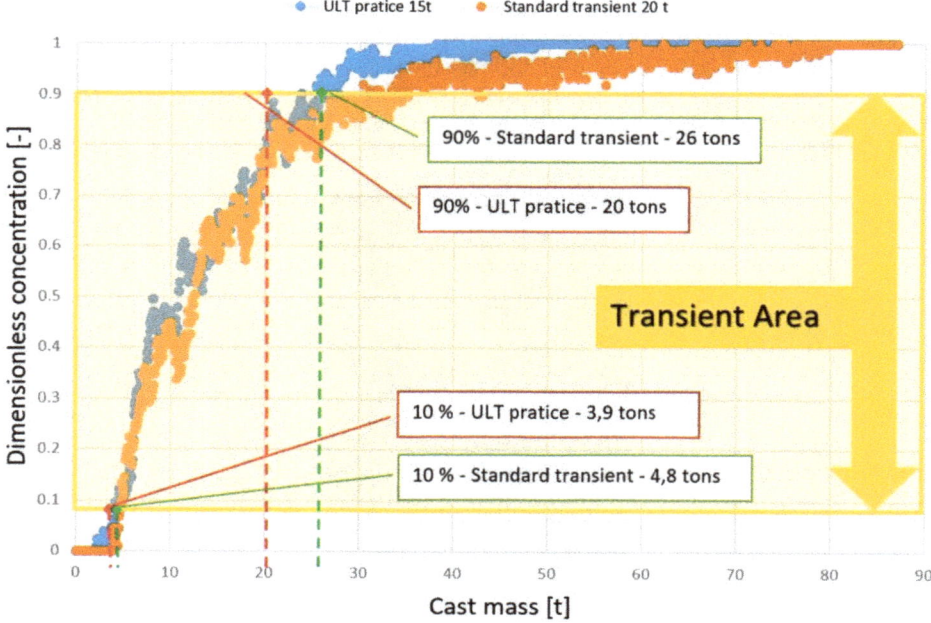

Figure 4. F-curves for standard transition and for ULT practice (physical simulation).

Table 1. Parameters for the physical simulations. ULT—ultra-low tundish.

Mold Width (mm)	Casting Speed (m·min^{-1})	Theoretical Hourly Production of CCM (tons·h^{-1})	Tundish Width-to-Length Ratio	Operating Level of Water in Tundish (mm)	Initial Level of Water in Tundish for Simulation of 20-Ton Standard Transition	Initial Level of Water in Tundish for Simulation of 15-Ton ULT Practice
1250	1	257.4	0.16	420	200	160

According to [6], we can consider the examined tundish with a width-to-length ratio of 0.16 to be a narrow tundish. Consequently, the walls of the narrow tundish are assumed to have a significant effect on the flow of liquid steel, which is revealed in the F-curves in Figure 4 by an uneven course of change in tracer concentration when the steel is reflected from these walls [16–18].

An in-depth analysis of the effect of the ULT practice on steel cleanliness was performed by means of operating tests. Samples of the steel from the mold were taken at specified time intervals during the Intermix transition. To compare the results, each transition was separately mapped in more detail for standard transient casting and using the ULT practice.

An important step, apart from the preparation of the operating tests themselves, was also the choice of the most appropriate method for evaluating the results obtained, regarding the change in chemical composition in the transient area of the slab for the given conditions. This choice was based on our previous knowledge and experience in this area of research, and consequently, the methodology of determining the dependence of the change in the so-called normalized composition by considering the mass of the cast steel in each mold from the moment the ladle was opened with a different steel grade was used [12,13].

The main advantage of this methodology is the ability to compare results from several realized transitions on the same basis, regardless of the variations in steel grade or casting conditions. The term "normalization of chemical composition of the sample" (NC) can be defined as the conversion of the chemical composition (percentage content) of the transition elements to be monitored, based on the melting analysis of the first and second melt, as well as the content of the individual samples taken during the given transition. Using the simple calculation in Equation (2), the percentage content of any chemical element with any value is transformed into the same interval of values from 0 to 1.

$$NC = \frac{(\% \text{ of element in transient sample} - \% \text{ of that element in first melt})}{(\% \text{ of the element in second melt} - \% \text{ of the element in first melt})} \qquad (2)$$

In order to study the effect of the joining conditions of different melts during the Intermix, a detailed, fully automatic analysis of the metallurgical cleanliness was carried out from the collected steel samples after their chemical analysis. For this purpose, one of the most modern AZtecFeature (Oxford Instruments, Abingdon, UK) tools was used, which is part of a VEGA3 scanning electron microscope from TESCAN ORSAY HOLDING a.s. (TESCAN ORSAY HOLDING, Brno, Czech Republic). The microscope itself is based on a hot cathode for use in both high and low vacuums. The microscope is equipped with an Oxford Instruments X-Max® 80 EDS analyzer (Oxford Instruments, Abingdon, UK), which is fully controlled by the latest AZtec software (Oxford Instruments, Abingdon, UK), also from Oxford Instruments. This tool makes it possible to determine the type, size, distribution, and shape, as well as the chemical composition, of individual types of non-metal inclusions in the analyzed samples, enabling their direct categorization into individual groups, as well as the export of the obtained information into standard available programs and formats for further processing.

Operational testing was carried out on the joining of selected A and B steel grades. Their basic chemical compositions are given in Table 2. The main monitored transition elements for this joining were silicon and aluminum.

Table 2. Basic chemical composition of steel grades A and B.

Sample	C (%)	Mn (%)	Si (%)	P (%)	S (%)	Al (%)	N (%)
Steel grade A	Max. 0.0105	0.295–0.405	0.895–1.054	0.08–0.1	Max. 0.01	0.095–0.165	Max. 0.007
Steel grade B	0.003–0.0105	0.146–0.305	0.595–0.705	0.1–0.13	Max. 0.009	Max. 0.005	Max. 0.005

3. Results

The results of the research in terms of Intermix evaluation are presented graphically in Figure 5, which presents a direct comparison of the standard transient conditions in the tundish at 20 tons, as well as with the new alternative tested, with an initial steel level of 15 tons in the tundish, i.e., ULT.

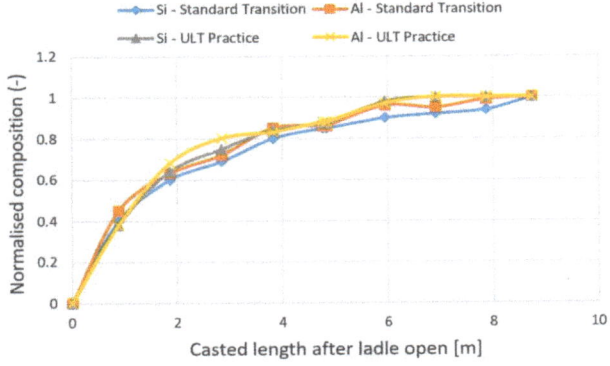

Figure 5. Comparison of transition curves for standard transition and ULT practice.

On the basis of a more detailed assessment of the progress of these transition curves, it can be concluded that the ULT practice produced a more rapid change in the chemical composition within the transition zone, especially after casting approximately six tons of steel from the tundish on a given strand. In this ULT practice, from the point of view of chemical composition, the transition zone was intensified and accelerated, and thus, the overall period of the Intermix was shortened.

The trends in chemical composition changes with the number of tons of cast steel, comparing the two procedures with different initial steel levels in the tundish, are presented in detail in Tables 3 and 4. Each line in this table represents the moment of sampling of the steel from the mold during the particular type of transition. Samples were taken approximately from every two meters of cast slab. From the time data (in seconds) between sampling, the corresponding amount of steel cast (from the moment of opening of the second ladle) was calculated. Subsequently, the corresponding chemical composition and the calculated values of the change in the normalized composition were ascribed to the individual samples.

Due to this being a standard Intermix evaluation methodology [12–15], the limit of the standardized composition of the main transition element is generally considered to be 0.9 at the end of the transition zone under the stricter assessment criterion (10/90). The area for achieving this limit value is shown in Tables 3 and 4, and for the standard practice in samples 5 and 6, and for the ULT practice in samples 4 and 5.

Table 3. Detailed course of the change in normalized composition for steel grades A and B (20 tons).

Standard Practice of Joining Steel Grades A and B (20 tons)			
No. of Sample	Cast Mass during Intermix (tons)	Si Normalized	Al Normalized
1	3.78	0.430	0.450
2	8.5	0.604	0.634
3	12.58	0.683	0.733
4	16.65	0.795	0.847
5	20.73	0.850	0.885
6	25.49	0.904	0.962
7	29.53	0.915	0.962
8	33.69	0.925	0.992
9	37.73	1	1

Table 4. Detailed course of the change in normalized composition for steel grades A and B (15 tons).

ULT Practice of Joining Steel Grades A and B (15 tons)			
No. of Sample	Cast Mass during Intermix (tons)	Si Normalized	Al Normalized
1	3.45	0.372	0.372
2	8.15	0.661	0.669
3	12.44	0.775	0.804
4	16.27	0.846	0.831
5	19.95	0.913	0.908
6	23.44	0.966	0.953
7	27.76	1	1
8	31.59	1	1
9	35.47	1	1

It is clear from these detailed areas that, in the standard procedure, the value of the normalized composition of the main transition element was 0.904% Si for the volume of cast steel at 25.49 tons, while the ULT procedure already achieved 0.913% Si with the volume of cast steel at 19.95 tons.

Despite the fact that the qualitative results of the processed slabs from the ULT practice testing did not confirm any possible negative trend or problems with their processing, detailed analysis of the standard of steel cleanliness was also performed on the taken samples. As can be seen from the results of this analysis presented in Figures 6 and 7, in this context, the most highly represented non-metallic inclusions (Si–O and Al–Ca–O complex inclusions) were classified and evaluated in more detail. The analyzed area in each sample was 100 mm^2. In our assessment of the achieved level of steel cleanliness, the first samples (No. 1) were taken at the moment of opening of the second ladle within the given transition, while subsequent samples were taken after casting approximately 2 m of slab. As in the case of the overall Intermix range evaluation, the two procedures (standard and ULT) were compared.

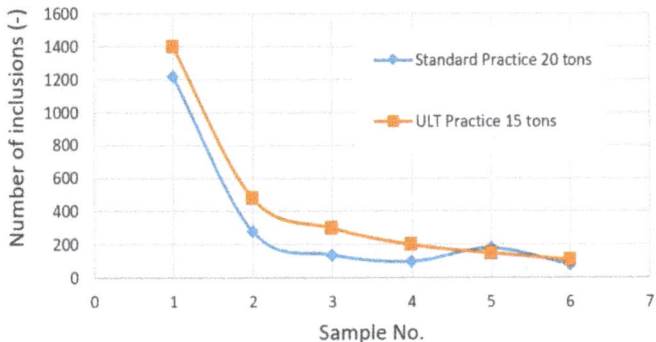

Figure 6. Comparison of the course of total Si–O inclusions in the different A and B grade joining conditions (standard vs. ULT practice).

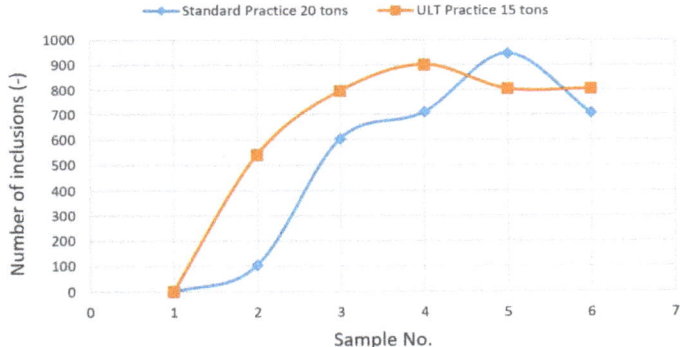

Figure 7. Comparison of the course of total Al–Ca–O inclusions in the different A and B grade joining conditions (standard vs. ULT practice).

4. Discussion

Based on these results from our detailed analysis of steel cleanliness, it can be stated that the significantly different trends within the monitored categories of non-metal inclusions were due to the specific chemical concept and the process of production of specific steel grades.

The most important result of this research is the finding that increased inclusion content was recorded using the ULT practice compared to the standard transition process. Under the given casting conditions, the most significant decrease in steel cleanliness was recorded in the first four to five samples taken and analyzed (roughly equivalent to the first 6–8 m cast per strand from the moment of opening of the second ladle). Thus, the significantly affected section of slab (from the point of view of steel cleanliness) corresponded, in fact, to the range of the first transition slab cast on the strand within the given joining, which, in any case, in view of the significant differences in its chemical composition over its length, is generally regarded as problematic in terms of quality. Thus, it can be concluded that the steel cleanliness outside the transition zone was not affected by use of the ULT practice.

5. Conclusions

Based on our research results, it can be concluded that the ultra-low tundish practice produces savings in terms of shorter transition zones, thereby increasing steel yields in the continuous-casting process. Under the given conditions, the ULT practice enabled a 20% shortening of the transition zone. The area of higher inclusion concentration during the ULT practice was located in the transition part of the slab, and therefore, had no negative effect on the final quality of the cast steel.

Author Contributions: B.B. performed the physical simulations, wrote the majority of the manuscript, and arranged the funding; M.M. carried out the experiments in the steel plant, performed the analyses of slab samples, evaluated the influence of ULT practice on steel cleanliness, and wrote the corresponding part of the manuscript; P.D. evaluated the results of the physical simulations; D.B. designed the experiments performed the water model of CCM; P.F. provided the graphical evaluation of the experiments; and A.P. revised the original manuscript.

Funding: This research was funded by VEGA MŠ SR a SAV grant number 1/0868/17.

Conflicts of Interest: The authors declare no conflicts of interest.

References

1. Cho, M.J.; Kim, I.C. Simple Tundish Mixing Model of Continuous Casting during a Grade Transition. *ISIJ Int.* **2006**, *46*, 1416–1420. [CrossRef]
2. Alizdeh, M.; Edris, H.; Pishevar, A.R. Behavior of Mixed Grade during the Grade Transition for Different Conditions in the Slab Continuous Casting. *ISIJ Int.* **2008**, *48*, 28–37. [CrossRef]
3. Braga, B.M.; Tavares, R. Additional Information on "Simple Tundish Mixing Model of Continuous Casting during a Grade Transition" by Cho and Kim. *ISIJ Int.* **2018**, *58*, 1178–1180. [CrossRef]
4. Huang, X.; Thomas, B.G. Modeling of Steel Grade Transition in Continuous Slab Casting Process. *Metall. Mater. Trans. B* **1993**, *24*, 379–393. [CrossRef]
5. Wang, Y.; Zhang, L. Transient Fluid Flow Phenomena During Continuous Casting: Part II—Cast Speed Change, Temperature Fluctuation and Steel Grades Mixing. *ISIJ Int.* **2010**, *50*, 1783–1791. [CrossRef]
6. Wang, Y.; Wen, X.; Qi, X.; Tang, X. Design of Bottom Height and Flow Control Device of Tundish for Slab Continuous Caster. *J. Iron Steel Res.* **2006**, *13*, 274–278.
7. Morávka, J.; Mrajca, V.; Pindor, J. Possible ways of processing results of transition phenomena occuring in the tundish using approximation methods and their implementaion in field practice. In Proceedings of the 1st International Conference STEELSIM, Brno, Czech Republic, 25–27 October 2005.
8. Mazumdar, K.D. Transient, Multiphase Simulation of Grade Intermixing in a Tundish under Constant Casting Rate and Validation Against Physical Modeling. *JOM* **2018**, *70*, 2139–2147. [CrossRef]
9. Yeh, J.-L.; Hwang, W.-S.; Chou, C.-L. The Development of a Mathematical Model to Predict Composition Distribution in Casting Slab and Intermix Slab Length during Ladle Changeover Period and Its Verification by Physical Model. *ISIJ Int.* **1993**, *33*, 588–594. [CrossRef]
10. Michalek, K. *Využití Fyzikálního a Numerického Modelování pro Optimalizaci Metalurgických Procesů (Utilization of Physical and Numerical Simulation for Optimization of Metallurgical Processes)*; VŠB—Technical University of Ostrava: Ostrava, Czech Republic, 2001; ISBN 80-7078-861-5. (In Czech)
11. Michalek, K.; Sawová, M.; Střasák, P. Modelling of transition phenomena in tundish during grade change of CC-Steel. In Proceedings of the 1st International Conference STEELSIM, Brno, Czech Republic, 25–27 October 2005.
12. Buľko, B.; Kijac, J. Optimization of Tundish Equipment. *Acta Metall. Slovaca* **2010**, *16*, 76–83.
13. Pieprzyca, J.; Merder, T.; Jowsa, J. Method for determining the time constants characterizing the intensity of steel mixing in continuous casting tundish. *Arch. Metall. Mater.* **2015**, *60*, 245–249. [CrossRef]
14. Michalek, K.; Gryc, K.; Tkadlečková, M.; Bocek, D. Model study of tundish steel intermixing and operational verification. *Arch. Metall. Mater.* **2012**, *57*, 291–296. [CrossRef]
15. Jha, P.K.; Kant, S.; Kumar, P.; Kumar, A. *Time Zone Analysis of F-Curve for Intermixing during Ladle Change-Over*; The Minerals, Metals, & Materials Society: Orlando, FL, USA, 2012. [CrossRef]
16. Väyrynen, P.J.; Vapalahti, S.K.; Louhenkilpi, S. On Validation of Mathematical Fluid Flow Models for Simulation of Tundish Water Models and Industrial Examples. In Proceedings of the 2008 AISTech, Pittsburgh, PA, USA, 5–8 May 2008; Volume 2, pp. 41–50.

17. Braun, A.; Pfeifer, H. Investigations of Non-isothermal Flow Conditions in a Two Strand Tundish Water Model using DPIV and PLIF-technique. In Proceedings of the SteelSim 2007—2nd International Conference of Simulation and Modeling of Metallurgical Processes in Steelmaking, Seggauberg, Austria, 12–14 September 2007; pp. 141–146.
18. Singh, S.; Koria, S.C. Model Study of the Dynamics of Flow of Steel Melt in the Tundish. *ISIJ Int.* **1993**, *33*, 1228–1237. [CrossRef]

© 2018 by the authors. Licensee MDPI, Basel, Switzerland. This article is an open access article distributed under the terms and conditions of the Creative Commons Attribution (CC BY) license (http://creativecommons.org/licenses/by/4.0/).

Article

Reoxidation of Al-Killed Steel by Cr_2O_3 from Tundish Cover Flux

Feng Wang [1], Daoxu Liu [2], Wei Liu [2], Shufeng Yang [2,*] and Jingshe Li [1,2,*]

[1] Engineering Research Institute, University of Science and Technology Beijing, Beijing 100083, China; b20150439@xs.ustb.edu.cn
[2] School of Metallurgical and Ecological Engineering, University of Science and Technology Beijing, Beijing 100083, China; liudaoxuustb@163.com (D.L.); youiithe@foxmail.com (W.L.)
* Correspondence: yangshufeng@ustb.edu.cn (S.Y.); lijingshe@ustb.edu.cn (J.L.); Tel.: +861062334277 (S.Y.)

Received: 2 April 2019; Accepted: 4 May 2019; Published: 12 May 2019

Abstract: Reoxidation has long been a problem when casting ultra-low oxygen liquid steel. An experimental study of the reoxidation phenomenon caused by Cr_2O_3-bearing cover flux of Al-killed steel is presented here. MgO-CaO-SiO_2-Al_2O_3-Cr_2O_3 tundish cover flux with various Cr_2O_3 contents were used to study the effects of Cr_2O_3 on total oxygen content (T[O]) and alumina and silicone loss of Al-killed steel at 1923 K (1650 °C). It was found that Cr_2O_3 can be reduced by Al to cause reoxidation, and the reaction occurs mainly within 2 to 3 min after the addition of the tundish cover flux with 5% and 10% Cr_2O_3 concentration. T[O] and Al loss increase with higher Cr_2O_3 concentration flux. Two controlled experiments were also made to investigate the oxygen transported to the steel by the decomposition of Cr_2O_3. It was calculated that when Al is present in steel, more than 90% of the reoxidation of Cr_2O_3 is caused by Al, and the rest is caused by decomposition.

Keywords: reoxidation; tundish cover flux; Cr_2O_3; decomposition reaction

1. Introduction

The total oxygen content (T[O]) of high-grade steel has been required to be as low as possible in recent years [1,2]. For example, line pipe steel requires sulfur, phosphorus, and T[O] all to be less than 30 ppm, and bearing steel requires T[O] to be less than 10 ppm [3]. This low oxygen liquid steel is very hard to cast due to reoxidation during the process. The tundish used in continuous casting is the last vessel in contact with molten steel during steel production, with this final step being the most important in protecting the steel from oxygen and maintaining cleanliness [4,5]. However, due to the extremely low oxygen content in liquid steel, it can be easily reoxidized by the oxygen from the refractory, the slag, and the air environment [6,7]. The reoxidation phenomenon has been a problem for tundishes and is the obstacle to producing clean steel [8–10].

The role of tundish cover flux is to protect the molten steel from the air. However, in recent years, reoxidation of steel by the cover flux has been observed and widely investigated. Researchers have found that the reoxidation capacity of the slag was related to the oxygen potential in the slag, such as the FeO, SiO_2, and MnO content in the slag phase [11–13]. The high oxygen potential slag easily reacts with deoxidizing agents such as the Al and Ti contained in liquid steel. Some researchers [14] studied the effect of refining slag components on the reoxidation of molten steel through the double-membrane theory and found that reducing the CaO/Al_2O_3 ratio and FeO content in the slag can reduce the reoxidation of steel by refined slag.

Recently, Cr_2O_3 has been found in tundish cover flux [15]. Cr is usually used as an alloying element for stainless steel, working as a protecting element that generates Cr_2O_3 on the metal surface to prevent further oxidation [16]. Due to the steel-slag equilibrium reaction, some Cr_2O_3 shows up in the refining slag of stainless steel [17,18]. The reduction of Cr_2O_3 during the refining process will increase T[O] and thus will do harm to steel cleanliness [19]. It has been reported that the amount of Cr_2O_3 forming in slag could be reduced by lowering the basicity of the slag or by adding Al_2O_3, the reason for this being the increase in the liquid phase fraction of the slag [19].

When using cover flux containing Cr_2O_3, Cr_2O_3 may work as a medium carrying oxygen from the air that reoxidizes the liquid steel instead of protecting it. Cover flux is a semi-molten slag that, when compared to other totally melted refining slags, makes the reduction of Cr_2O_3 in the flux special and worth studying. Reoxidation is related to the Si and Al content in steel and the slag composition. Therefore, in the present work, reoxidation of steel, with and without Al, by Cr_2O_3 in the tundish cover flux was investigated using experimental methods. The effects of the Cr_2O_3 content of flux on the T[O] and the loss of Al and Si in steel was investigated, and the different results of reoxidation by reduction and decomposition were compared and discussed.

2. Experimental Procedures

A tube resistance furnace (Baotou Yunjie Furnace Ltd., Baotou, China) was used in the experiments to melt the steel and the tundish cover flux. The schematic diagram of the tube resistance furnace is shown in Figure 1. Four hundred grams of pure industrial iron (the composition is shown in Table 1) was put into a magnesia crucible (inner diameter: 42 mm, outside diameter: 51 mm, height: 101 mm, Boshan Refratory Material Ltd., Zibo, China) and then melted in an argon gas atmosphere at 1873 K (1600 °C). Al was added to the molten steel at a concentration of 0.2 mass% to make Al-killed steel. Then, after adjusting the molten steel temperature to 1923 K (1650 °C) and holding for 30 min, 20 g tundish cover flux of different Cr_2O_3 concentrations were added wrapped with iron foil. The crucible was held at 1923 K (1650 °C) for another 30 min to ensure a uniform covering of the molten steel by slag. The composition of the cover flux is presented in Table 2.

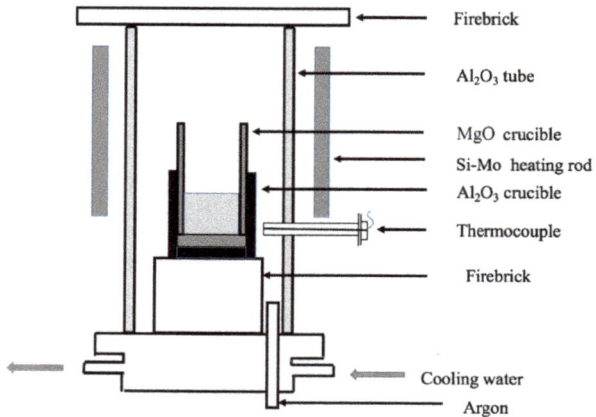

Figure 1. Schematic diagram of experimental apparatus.

Table 1. Chemical composition of YT01 (wt. %).

C	Si	Mn	P	S	Al	Ni	Ti	N	Ca	Fe
0.0016	0.0033	0.01	0.0053	0.0017	0.003	0.0038	0.001	0.002	-	99.95

Table 2. Compositions of slags (wt. %).

Experiment	CaO	Al$_2$O$_3$	MgO	SiO$_2$	Cr$_2$O$_3$
1	50	30	10	10	0
2	48	27	10	10	5
3	46	24	10	10	10
4	48	27	10	10	5
5	46	24	10	10	10

During the experiments, steel samples were taken as follows: (a) before adding the tundish cover flux, and (b) 5, 10, and 15 min after adding the tundish cover flux. Tundish cover flux samples were collected after the experiments.

Oxygen content was determined using the infrared method. The calculated aluminum, silicon, and chromium contents of the steel were determined using ICP-AES (NCS plasma 2000, NCS Testing Technology Co., Ltd., Beijing, China). FeO content in the tundish cover flux was analyzed using XRF (ZSX Primus II, Rigaku Corporation, Tokyo, Japan).

3. Results and Discussion

3.1. Reoxidation of Al-Killed Steel by Cr$_2$O$_3$

The changes in Al and T[O] content in Experiments 1, 2, and 3 are shown in Figures 2 and 3. It can be seen that the Al content in the steel dropped sharply and the O content rose sharply when flux with Cr$_2$O$_3$ was added. This indicated that Cr$_2$O$_3$ in the tundish cover flux can be reduced by Al in the molten steel and lead to the reoxidation. The changes in Al and O contents mainly occurred in the first 5 min after the addition of the tundish cover flux. The reduction of Cr$_2$O$_3$ by Al occurred immediately after the addition of the tundish cover flux, and it was very quick. The increase in Cr$_2$O$_3$ from 5% to 10% showed no obvious difference on the Al and O contents. This indicates that for the slag-steel system used in this study, as the content of Cr$_2$O$_3$ increased, the transport of O from slag to steel showed no increase, and that the 5% Cr$_2$O$_3$ flux had already reached the limitation. After 5 min, T[O] began to drop sharply, suggesting that the reduction reaction was very fast and reached the equilibrium quickly. Thus, after the first 5 min of the reaction, Al$_2$O$_3$ inclusions floated up and were removed, resulting in the drop in T[O] [11]. However, compared with no Cr$_2$O$_3$ flux, the T[O] at 15 min was much higher in the other two experiments, indicating the reoxidation of steel by the Cr$_2$O$_3$ flux.

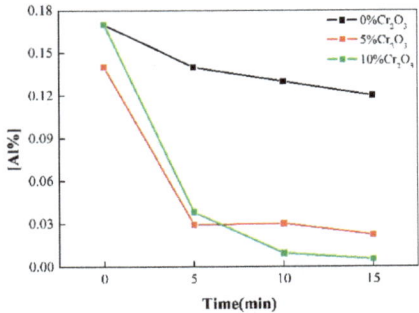

Figure 2. Variations in Al concentrations of steel with time with different concentrations of Cr$_2$O$_3$ cover flux added.

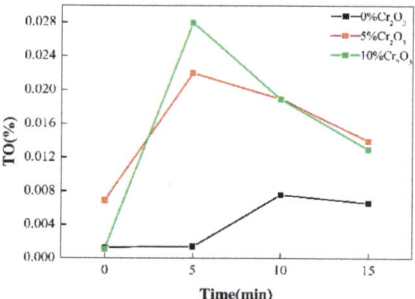

Figure 3. Variations in total oxygen content (T[O]) of steel with different concentrations of Cr_2O_3 cover flux added.

In order to determine the speed of the reduction reaction, another experiment with more frequent sampling was designed. The sampling time was 1, 2, 3, 4, and 5 min after the tundish cover flux was added, as shown in Figure 4. It can be seen that the reaction of Al reducing Cr_2O_3 occurred mainly within 2 min after adding the cover flux for the 5% Cr_2O_3 flux and within 3 min for the 10% flux. Moreover, the reaction rates were almost the same in the first minute after adding the flux, and the Cr content rose to 0.08% in the first minute for both. Between the first and second minute, the rate remained at 0.08%/min for the 5% Cr_2O_3 flux, while the 10% flux reached the highest rate of 0.16%/min. Then at 3 min, the reduction of Cr_2O_3 stopped for the 5% Cr_2O_3 flux, while the reduction of the 10% flux lasted another minute at a rate of 0.08%/min. The reduction reaction in this study happened at a very high rate and required little time. The increase of the equilibrium of Cr in steel was caused by the change in Cr_2O_3 content of the flux. The highest reaction rate occurred with the higher Cr_2O_3 concentration flux, and this may be explained by the doubling of the concentration of Cr_2O_3 in the slag being the driving force behind the transport of Cr_2O_3 from the slag to the interface [19].

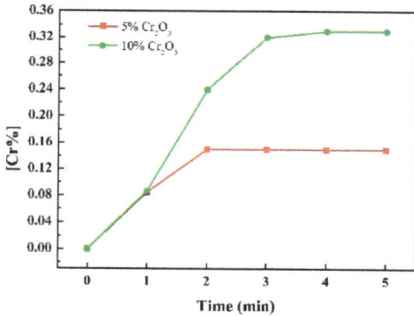

Figure 4. Variations in Cr concentrations of steel with time for Experiments 2 and 3.

The changes in Si and Cr in Experiments 1, 2, and 3 are shown in Figures 5 and 6. It can be seen from Figure 5 that before the addition of flux with Cr_2O_3, the Si content in the steel increased first and then decreased, and after the flux was added, the Si content in the steel was significantly reduced. Furthermore, when adding flux with 10% Cr_2O_3, the Si content in the steel first decreased and then increased. It is known that Al in steel preferentially reacts with Cr_2O_3 in the cover flux. This means Cr_2O_3 is superior to SiO_2 when reacting with Al in steel. Thus, it is assumed that Reaction (1) will be weakened by Reaction (2) [20,21] because of the limited oxygen supplement [17,18]. The dashed line in Figure 6 represents the ideal value of Cr content that can be reduced in the flux according to a mass balance calculation. It was found that almost all of the Cr_2O_3 in the cover flux was reduced, which made Cr close to the ideal value. When increasing the Cr_2O_3 content from 5% to 10%, there was even less Si due to the oxidation of the Si in the steel. This indicates that if not enough oxygen is provided by the flux, the reoxidation of molten steel by Cr_2O_3 will first involve Al, while the reaction will involve both Al and Si when sufficient oxygen is supplied by the flux.

$$4[Al] + 3(SiO_2) \rightarrow 3[Si] + 2(Al_2O_3) \quad (1)$$

$$[Al] + (Cr_2O_3) \rightarrow [Cr] + (Al_2O_3) \quad (2)$$

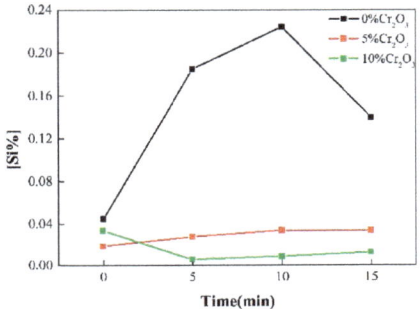

Figure 5. Variations in Si concentrations of steel with time for Experiments 1, 2, and 3.

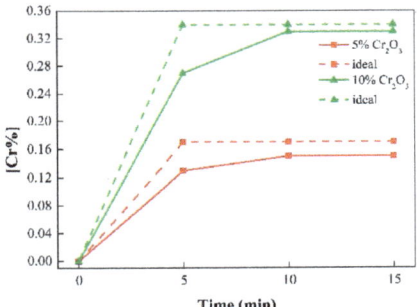

Figure 6. Variations in Cr concentrations of steel with time for Experiments 2 and 3.

3.2. Reoxidation of Non-Al Steel by Cr_2O_3

Compared to Experiments 2 and 3, no Al was added to the molten steel in Experiments 4 and 5. The changes in the Cr and Si contents in Experiments 4 and 5 are shown in Figures 7 and 8, respectively. It can be seen from these figures that even if Al is not added to the steel, the Cr content in the steel will still rise dominantly. In Experiment 4, the Si content of the steel first increased and then decreased, which indicates that Cr_2O_3 and SiO_2 in the tundish cover flux can transport oxygen to the steel by decomposition [22]. By comparing the magnitude of the rise, it can be seen that the decomposition of Cr_2O_3 dominates, and that the Si content in the steel begins to decrease due to the increase in the oxygen potential of the steel caused by the continued decomposition of Cr_2O_3, which causes the Si in the steel to be oxidized. The Si content in Experiment 5 first decreased and then remained unchanged. This was due to the Cr_2O_3 content in the tundish cover flux being higher, causing the decomposition of Cr_2O_3 to be very intense from the beginning. Therefore, the oxidation rate of the Si in the steel was greater than the decomposition rate of SiO_2 in the tundish cover flux, which led to a decrease in the Si content of the steel. The subsequent decomposition of the Cr_2O_3 further reduced the Si content in the steel until it reached a very low value, after which the Cr in the steel began to be oxidized (as shown in Figure 8). Combining the change in Al content of the Experiment 3 steel and the stoichiometry of Reaction (2), it can be calculated that when Al is present in steel, the contribution of Al reduction to the increase in Cr content of steel is more than 90%, and that of Cr_2O_3 decomposition is less than 10%. Figures 9 and 10 compare the T[O] changes with time between Experiments 2 and 4 and Experiments 3 and 5, respectively. It can be seen from Figures 9 and 10 that the oxygen content in Experiments 4 and 5 was significantly greater than that in Experiments 2 and 3. The main reason for this was that Al was not added, and thus the deoxidation element was mainly Si in Experiments 4 and 5.

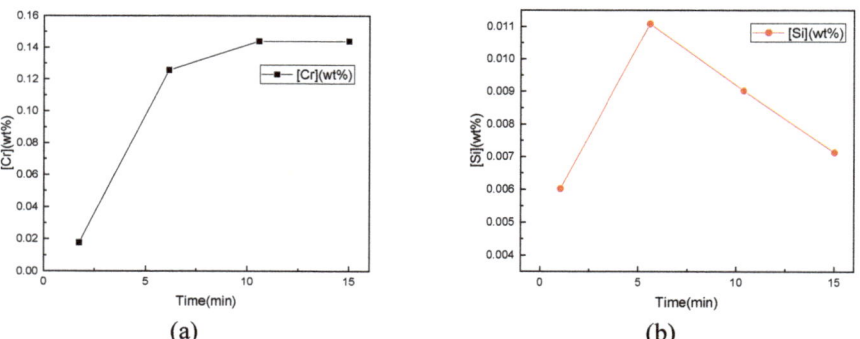

Figure 7. Variations in Cr and Si concentrations of steel with time for Experiment 4: (**a**) Cr, (**b**) Si.

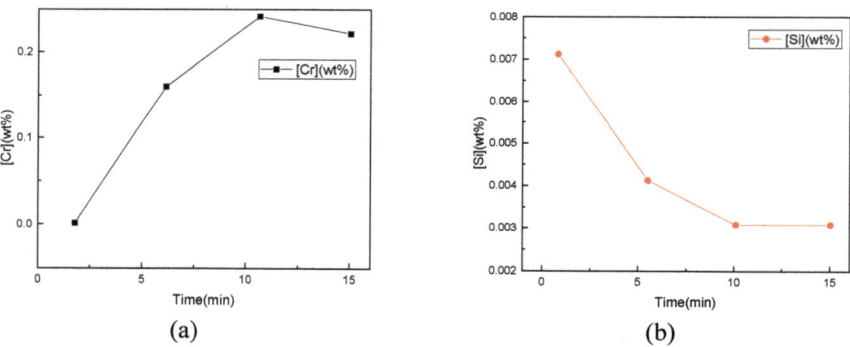

Figure 8. Variations in Cr and Si concentrations of steel with time for Experiment 5: (**a**) [Cr], (**b**) [Si].

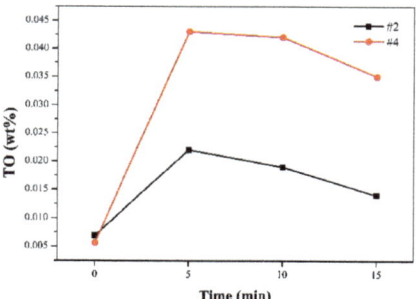

Figure 9. Variations in T[O] of steel with time for Experiments 2 and 4.

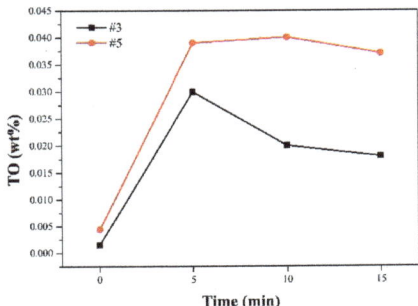

Figure 10. Variation in T[O] of steel with time for Experiments 3 and 5.

4. Conclusions

In this work, the unwanted reoxidation of steel by Cr_2O_3-bearing tundish cover flux was investigated through high temperature experiments. It was found that Cr_2O_3 can be reduced by Al to cause the reoxidation of steel. Moreover, as the Cr_2O_3 concentration increased, the reoxidation of the steel became more and more serious, and the Al loss also increased. It was also found that the reduction of Cr_2O_3 by Al mainly occurred within 2 to 3 min after the addition of the tundish cover flux with Cr_2O_3 concentrations of 5% and 10%. During the reduction of Cr_2O_3 by Al, the reaction rate reached a maximum value 2 min after the addition of the tundish cover flux with 10% Cr_2O_3, which may be related to kinetic conditions, and requires further investigation. Moreover, the reoxidation of molten steel by Cr_2O_3 caused the oxygen potential of the molten steel to rise, thereby suppressing the reoxidation of SiO_2. It was concluded from two controlled experiments that Cr_2O_3 can also transport oxygen to the steel through decomposition. Furthermore, it was calculated that when Al is present in steel, more than 90% the reoxidation of Cr_2O_3 is caused by Al, and the rest is caused by decomposition.

Author Contributions: S.Y. and F.W. conceived and designed the experiments and interpreted the data; D.L. and W.L. wrote the paper; F.W. and J.L. analyzed the data and collected the literature; D.L. and F.W. performed the experiments.

Funding: This research was funded by National Natural Science Foundation of China, grant number 51734003 and 51674023.

Conflicts of Interest: The authors declare no conflict of interest.

References

1. Yoon, B.H.; Heo, K.H.; Kim, J.S.; Sohn, H.S. Improvement of steel cleanliness by controlling slag composition. *Ironmak. Steelmak.* **2002**, *29*, 214–217. [CrossRef]
2. Deng, Z.; Zhu, M. A new double calcium treatment method for clean steel refining. *Steel Res. Int.* **2013**, *84*, 519–525. [CrossRef]
3. Zhang, L.; Zhi, J.; Mei, F.; Zhu, L.; Jiang, X.; Shen, J.; Thomas, B.G. Basic oxygen furnace based steelmaking processes and cleanliness control at Baosteel. *Ironmak. Steelmak.* **2006**, *33*, 129–139. [CrossRef]
4. Yan, P.; Van Ende, M.A.; Zinngrebe, E.; Laan, S.; Blanpain, B.; Guo, M. Interaction between steel and distinct gunning materials in the tundish. *ISIJ Int.* **2014**, *54*, 2551–2558. [CrossRef]
5. Mantovani, M.C.; Moraes, L.R.; Silva, R.L.; Cabral, E.F.; Possente, E.A.; Barbosa, C.A.; Ramos, B.P. Interaction between molten steel and different kinds of MgO based tundish linings. *Ironmak. Steelmak.* **2013**, *40*, 319–325. [CrossRef]
6. SchÜRmann, E.; Braun, U.; Pluschkell, W. Investigations on the equilibria between Al-Ca-O containing iron melts and CaO-Al_2O_3-FeO_n slags. *Steel Res.* **1998**, *69*, 355–361. [CrossRef]
7. Katsuhiro, S.; Yoshimasa, M. Reoxidation behavior of molten steel in tundish. *ISIJ Int.* **2000**, *40*, 40–47.
8. Wang, C.; Verma, N.; Kwon, Y.; Tiekink, W.; Kikuchi, N.; Sridhar, S. A study on the transient inclusion evolution during reoxidation of a Fe-Al-Ti-O melt. *ISIJ Int.* **2011**, *51*, 375–381. [CrossRef]

9. Yan, P.; Arnout, S.; Ende, M.A.V.; Zinngrebe, E.; Jones, T.; Blanpain, B.; Guo, M.X. Steel Reoxidation by Gunning Mass and Tundish Slag. *Metall. Mater. Trans. B* **2015**, *46*, 1242–1251. [CrossRef]
10. Yang, G.; Wang, X.; Huang, F.; Wang, W.; Yin, Y.; Tang, C. Influence of reoxidation in tundish on inclusion for Ca-treated Al-killed steel. *Steel Res. Int.* **2014**, *85*, 784–792. [CrossRef]
11. Kim, S.H.; Song, B. Thermodynamic aspects of steel reoxidation behavior by the ladle slag system of CaO-MgO-SiO$_2$-Al$_2$O$_3$-FetO-MnO-P$_2$O$_5$. *Metall. Mater. Trans. B* **1999**, *30*, 435–442. [CrossRef]
12. Brooks, G.A.; Rhamdhani, M.A.; Coley, K.S.; Subagyo; Pan, Y. Transient kinetics of slag metal reactions. *Metall. Mater. Trans. B* **2009**, *40*, 353–362. [CrossRef]
13. Jiang, F.; Liu, Y.; Xie, Y.; Cheng, G. Reoxidation of Al-Killed Steel by Ca(OH)$_2$ in the High Basicity Slag. *Steel Res. Int.* **2012**, *83*, 892–895. [CrossRef]
14. Qin, Y.; Wang, X.; Li, L.; Huang, F. Effect of oxidizing slag on cleanliness of IF steel during ladle holding process. *Steel Res. Int.* **2015**, *86*, 1037–1045. [CrossRef]
15. Ludlow, V.; Harris, B.; Riaz, S.; Normanton, A. Continuous casting mould powder and casting process interaction: Why powders do not always work as expected. *Ironmak. Steelmak.* **2005**, *32*, 120–126. [CrossRef]
16. Sabioni, A.C.S.; Huntz, A.M.; Silva, F.D.; Jomard, F. Diffusion of iron in Cr$_2$O$_3$: Polycrystals and thin films. *Mater. Sci. Eng. A* **2005**, *392*, 254–261. [CrossRef]
17. Pretorius, E.B.; Nunnington, R.C. Stainless steel slag fundamentals: From furnace to tundish. *Ironmak. Steelmak.* **2002**, *29*, 133–139. [CrossRef]
18. Albertsson, G.; Teng, L.; Björkman, B.; Seetharaman, S.; Engström, F. Effect of low oxygen partial pressure on the chromium partition in CaO–MgO–SiO$_2$–Cr$_2$O$_3$–Al$_2$O$_3$ synthetic slag at elevated temperatures. *Steel Res. Int.* **2013**, *84*, 670–679. [CrossRef]
19. Jo, S.K.; Kim, S.H. Thermodynamic assessment of CaO-SiO$_2$-Al$_2$O$_3$-MgO-Cr$_2$O$_3$-MnO-FetO slags for refining chromium-containing steels. *Steel Res. Int.* **2000**, *71*, 281–287. [CrossRef]
20. Nakasuga, T.; Nakashima, K.; Mori, K. Recovery rate of chromium from stainless slag by iron melts. *ISIJ Int.* **2004**, *44*, 665–672. [CrossRef]
21. Jiang, F.; Cheng, G.; Xie, Y.; Qian, G.; Rui, Q.; Song, Y. Reoxidation of Al-killed molten steel by Fe$_2$O$_3$ and Cr$_2$O$_3$ in the magnesia-chromite refractory. *Steel Res. Int.* **2013**, *84*, 1098–1103. [CrossRef]
22. Park, D.C.; Jung, I.H.; Rhee, P.C.; Lee, H.G. Reoxidation of Al-Ti containing steels by CaO-Al$_2$O$_3$-MgO-SiO$_2$ slag. *ISIJ Int.* **2004**, *44*, 1669–1678. [CrossRef]

© 2019 by the authors. Licensee MDPI, Basel, Switzerland. This article is an open access article distributed under the terms and conditions of the Creative Commons Attribution (CC BY) license (http://creativecommons.org/licenses/by/4.0/).

Article

Effect of Immersion Depth of a Swirling Flow Tundish SEN on Multiphase Flow and Heat Transfer in Mold

Peiyuan Ni [1,2,*], Mikael Ersson [3], Lage T. I. Jonsson [3], Ting-An Zhang [1] and Pär Göran Jönsson [3]

1. Key Laboratory of Ecological Metallurgy of Multi-Metal Intergrown Ores of Education Ministry, School of Metallurgy, Northeastern University, Shenyang 110819, China; zta2000@163.net
2. Department of Materials and Manufacturing Science, Graduate School of Engineering, Osaka University, 2-1, Yamadaoka, Suita, Osaka 565-0871, Japan
3. Department of Materials Science and Engineering, KTH Royal Institute of Technology, SE-100 44 Stockholm, Sweden; bergsman@kth.se (M.E.); lage@kth.se (L.T.I.J.); parj@kth.se (P.G.J.)
* Correspondence: peiyuan_ni@163.com; Tel.: +86-024-83686283

Received: 15 October 2018; Accepted: 5 November 2018; Published: 6 November 2018

Abstract: The effect of the immersion depth of a new swirling flow tundish SEN (Submerged Entry Nozzle) on the multiphase flow and heat transfer in a mold was studied using numerical simulation. The RSM (Reynolds Stress Model) and the VOF (Volume of Fluid) model were used to solve the steel and slag flow phenomena. The results show that the SEN immersion depth can significantly influence the steel flow near the meniscus. Specifically, an increase of the SEN immersion depth decreases the interfacial velocity, and this reduces the risk for the slag entrainment. The calculated Weber Number decreases from 0.8 to 0.2 when the SEN immersion depth increases from 15 cm to 25 cm. With a large SEN immersion depth, the steel flow velocity near the solidification front, which is below the mold level of SEN outlet, was increased. The temperature distribution has a similar distribution characteristic for different SEN immersion depths. The high temperature region is located near the solidification front. Temperature near the meniscus was slightly decreased when the SEN immersion depth was increased, due to an increased steel moving distance from the SEN outlet to the meniscus.

Keywords: swirling flow tundish; SEN immersion depth; multiphase flow; heat transfer; continuous casting mold

---

1. Introduction

Multiphase flow and heat transfer are very important phenomena in the continuous casting mold. These phenomena include steel-slag flow, inclusion motion, solidification, and so on. They can significantly influence the quality of the semifinal steel product. The basis for a good control on multiphase flow and heat transfer is a desirable steel flow in mold.

In the past, many studies have been carried out to optimize the multiphase flow and heat transfer in mold. The optimization investigations firstly focused on the structure of the SEN (Submerged Entry Nozzle), such as the SEN type (straight or bifurcated) [1,2], SEN port design (shape, angle, thickness) [3–10], and SEN immersion depth [3,4,6,11]. Argon injection in SEN [11] was also a widely investigated method to improve the continuous casting process, with the aim to reduce the nozzle clogging, reduce the steel reoxidation and increase the inclusion floatation in mold. In addition to these, EMBr (Electromagnetic Braking) [12,13] and M-EMS (Mold Electromagnetic Stirring) [8–10] have been vastly investigated to optimize the steel flow in molds. All the above investigations aim to reduce the steel flow velocity and the flow fluctuation near the meniscus, increase superheat removal, optimize the temperature distribution, and so on. Recently, swirling flow SEN has been considered to be a promising method to further modify the steel flow in mold. The significant improvement with this method is that it can directly change the steel flow characteristics before the steel flows into the mold—for example, the prevention of an impingement jet flow from a straight SEN. It was found that the heat and mass transfer near the meniscus can be remarkably activated [14–17], and a uniform velocity distribution can be obtained within a short distance from the SEN outlet [14–16]. Furthermore, the penetration depth of the SEN outlet flow is remarkably decreased in a billet mold [14,17]. Industrial trial results [18] show that the swirling flow SEN effectively improved the steel product quality and reduced the clogging problem of the SEN side ports.

In order to produce a swirling flow inside the SEN, the swirl blade method [14–18] and electromagnetic stirring method [19–22] were investigated in many studies. The lifespan of the swirl blade and the inclusion deposition on its surface, which may lead to nozzle clogging, restrict its application for longer casting times. Therefore, it has still not been used in industry since it was developed in 1994. The electromagnetic stirring method is associated with an equipment cost and an electricity cost, and thus its application will increase the steel production cost. Recently, Ni et al. [23–26] proposed a new method to produce a swirling flow in an SEN simply using a cylindrical tundish design. Its effectiveness has been confirmed, both by water model experiments and also by numerical simulations [25]. In addition, the steel flow characteristics and temperature distribution in mold were found to be improved using this new tundish design [26].

Previously, the influence of a swirling flow tundish design on the steel-slag flow, temperature distribution, and the steel flow in the vicinity of the solidified shell in mold was studied [26]. In this study, the effect of the immersion depth of the swirling flow tundish SEN on the multiphase flow and heat transfer in a billet mold were further studied based on the previous investigation [26]. This aims to understand the flow characteristic change under different conditions as a basis for its future application. The VOF (Volume of Fluid) method was used to capture the steel-slag interface, and the energy equation was solved to study the temperature distribution in the mold. The changes of the steel flow characteristics, steel-slag interface velocity, mold fluctuation, and temperature distribution in the mold induced by the SEN immersion depth were investigated.

2. Model Description

A three-dimensional mathematical model has been developed to describe the multiphase flow and heat transfer in a billet mold during the continuous casting of steel. The geometry and the dimension of the billet mold model is shown in Figure 1.

2.1. Model Assumption

The numerical model is based on the following assumptions:

(1) Steel and slag behave as incompressible Newtonian fluids;
(2) Solidification in the mold is not considered;
(3) A constant molecular viscosity for steel and slag was assumed. This is due to the fact that the maximum temperature difference in the mold is only 30 K between 1788 K and 1818 K as the superheat of the steel. The viscosity change in this temperature range is not significant, and this can be seen from a previous study [10];
(4) A constant steel and slag density was used. The temperature influence on the steel density change was accounted for in the source term of the momentum equation;
(5) The SEN wall was assumed to be a smooth wall.

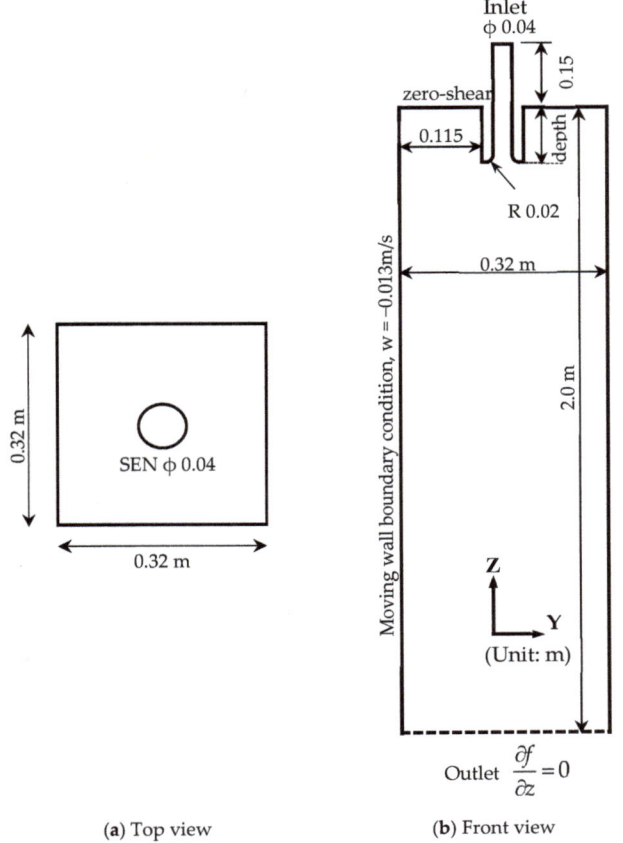

Figure 1. Geometry of the casting mold, (a) top view of the billet mold and (b) front view of the mold.

2.2. Transport Equations

The conservation of a general variable ϕ within a finite control volume can be expressed as a balance among the various processes, which tends to increase or decrease the variable values. The conservation equations, for example, continuity, volume fraction, momentum, turbulence equations, and energy equation, can be expressed by the following general equation [27]:

$$\frac{\partial}{\partial t}(\rho\phi) + \frac{\partial}{\partial x_i}(\rho\phi u_i) = \frac{\partial}{\partial x_i}\left(\Gamma_\phi \frac{\partial \phi}{\partial x_i}\right) + S_\phi, \tag{1}$$

where the first term on the left-hand side is the instantaneous change of ϕ with time, the second term on the left-hand side represents the transport due to convection, and the first term on the right-hand side expresses the transport due to diffusion where Γ_ϕ is the diffusion coefficient with different values for different turbulence models, or the effective thermal conductivity. Furthermore, the second term on the right-hand side is the source term.

The steel-slag interface was tracked by the VOF model [28]. The sum of the slag phase fraction α_{slag} and the steel phase fraction α_{steel} is equal to 1. The mixed material properties in the grid cell, where the interface exists, are required by the momentum equation and can be calculated by the following equations:

$$\rho_{mix} = \alpha_{steel}\rho_{steel} + \alpha_{slag}\rho_{slag} \qquad (2)$$

$$\mu_{mix} = \alpha_{steel}\mu_{steel} + \alpha_{slag}\mu_{slag}. \qquad (3)$$

The realizable k-ε turbulence model, coupled with the Enhanced Wall Treatment model, was first used to produce an initial flow field [28]. Then, the Reynolds stress model (RSM) model combined with the Stress-Omega submodel was used to simulate the steel flow. The Stress-Omega submodel is good for modeling flows over the curved surfaces and swirling flows [28]. The Reynolds stress terms emerging from the Reynolds averaging of Navier-Stockes equations are directly solved to account for the possible anisotropic fluctuation in a swirling flow.

The temperature distribution in mold was obtained by solving the following energy equation [28]:

$$\frac{\partial}{\partial t}(\rho E) + \frac{\partial}{\partial x_i}(\mu_i(\rho E + p)) = \frac{\partial}{\partial x_i}\left(\left(k + \frac{c_p \mu_t}{Pr_t}\right)\frac{\partial \phi}{\partial x_i}\right), \qquad (4)$$

where E is energy in the unit of J, k is the thermal conductivity with the unit of W/(m·K), c_p is the specific heat capacity in J/(kg·K), μ_t is the turbulent viscosity, Pr_t is the turbulent Prandtl Number, ρ is fluid density in kg/m^3, p is pressure in Pa, and T is temperature in K. The steel density change and subsequent natural convection due to temperature variance was accounted for by the Boussinesq Model [29]. This model treats density as a constant value in all solved equations, except for the buoyancy in the momentum equation (it is normally put in source term) as follows:

$$(\rho - \rho_0)g \approx -\rho_0\beta(T - T_0)g, \qquad (5)$$

where ρ_0 is the (constant) density of the liquid steel with the unit of kg/m^3, T_0 is the operating temperature in K, and β is the thermal expansion coefficient of the liquid steel. The thermal properties of the fluids and some parameters are shown in Table 1.

Table 1. Thermal properties of the steel and slag.

Parameters	Symbols	Steel	Slag
Density, kg/m^3	ρ_0	7000	2600
Viscosity, kg/(m·s)	μ	0.0064	0.09
Thermal conductivity, w/(m·K)	k	35	1.1
Specific heat, J/(kg·K)	c_p	628	1200
Thermal expansion Coefficient, 1/K	β	10^{-4}	-
Interface tension, N/m	σ	1.6	
Operating Temperature, K	T_o	1788	
Turbulent Prandtl Number	Pr_t	0.85	

2.3. Boundary Conditions

The velocity profile on the cross section of the cylindrical tundish SEN, which has been solved in a previous study [25], was used as the inlet boundary condition for the current simulation of the mold flow. This steel flow velocity at the inlet in Figure 1b has been presented in a previous study [26] and, thus, it is not repeated here. A nonslip boundary condition was imposed on the SEN wall. A zero-shear slip wall boundary condition was used at the mold surface. For the mold wall, a moving wall boundary condition with the velocity of 0.013 m/s in Z or downwards direction was used to account for the movement of the solidified shell in a real casting process. A fully developed flow condition is adopted at the mold outlet, where the normal gradients of all variables are set to zero. A constant steel temperature of 1818 K was used at the inlet, with a superheat of 30 K. A constant temperature of 1788 K was imposed on the solidified shell. An adiabatic condition was used both at the SEN wall and at the free surface.

2.4. Solution Method

The numerical model was solved using the commercial software ANSYS FLUENT 18.0® (ANSYS, Canonsburg, PA, USA). The numerical simulations were carried out based on 1.4 million grid cells to guarantee the grid-independent solution. A fine grid was used in the near-wall region, with the y⁺ value of the first grid layer around 1. The PISO (Pressure-Implicit with Splitting of Operators) scheme was used for the pressure-velocity coupling. Furthermore, the PRESTO (PREssure STaggering Option) method was adopted to discretize the pressure. The governing equations were discretized using a second order upwind scheme. The convergence criteria were as follow: The residuals of all dependent variables were smaller than 1×10^{-3} at each time step.

3. Results and Discussion

The multiphase flow and heat transfer in the mold with different SEN immersion depths were firstly solved by the realizable k-ε model with an Enhanced Wall Treatment for the first 75 s. After that, this solution was used as an initial condition for the RSM model calculation to 125 s for a developed flow field. The multiphysics in the mold with different SEN immersion depths were analyzed and compared in the following.

3.1. Steel Flow Phenomena

Figure 2 shows the steel flow path in the mold with different SEN immersion depths. It can be observed that the steel flow pattern in mold was similar for different SEN depths. It delivers the steel into the mold along the periphery of the SEN, which is in 360°. The SEN outlet flow moves towards the solidified shell after it flows out from the straight SEN due to the swirling flow effect, inducing a rotational steel flow momentum. After the steel stream reaches the solidified shell, a part of the steel flows downwards along the solidified shell with a horizontally rotational flow momentum, and another part of the steel moves upwards and towards the meniscus. Due to the difference in SEN immersion depth, the top rotational flow region near the meniscus was large when a large immersion depth of SEN was used. This should be beneficial for the decrease of the steel flow velocity, since the steel from SEN outlet needs a long distance to reach the steel-slag interface. Therefore, the current swirling flow tundish SEN can deliver high temperature steel uniformly distributed towards the solidified shell, no matter the change of the SEN depth.

Figure 3 shows the velocity on the vertical plane located at the middle of the mold for different SEN immersion depths. It can be seen that the high velocity region was located at the solidification front in the mold. Steel moves downwards at the region near the solidified shell and it flows upwards in the center of the mold. The effect of the SEN depth is mainly on the steel flow velocity at the top of the mold. It can be seen that the region with a high steel flow velocity was reduced when a large SEN immersion depth was used. This is expected to reduce the risk of the slag entrainment at the steel–slag interface. When a large SEN immersion depth was used, the length of the SEN was increased. The dissipation of the rotational momentum was expected due to the friction of the SEN wall. However, it did not show significant influence on the steel flow in the mold below the height of the SEN outlet. Figure 4 shows a comparison of vertical velocity distributions along the line, with the mold depth of 1.5 m, for different SEN immersion depths. It can be seen that a large velocity with a magnitude of 0.03 m/s exists in the solidification front. This may be helpful to shear off the dendrites from the solidification interface and promotes the nucleate, which results in an enhancement of the transition from a columnar to equiaxed solidification [30]. Although the immersion depth of the SEN is different, the vertical velocity shows a similar distribution and magnitude. This means that the 10 cm difference in the immersion depth does not lead to a large change of the steel flow below the level of SEN outlet in mold.

Figure 5 shows that the velocities of the steel flow at the different cross sections of for different SEN immersion depths. It can be seen from Figure 5a,b that the rotational flow developed very well in the mold depth of 1.0 m. A large tangential velocity was located at the solidification front, and the maximum tangential velocity can reach around 0.014 m/s near the solidified shell, which is a similar value for different SEN immersion depths. However, the rotational steel flow near the meniscus (Mold depth of 0.05 m) was not observed for these two SEN immersion depths. The steel flow magnitude in Figure 5c for the SEN immersion depth of 25 cm is smaller, compared to the immersion depth of 15 cm. It can be concluded that the rotational steel flow mainly develops in the deep mold rather than near the meniscus. Figure 6 shows the magnitude of the tangential velocity along different horizontal lines in different depths. The top view of the locations of these horizontal lines is shown in Figure 5a. It can be seen that for both cases, the maximum tangential velocity decreases when the steel moves downwards. The large tangential velocity region is located near the solidification front. In addition, at the location with the same distance from the SEN outlet ($Z = -0.5$ m and -0.6 m for SEN immersion depth of 15 cm and 25 cm, respectively), the maximum tangential velocity decreases with an increased SEN immersion depth. This is due to the fact that the swirling steel flow passed a larger SEN length when a larger immersion depth was used and the dissipation of SEN wall on the rotational steel flow momentum was avoided.

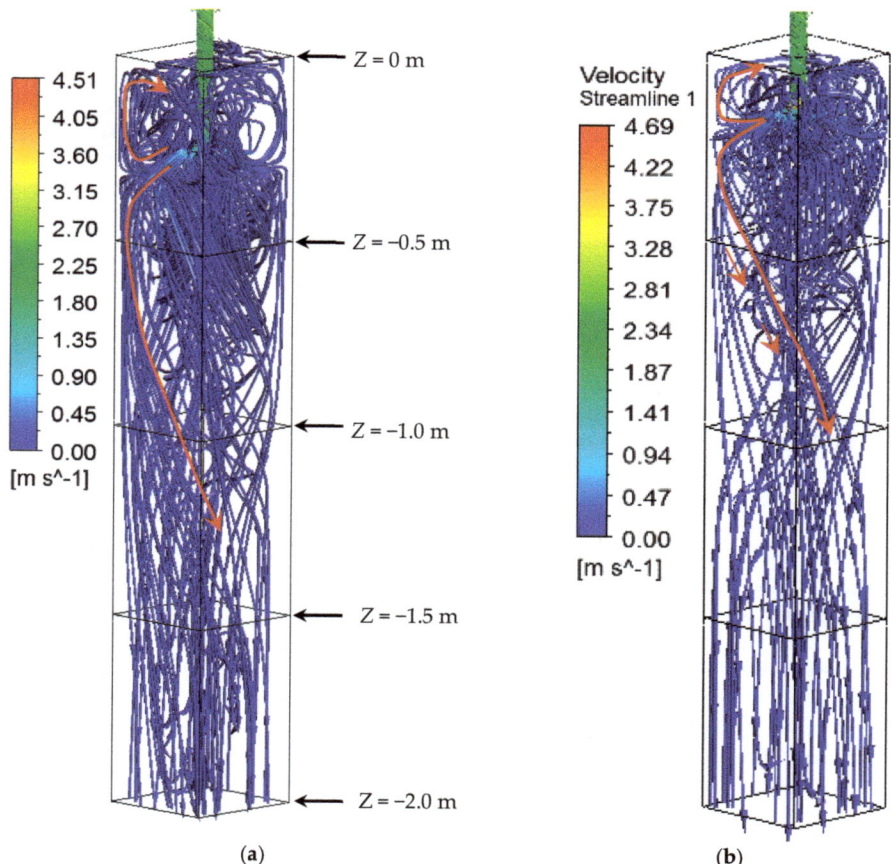

Figure 2. Comparison of steel flow paths in mold, (**a**) SEN immersion depth of 25 cm and (**b**) SEN immersion depth of 15 cm [26].

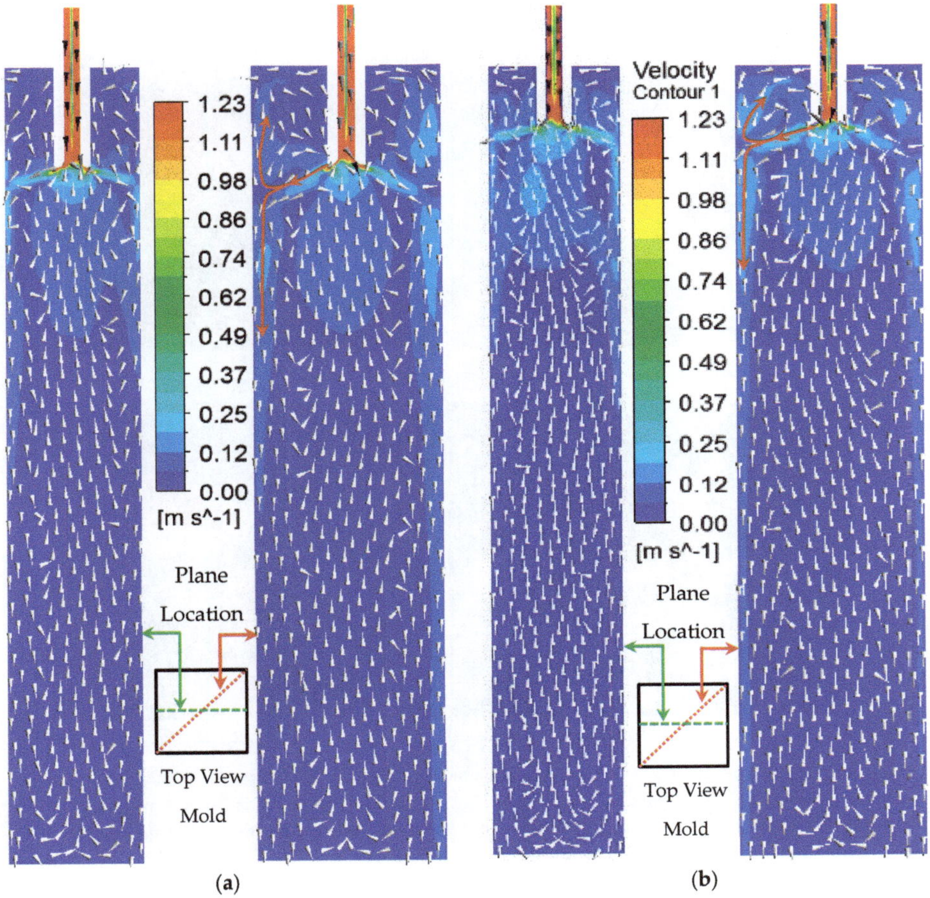

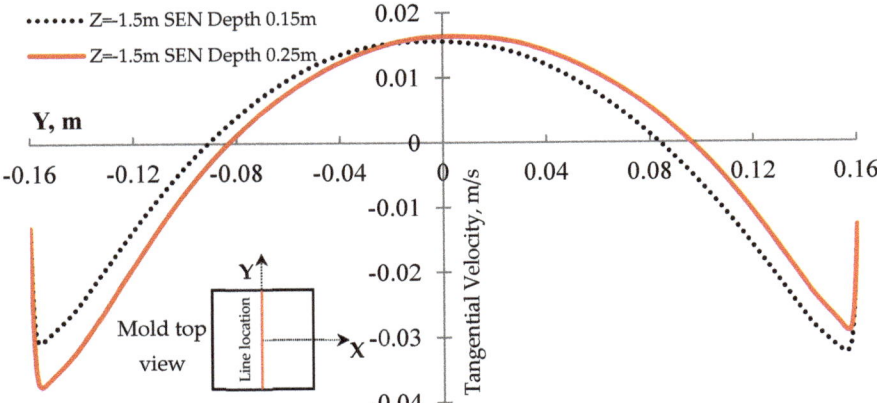

Figure 3. Steel flow velocity in the vertical middle plane of the mold, (**a**) SEN immersion depth of 0.25 m and (**b**) SEN immersion depth of 0.15 m [26]. (Arrows are the steel flow directions).

Figure 4. Vertical steel flow velocity along horizontal lines in different mold depths.

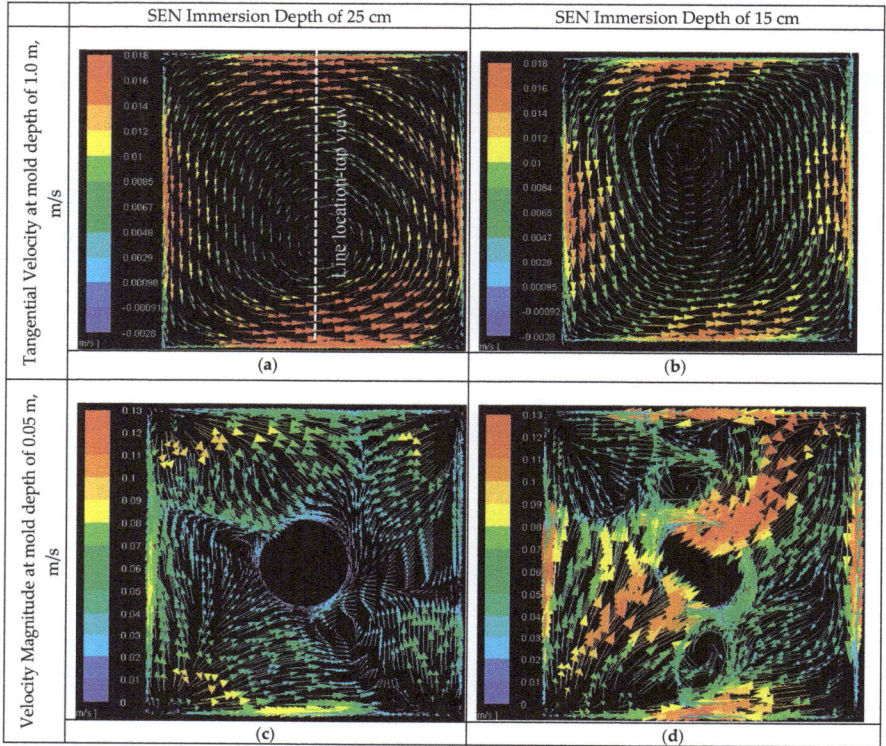

Figure 5. Velocities on different cross sections of the mold at different mold depths and SEN immersion depths, (**a**) tangential velocity for SEN depth of 0.25 m and mold depth of 1.0 m, (**b**) tangential velocity for the SEN depth of 0.15 m and mold depth of 1.0 m, (**c**) velocity magnitude for the SEN depth of 0.25 m and mold depth of 0.05 m and (**d**) velocity magnitude for the SEN depth of 0.15 m and mold depth of 0.05 m.

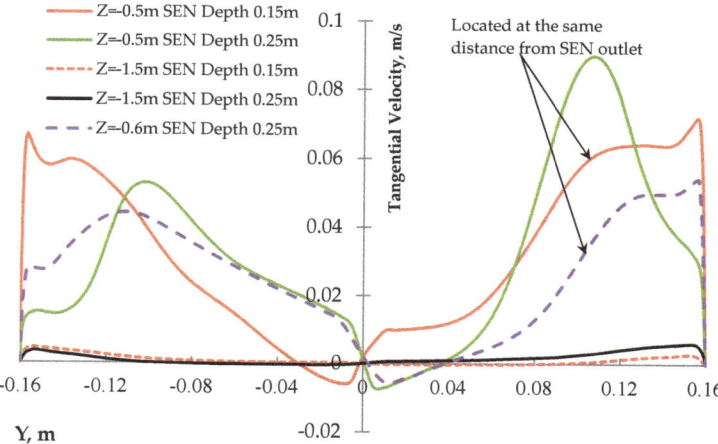

Figure 6. Tangential velocity distribution along different horizontal lines in different mold depths.

Figure 7 shows the velocity magnitude distribution along different lines in mold depth direction. Figure 7a is the velocity distribution at Location A with 1 cm away from the solidification shell. At the top of the mold, it can be seen that the velocity magnitude with a large SEN immersion depth is smaller than that with a small SEN immersion depth. This is helpful to reduce the risk of the slag entrainment. In the low part of the mold, the velocity near the solidification front is larger with a larger SEN immersion depth, and this is helpful for the formation of equiaxed crystals. In Figure 7b, the velocity distribution at Location B, which is close to the mold center, was presented. It can be seen that the major difference exists at the top of the mold, with a smaller velocity when a larger SEN depth was used. Furthermore, the velocity was similar at the location in deep mold. In summary, the general trend of the flow change when the SEN immersion depth was increased is that the velocity in the top mold decreased while the velocity at the low part of the mold increased.

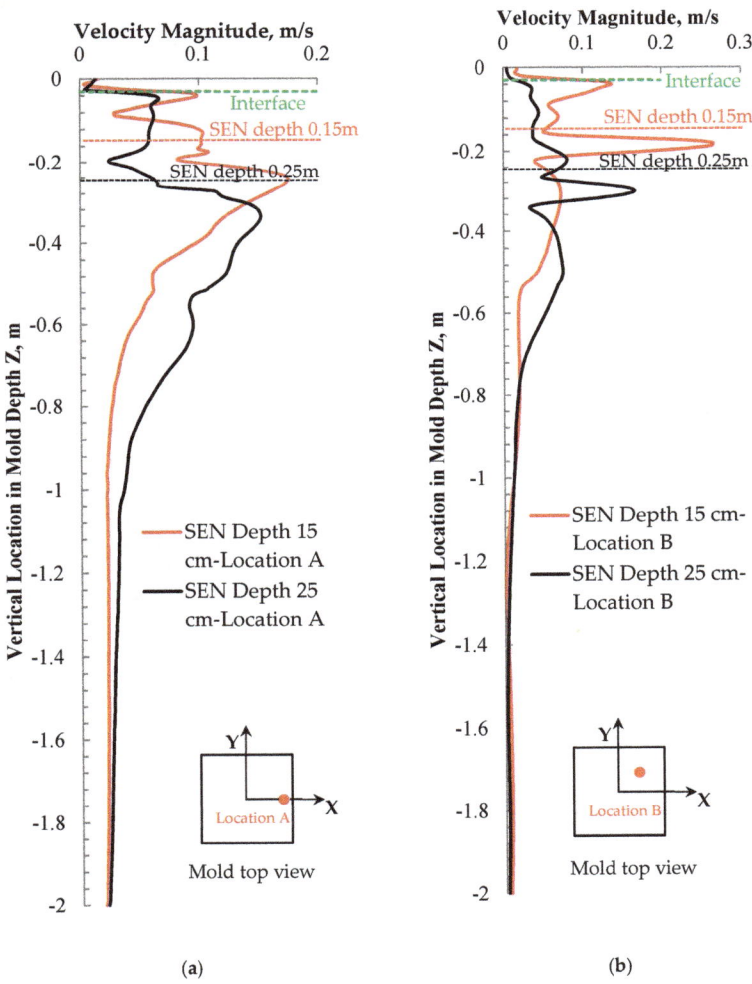

Figure 7. Total velocity distribution along different lines in mold depth direction, (**a**) Location A with 1 cm away from the wall and (**b**) Location B in the center of the top-right one quarter cross section.

3.2. Steel-Slag Interface Phenomena

One of the most important concerns about the swirling flow SEN is about the steel flow and heat transfer near the meniscus. Due to the existence of the swirling flow, the impingement jet flow in a conventional tundish casting disappeared [26]. The steel flow moves towards the solidification front, as shown in Figure 3. The induced steel flow in the meniscus region was increased, and this led to the heat transfer near the meniscus accelerating [26]. However, a large steel flow velocity near the meniscus region also illustrates a high risk of the slag entrainment. Therefore, it is very important to investigate the effect of the SEN immersion depth on the steel-slag interface behavior.

Figure 8 shows the steel-slag interface and the flow pattern in mold for different SEN immersion depths. It can be seen that the thickness of the steel-slag interface region was larger when a small immersion depth was used. This can also be seen from the iso-surface of the slag where the density of 2601 kg/m^3 was plotted (the pure slag has the density of 2600 kg/m^3). Above this iso-surface, almost pure slag existed. It can be seen that the steel flow here is strong to push the slag upwards. With a large SEN immersion depth, this phenomenon was alleviated. The difference comes from the upwards steel flow, indicated with the arrow in Figure 8, where the upwards steel flow is caused by the rotational steel flow from the swirling flow SEN. The VOF model cannot predict the sharp interface of two phases. Instead, a mixing region was predicted, and this mixing region does not mean that steel and slag are well mixed in reality. However, the thickness of the mixing region reflects the intensity of a steel-slag mixing. Figure 9 shows the enlarged view of the thickness of the steel-slag mixing region. It can be seen that the predicted thickness of the mixing region was much smaller when a larger SEN immersion depth was used. This means that the interface became stable. Figure 10 shows the distributions of the velocity magnitude and turbulent kinetic energy along the steel-slag interface for different SEN immersion depths. The location of the Line plot is shown in Figure 8. It can be seen that the velocity magnitude at the interface is decreased with an increased SEN immersion depth, with the maximum value decreasing from 0.05 m/s to 0.025 m/s. The maximum turbulent kinetic energy is similar for different SEN immersion depths, while its distribution shows a large difference. This may be due to the fact that the upwards steel flow arriving at the interface became more uniform when a large SEN immersion depth was used. In the case with M-EMS, the level fluctuation was also found to be increased [10,30]. The meniscus surface has a swirl flow and the meniscus level rises near the bloom strand wall and sinks around the SEN wall, which shows an inclined steel-slag interface [30]. Sometimes, a vortex formation near the SEN wall was found with M-EMS [31]. Therefore, the mold level fluctuation should be considered to make it as low as possible, both for M-EMS applications and for the use of swirling flow SEN. For the current swirling flow SEN, the interface flow can be controlled by the SEN immersion depth.

According to previous research [32], the slag entrainment into liquid steel may occur when the Weber number is greater than 12.3. The Weber number can be defined as:

$$We = \frac{\mu_l^2 \rho_l}{\sqrt{\sigma g (\rho_l - \rho_s)}} \tag{6}$$

where μ_l is the radial steel velocity, g is gravitational acceleration, and σ is the interfacial tension between steel and slag. A slag density value of 2600 kg/m^3 was used, and the value of interfacial tension between the steel and the slag was set to 1.16 N/m. The maximum total velocity at the steel-slag interface in the mold, 0.05 m/s for immersion depth of 15 cm and 0.025 m/s for immersion depth of 25 cm, was used to calculate the Weber Number. The calculated maximum Weber number is around 0.8 and 0.2 for the small and large immersion depth, respectively. Therefore, the Weber number is still much smaller than 12.3, which means a small risk for the slag entrainment. Furthermore, 25 cm immersion depth can further decrease this risk. However, slag entrainment should be experimentally investigated in the future.

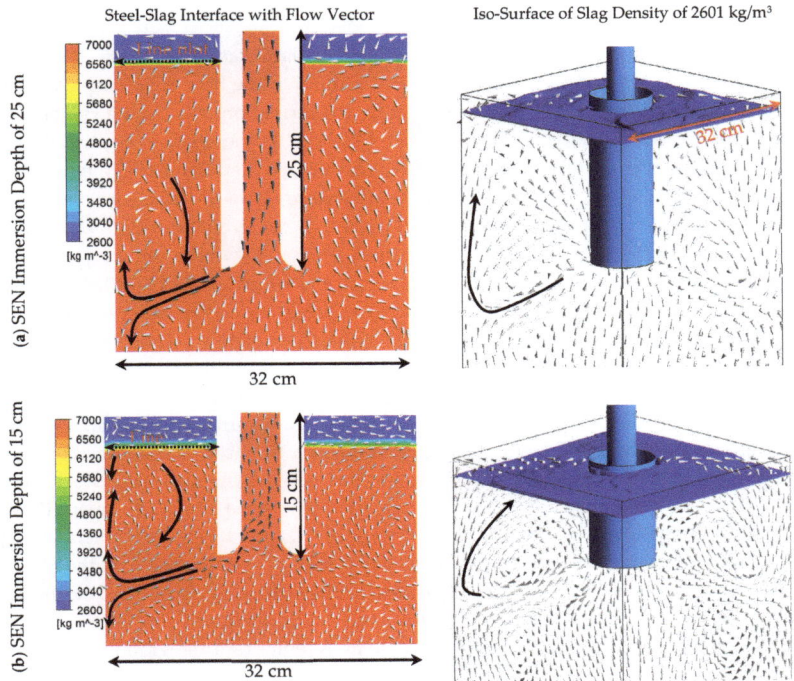

Figure 8. Steel-slag interface with steel flow vectors.

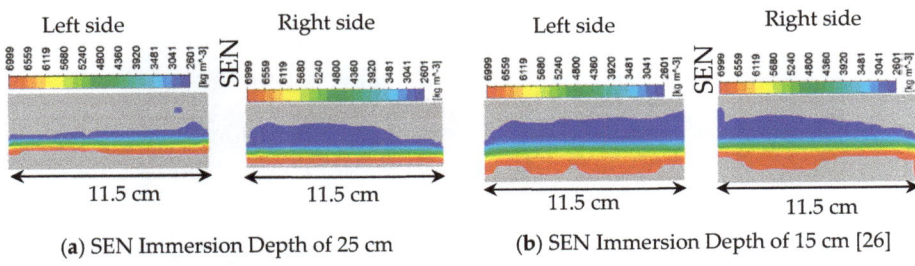

Figure 9. Enlarged interface region in Figure 8, with a density range from pure slag to pure steel.

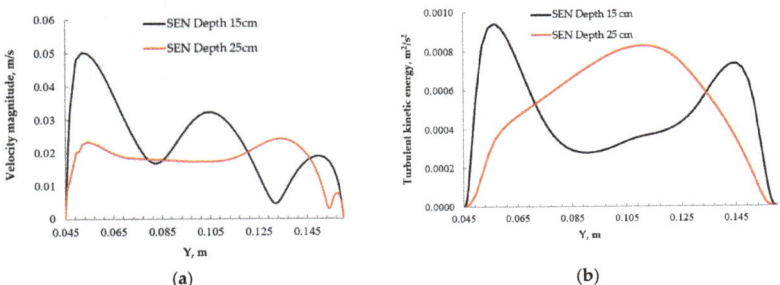

Figure 10. (**a**) Velocity magnitude at the steel-slag interface and (**b**) turbulent kinetic energy at the steel-slag interface.

3.3. Temperature Field

Steel temperature in the mold is very important, since it significantly influences the solidification structure, which in turn determines the product quality. The swirling flow SEN has proven that it can accelerate steel superheat removal [26]. This is good for the formation of equiaxed crystals. Figure 11 shows the temperature distribution in the mold with different SEN immersion depths. It can be seen that similar temperature distribution characteristics were observed for different SEN immersion depths. Due to the swirling flow effect, steel with a high temperature flows towards the solidified shell. It increased the temperature near the solidified shell as well as the temperature gradient there, while the core temperature of the billet was low. On the cross section at a depth of 0.5 m in the mold, the maximum temperatures for the immersion depths of 25 cm and 15 cm are 1806 and 1804 K, respectively. It can be seen that the high temperature region is not located in the center of the mold. These values decrease to 1796 K and 1795 K at the mold depth of 1.5 m, respectively. Here, the high temperature region was located in the mold center. This is due to the fact that the superheat of the steel near the solidification front can be removed fast, while that in the mold center cannot be easily dissipated. In addition, there are some differences induced by the increase of the immersion depth. The first issue is about the temperature near the meniscus, where a low temperature was observed when a large immersion depth was used. This is obviously due to the fact that the upwards steel flow near the solidified shell (as shown in Figure 3a) with a high temperature needs a long way to reach the meniscus and the superheat was dissipated along the way. Furthermore, due to a large immersion depth, the high temperature region moves downwards around 25 cm, compared to that with the SEN depth of 15 cm. This directly leads to the thickness of the low temperature region in the solidification front decreasing, as shown in Figure 11. Therefore, an increase in the SEN depth only slightly decreases the steel temperature near the meniscus and it makes the high temperature region in the mold move a bit downwards.

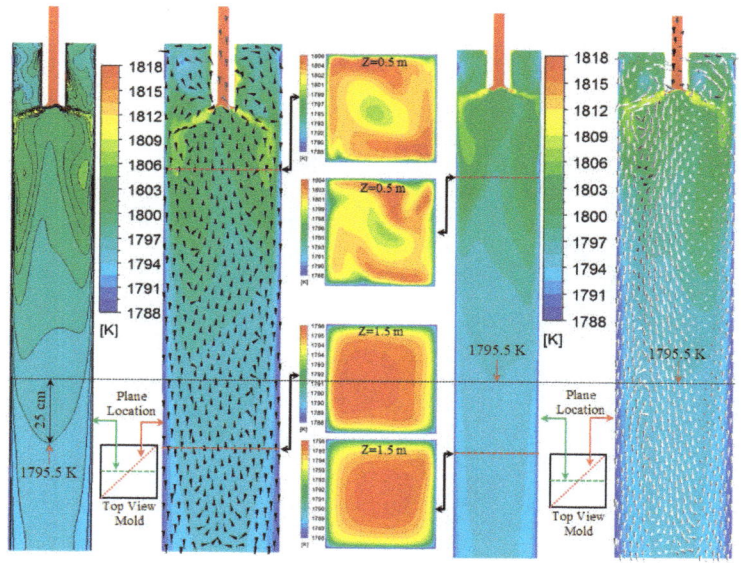

(a) SEN Immersion Depth of 25 cm   (b) SEN Immersion Depth of 15 cm [26]

Figure 11. Temperature distribution in mold for different SEN immersion depths.

4. Conclusions

The effects of the SEN immersion depth on the multiphase flow and heat transfer in a mold with a new cylindrical tundish design for continuous casting were investigated using numerical simulations. The main conclusions were the following:

1. Steel flow patterns are similar for different SEN immersion depths, with the flow direction towards the solidification front.
2. An increase in the SEN immersion depth decreases the interfacial velocity and this reduces the risk of slag entrainment. The calculated Weber Number is 0.8 and 0.2 for the SEN depth of 15 cm and 25 cm, respectively. The steel flow velocity near the solidification front below the SEN outlet is increased with a large SEN immersion depth.
3. The temperature distribution has a similar distribution characteristic. The high temperature region is located near the solidification front. Temperature near the meniscus was slightly decreased when the SEN immersion depth was increased.
4. A large SEN immersion depth was recommended in order to reduce the slag entrainment. This will not reduce the steel flow velocity near the solidification front, nor will it significantly reduce the temperature near the meniscus.

Author Contributions: P.N. and L.T.I.J. designed the paper; P.N. and M.E. did the numerical simulation; all the authors analyzed and discussed the results; P.N. wrote the paper; L.T.I.J., T.Z. and P.G.J. revised the paper.

Funding: The authors want to thank the National Natural Science Foundation of China (Grant No. 51704062) for the support on this work.

Conflicts of Interest: The authors declare no conflict of interest.

References

1. Szekely, J.; Yadoya, R.T. The physical and mathematical modelling of the flow field in the mold region of continuous casting systems. Part II. The mathematical representation of the turbulence flow field. *Metall. Mater. Trans.* **1973**, *4*, 1379–1388. [CrossRef]
2. Xu, M.; Zhu, M. Transport phenomena in a Beam-Blank continuous casting mold with two types of submerged entry nozzle. *ISIJ Int.* **2015**, *55*, 791–798. [CrossRef]
3. Thomas, B.G.; Mika, L.J.; Najjar, F.M. Simulation of fluid flow inside a continuous slab-casting machine. *Metall. Mater. Trans. B* **1990**, *21*, 387–400. [CrossRef]
4. Calderon-Ramos, I.; Morales, R.D.; Garcia-Hernandez, S.; Ceballos-Huerta, A. Effects of immersion depth on flow turbulence of liquid steel in a slab mold using a nozzle with upward angle rectangular ports. *ISIJ Int.* **2014**, *54*, 1797–1806. [CrossRef]
5. Calderon-Ramos, I.; Morales, R.D.; Salazar-Campoy, M. Modeling flow turbulence in a continuous casting slab mold comparing the use of two bifurcated nozzles with square and circular ports. *Steel Res. Int.* **2015**, *86*, 1610–1621. [CrossRef]
6. Calderon-Ramos, I.; Morales, R.D. The role of submerged entry nozzle port shape on fluid flow turbulence in a slab mold. *Metall. Mater. Trans. B* **2015**, *46*, 1314–1325. [CrossRef]
7. Salazar-Campoy, M.; Morales, R.D.; Najera-Bastida, A.; Cedillo-Hernandez, V.; Delgado-Pureco, J.C. A physical model to study the effects of nozzle design on dense two-phase flows in a slab mold casting ultra-low carbon steels. *Metall. Mater. Trans. B* **2017**, *48*, 1376–1389. [CrossRef]
8. Sun, H.; Zhang, J. Macrosegregation improvement by swirling flow nozzle for bloom continuous castings. *Metall. Mater. Trans. B* **2014**, *45*, 936–946. [CrossRef]
9. Sun, H.; Li, L. Application of swirling flow nozzle and investigation of superheat dissipation casting for bloom continuous casing. *Ironmak. Steelmak.* **2016**, *43*, 228–233. [CrossRef]
10. Fang, Q.; Ni, H.; Zhang, H.; Wang, B.; Lv, Z. The effects of a submerged entry nozzle on flow and initial solidification in a continuous casting bloom mold with electromagnetic stirring. *Metals* **2017**, *7*, 146. [CrossRef]

11. Thomas, B.G.; Dennisov, A.; Bai, H. Behavior of Argon Bubbles during Continuous Casting of Steel. In Proceedings of the 80th Steelmaking Conference, Chicago, IL, USA, 13–16 April 1997.
12. Wang, Y.; Zhang, L. Fluid flow-related transport phenomena in steel slab continuous casting strands under electromagnetic brake. *Metall. Mater. Trans. B* **2011**, *42*, 1319–1351. [CrossRef]
13. Yu, H.; Zhu, M. Numerical simulation of the effects of electromagnetic brake and argon gas injection on the three-dimensional multiphase flow and heat transfer in slab continuous casting mold. *ISIJ Int.* **2008**, *48*, 584–591. [CrossRef]
14. Yokoya, S.; Takagi, S.; Iguchi, M.; Asako, Y.; Westoff, R.; Hara, S. Swirling effect in immersion nozzle on flow and heat transport in billet continuous casting mold. *ISIJ Int.* **1998**, *38*, 827–833. [CrossRef]
15. Yokoya, S.; Jönsson, P.G.; Sasaki, K.; Tada, K.; Takagi, S.; Iguchi, M. The effect of swirl flow in an immersion nozzle on the heat and fluid flow in a billet continuous casting mold. *Scand. J. Metall.* **2004**, *33*, 22–28. [CrossRef]
16. Kholmatov, S.; Takagi, S.; Jonsson, L.; Jönsson, P.; Yokoya, S. Development of flow field and temperature distribution during changing divergent angle of the nozzle when using swirl flow in a square continuous casting billet mold. *ISIJ Int.* **2007**, *47*, 80–87. [CrossRef]
17. Kholmatov, S.; Takagi, S.; Jonsson, L.; Jönsson, P.; Yokoya, S. Effect of nozzle angle on flow field and temperature distribution in a billet mold when using swirl flow. *Steel Res. Int.* **2008**, *79*, 31–39. [CrossRef]
18. Tsukaguchi, Y.; Hayashi, H.; Kurimoto, H.; Yokoya, S.; Marukawa, K.; Tanaka, T. Development of swirling-flow submerged entry nozzles for slab casting. *ISIJ Int.* **2010**, *50*, 721–729. [CrossRef]
19. Geng, D.; Lei, H.; He, J.; Liu, H. Effect of electromagnetic swirling flow in slide-gate SEN on flow field in square billet continuous casting mold. *Acta Metall. Sin.* **2012**, *25*, 347–356.
20. Wondrak, Th.; Eckert, S.; Galindo, V.; Gerbeth, G.; Stefani, F.; Timmel, K.; Peyton, A.J.; Yin, W.; Riaz, S. Liquid metal experiments with swirling flow submerged entry nozzle. *Ironmak. Steelmak.* **2012**, *39*, 1–9. [CrossRef]
21. Li, D.; Su, Z.; Chen, J.; Wang, Q.; Yang, Y.; Nakajima, K.; Marukaw, K.; He, J. Effects of electromagnetic swirling flow in submerged entry nozzle on square billet continuous casting of steel process. *ISIJ Int.* **2013**, *53*, 1187–1194. [CrossRef]
22. Yang, Y.; Jönsson, P.G.; Ersson, M.; Su, Z.; He, J.; Nakajima, K. The Influence of swirl flow on the flow field, temperature field and inclusion behavior when using a half type electromagnetic swirl flow generator in a submerged entry and mold. *Steel Res. Int.* **2015**, *86*, 1312–1327. [CrossRef]
23. Ni, P.; Jonsson, L.; Ersson, M.; Jönsson, P. A new tundish design to produce a swirling flow in the SEN during continuous casting of steel. *Steel Res. Int.* **2016**, *87*, 1356–1365. [CrossRef]
24. Ni, P.; Jonsson, L.; Ersson, M.; Jönsson, P. Non-Metallic inclusion behaviors in a new tundish and SEN design using a swirling flow during continuous casting of steel. *Steel Res. Int.* **2017**, *88*, 1600155. [CrossRef]
25. Ni, P.; Wang, D.; Jonsson, L.; Ersson, M.; Zhang, T.; Jönsson, P. Numerical and physical study on a cylindrical tundish design to produce a swirling flow in the SEN during continuous casting of steel. *Metall. Mater. Trans. B* **2017**, *48*, 2695–2706. [CrossRef]
26. Ni, P.; Ersson, M.; Jonsson, L.; Zhang, T.; Jönsson, P. Numerical study on the influence of a swirling flow tundish on multiphase flow and heat transfer in mold. *Metals* **2018**, *8*, 368. [CrossRef]
27. Patankar, S.V. *Numerical Heat Transfer and Fluid Flow*; Hemispere Publishing Corporation: New York, NY, USA, 1980.
28. *ANSYS Fluent Theory Guide, Release 18.0*; ANSYS: Canonsburg, PA, USA, 2017.
29. *ANSYS Fluent User's Guide, Release 18.0*; ANSYS: Canonsburg, PA, USA, 2017.
30. Liu, H.; Xu, M.; Qiu, S.; Zhang, H. Numerical simulation of fluid flow in a round bloom mold with In-Mold rotary electromagnetic stirring. *Metall. Mater. Trans. B* **2012**, *43*, 1657–1675. [CrossRef]
31. Willers, B.; Barna, M.; Reiter, J.; Eckert, S. Experimental investigations of rotary electromagnetic mould stirring in continuous casting using a cold liquid metal model. *ISIJ Int.* **2017**, *57*, 468–477. [CrossRef]
32. Jonsson, L.; Jönsson, P. Modeling of fluid flow conditions around the slag/metal interface in a gas-stirred ladle. *ISIJ Int.* **1996**, *36*, 1127–1134. [CrossRef]

© 2018 by the authors. Licensee MDPI, Basel, Switzerland. This article is an open access article distributed under the terms and conditions of the Creative Commons Attribution (CC BY) license (http://creativecommons.org/licenses/by/4.0/).

Article

Thermodynamic and Experimental Studies on Al Addition of 253MA Steel

Yandong Li [1], Tongsheng Zhang [2,*], Chengjun Liu [3] and Maofa Jiang [3]

1. Key Laboratory of Extraordinary Bond Engineering and Advanced Materials Technology, Yangtze Normal University, Chongqing 408000, China; andyydlee@gmail.com
2. School of Metallurgy and Environment, Central South University, Changsha 410083, China
3. Key Laboratory of Ecological Utilization of Multi-Metallic Mineral of Education Ministry, Northeastern University, Shenyang 110819, China; neu_zts@163.com (C.L.); anxin1984@126.com (M.J.)
* Correspondence: tongsheng.zhang@csu.edu.cn; Tel.: +86-139-7498-2473

Received: 27 March 2019; Accepted: 10 April 2019; Published: 12 April 2019

Abstract: To solve the nozzle clogging issue in the continuous casting process of 253MA steel, a method of modifying solid inclusions to liquid phases is proposed. The CALPHAD technique was employed to predict the liquid region of the Al_2O_3-SiO_2-Ce_2O_3 system. Then a thermodynamic package based on the extracted data during the phase diagram optimization process was developed. This package was then used to compute the appropriate aluminum addition, which was 0.01% in 253MA steel. The Si-Al alloy was chosen as the deoxidant according to the thermodynamic analysis. The solid inclusions were ultimately modified to liquid phases at 1500 °C when cerium was added through the equilibrium experiments in a $MoSi_2$ tube furnace.

Keywords: CALPHAD; thermodynamic model; inclusions; rare earth

1. Introduction

Rare earth elements (REs) are widely used in metallurgical, chemical, and advanced materials products [1–3]. REs were used to improve the quality of steel or slag. For example, the refining slag of CaO-AlO_3-MgO-SiO_2 systems containing Ce_2O_3 promotes the absorption of Al_2O_3 inclusions [4]. The 253MA steel is developed by adding 0.03–0.08% of cerium into the 21Cr-11Ni austenitic steel. The high temperature oxidation resistance of 253MA steel is superior to 310S stainless steel (25Cr-20Ni) with higher nickel addition [5,6]. However, the phase diagrams or thermodynamic data involving multicomponent-RE_2O_3 are missing, which restricts the further study and application of rare earth elements in metallurgy [7–9]. What's more, a large amount of fine and dispersed inclusions, such as RE_2O_3 and RE_2O_2S, are formed after the molten steel are alloyed by REs since the strong attaching power between O, S, and REs. These inclusions are easily attached to the inner wall and cause principle inducement of the nozzle clogging during the continuous casting, which deteriorate the productivity and the quality of production [10,11].

The clogging problems can usually be relieved by modifying the material or shape of submerged nozzle and calcium treatment [11–13]. Calcium treatment is the primary choice for dealing with the clogging issue of Al-killed steel by modifying Al_2O_3 to liquid phases at the casting temperature [14,15]. Kojola et al. demonstrated that the clogging frequency is remarkably reduced when 253MA steel is alloyed in the proper order of aluminum, cerium, and silicon [10]. The mechanism of Ca treatment for Al-killed steel might be similar to the Al treatment of Si-killed Ce-bearing steel, which is the generation of liquid inclusions, although the author gave the hypothesis that small inclusions might decompose after Si addition without sampling and analyzing the inclusions. To explain the declined clogging rate phenomenon in Kojola's experiments, the CALPHAD (CALculation of PHAse Diagrams) technique was introduced to obtain thermodynamic data of complex oxide systems containing REs. Then the thermodynamic model and package, as well as physical simulation at steelmaking temperature were employed to study the inclusions evolution behaviors in 253MA steel.

2. Research Methods

The technology routine is shown in Figure 1. The liquid or glass formation regions have been reported in Al_2O_3-SiO_2-Y_2O_3/La_2O_3/Sm_2O_3 systems. However, the liquid boundary of Al_2O_3-SiO_2 system involving Ce_2O_3 is unclear. For this reason, the Redlish–Kister polynomial expression and Kohler's extrapolation model in FactSage software were employed to optimize the phase diagram of Al_2O_3-SiO_2-Ce_2O_3 systems [16,17]. Then the interaction parameters and excess Gibbs free energy (G^E) were extracted during the optimization. Then the G^E were used to calculate the standard Gibbs free energy of liquid inclusions ($xCe_2O_3 \cdot yAl_2O_3 \cdot (1-x-y)SiO_2$, where $0 < x < 1, 0 < y < 1-x$) to represent the chemical equilibrium of every reactions based on Wagner's relations in the infinite dilute solution of molten steel, which is always adopted in steelmaking and is different from the minimum total Gibbs energy principle in FactSage software [18].

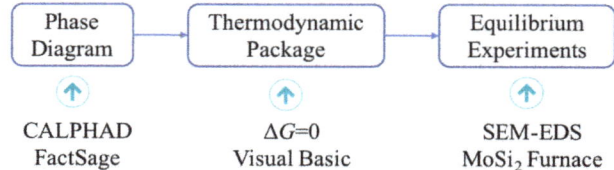

Figure 1. Research routine in this work.

The thermodynamic calculations were conducted by the Main.*exe* file compiled with the Visual Basic software (version 6.0, Microsoft Company, Redmond, WA, USA). The results were outputted in the format of *.txt* (as shown in the result module of Figure 2) and *.xlsx*. As shown in Figure 2, the initial input variants were temperature, calculation step, and original compositions of alloy elements.

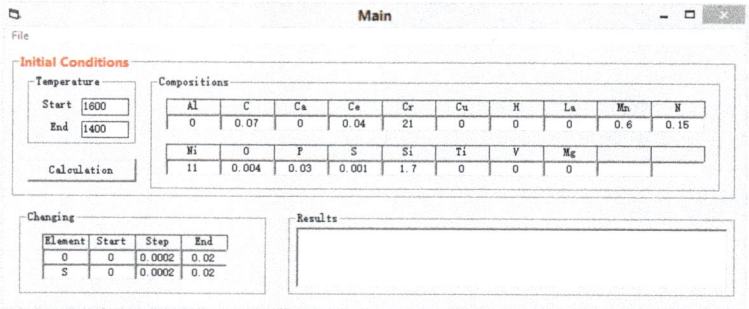

Figure 2. Main interface of the developed program.

According to the thermodynamic calculations, the Si-Al alloy was chosen and the 253MA steel was melt in the tube furnace heated with 8 $MoSi_2$ units (shown as Figure 3) at 1600 °C when the raw materials were collected in a MgO crucible placed in a graphite crucible in argon atmosphere. The materials used to melt the 253MA steel are listed in Table 1. Molten steel samples were extracted by a quartz tube at 1600 and 1500 °C and then quenched into the ice-water mixture to reserve the original morphologies of inclusions at the steelmaking and casting temperature, respectively. The quenched samples were polished and observed by the FE-SEM (field emission scanning electron microscope, JEOL, Tokyo, Japan) and EDS (energy dispersive spectrometer, JEOL, Tokyo, Japan).

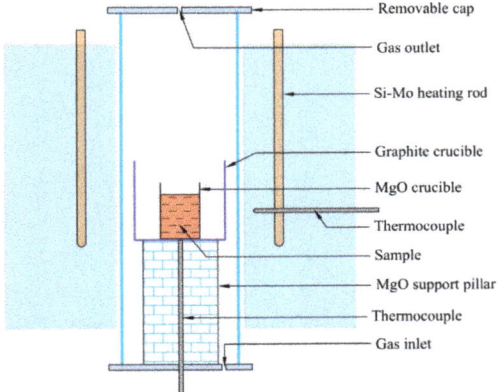

Figure 3. Schematic diagram of the tube furnace.

Table 1. Chemical compositions of melting alloys.

Raw Material	C	Si	Mn	P	S	N	Al	Cr	Ni	Ce	Fe
Fe	0.002	0.01	0.02	0.005	0.003	-	0.018	0.02	0.01	-	99.9
Ni	0.01	0.002	-	0.001	0.001	-	-	-	99.96	-	0.01
Cr	0.008	0.21	-	0.003	0.0015	0.05	0.21	99.1	-	-	0.16
Si-Al	0.068	75.7	0.24	0.026	0.005	-	1.3	-	-	-	22.61
Ce	0.08	0.045	-	-	-	-	-	-	-	99.4	-

3. Results and Discussion

Optimized liquid regions in the phase diagram of Al_2O_3-SiO_2-Ce_2O_3 (A-S-C) system by CALPHAD technology is given in Figure 4. The boundaries of full liquid (liquidus) in the A-S-C system have been compared to those in A-S-RE systems [19–21]. It can be seen that the liquid regions in the phase diagram of A-S-RE systems were almost located in similar sections, near the SiO_2 corner and symmetrically distributed on the isometric line of mole fraction ratio of RE_2O_3 and Al_2O_3 was 1. This is due to the similar physical and chemical properties of rare earth elements, especially the lanthanide series. The existence of the liquid regions implies it is possible to modify the fine solid inclusions to liquid phases.

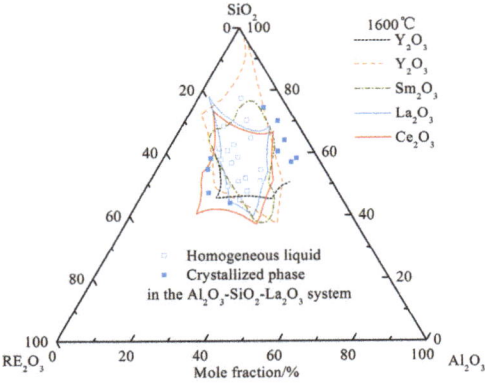

Figure 4. Liquid regions of Al_2O_3-SiO_2-RE_2O_3 systems.

During optimization of the of A-S-C phase diagram system, the excess Gibbs free energy of the complex liquid inclusions (xCe$_2$O$_3 \cdot y$Al$_2$O$_3 \cdot (1-x-y)$SiO$_2$) deviating from ideal mixture of Ce$_2$O$_3$, Al$_2$O$_3$ and SiO$_2$ was extracted as follow:

$$x\text{Ce}_2\text{O}_3(l) + y\text{Al}_2\text{O}_3(l) + (1-x-y)\text{SiO}_2(l) = x\text{Ce}_2\text{O}_3 \cdot y\text{Al}_2\text{O}_3 \cdot (1-x-y)\text{SiO}_2(l)$$
$$G^E = -825000xy(1-x-y)^2 + 42569.73xy + 19570.28y(1-y) - 140979.96x(1-x) + \frac{y(1-x-y)(1-x-2y)}{1-x}\left(14875.48 + 5640.02\frac{1-x-2y}{1-x}\right) + \frac{x(1-x-y)(1-2x-y)}{1-y}[215301.91 + 547791.07\frac{1-2x-y}{1-y} + 398115.65\left(\frac{1-2x-y}{1-y}\right)^2] + \{-98.65xy - 10.49y(1-y) + 78.91x(1-x) + \frac{y(1-x-y)(1-x-2y)}{1-x}\left(-0.71 + 1.21\frac{1-x-2y}{1-x}\right) + \frac{x(1-x-y)(1-2x-y)}{1-y}[23.6 - 202.71\frac{1-2x-y}{1-y} - 155.94\left(\frac{1-2x-y}{1-y}\right)^2]\}T$$

(1)

Equation (1) is used to calculate the Gibbs free energy of liquid inclusions generation. For other reactions of the solid inclusions generation, the equations are not listed in this work since they can be referenced from thermodynamic handbooks.

Based on the obtained thermodynamic data, we built the model and package of equilibrium calculation of multi-reactions including the liquid inclusions generation. The computed results are shown in Figure 5. As presented in Figure 5a, the liquid inclusions were stable in the region surrounded by the red dashed line as the mass fractions of aluminum and cerium were feasible. When the mass fraction of cerium was 0.02%, the inclusions in the 253MA steel transferred from Ce$_2$Si$_2$O$_7$ to liquid phases as the increase of aluminum addition, and cerium aluminates were formed as the aluminum content excess 0.017%, shown in Figure 5b. When the mass fraction of cerium was 0.03%, the inclusions first transferred from cerium silicates to liquid phase, and the amount of liquid inclusions began to decline as CeAlO$_3$ appeared, shown in Figure 5c. When the mass fraction of cerium increased to 0.04%, the liquid inclusions precipitated first and then disappeared as the aluminum addition reached 0.015%, Figure 5d. Considering the required cerium content of 0.03–0.04% in 253MA steel, the aluminum addition should be the key factor of restricting the inclusions to liquid, which was about 0.01%.

According to the above thermodynamic calculations, only 0.01% aluminum was needed to transfer the solid inclusions to liquid phases, then the Si-Al alloy, the chemical compositions are listed in Table 1, and were chosen as the deoxidant. The SEM observation and EDS analysis results of sampled inclusions from equilibrium experiments at steelmaking and casting temperature are present in Figure 6. It can be seen that, the inclusions were almost Al-Si-O system after Si-Al alloy was added for 60 min. The inclusions were mainly Ce$_2$O$_3$/Ce$_2$O$_2$S after cerium was added for 5 min. The morphologies and compositions changed during the following 25 min, and finally the spherical and liquid inclusions were formed at 1500 °C (the casting temperature). Altogether, the results showed that the solid inclusions were modified to liquid ones after aluminum was first added in the form of Si-Al alloy and then cerium was added. Nevertheless, it should take more than 30 min to finish the modifying process after cerium addition. The mechanism of the modifying process can be explained as: (1) the inclusions of Al-Si-O system are formed after Si-Al alloy is added; (2) a number of Ce$_2$O$_3$/Ce$_2$O$_2$S are formed immediately after cerium is added owing to the strong chemical reaction between cerium and oxygen solutes, and the oxygen activity in molten steel is sharply declined; (3) the early generated Al-Si-O inclusions decomposed to solutes of aluminum, silicon, and oxygen since the decline of oxygen activity; (4) the reactions between Ce$_2$O$_3$/Ce$_2$O$_2$S and silicon, aluminum, oxygen keep going on as the continuous diffusion of solutes in molten steel; and (5) the liquid inclusions are ultimately formed as the schematic diagram shown in Figure 6.

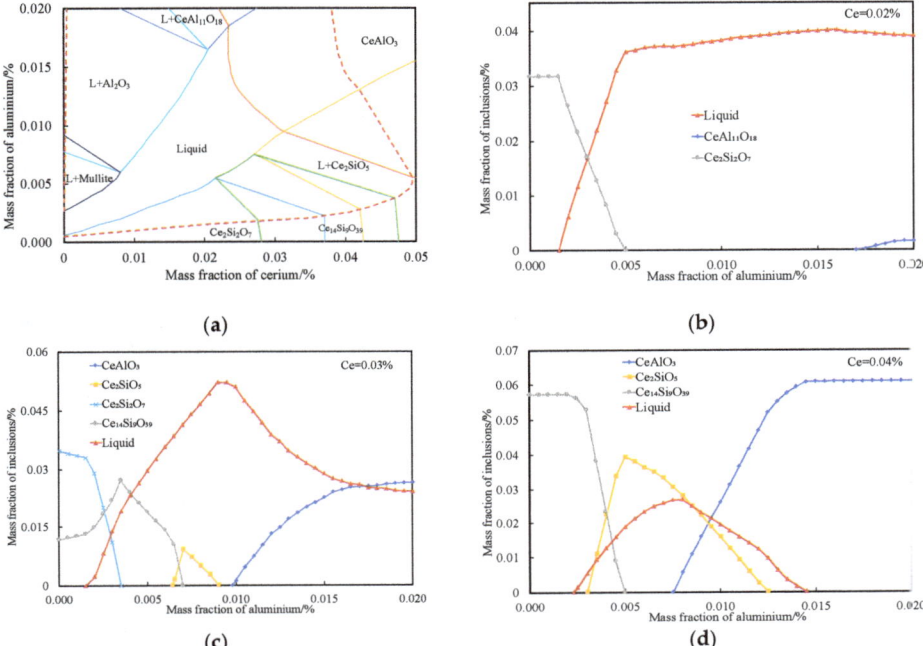

Figure 5. Calculation results by developed thermodynamic package. (**a**) Liquid regions of inclusions; (**b**) Effect of aluminum addition on inclusions, w_{Ce} = 0.02%; (**c**) Effect of aluminum addition on inclusions, w_{Ce} = 0.03%; (**d**) Effect of aluminum addition on inclusions, w_{Ce} = 0.04%.

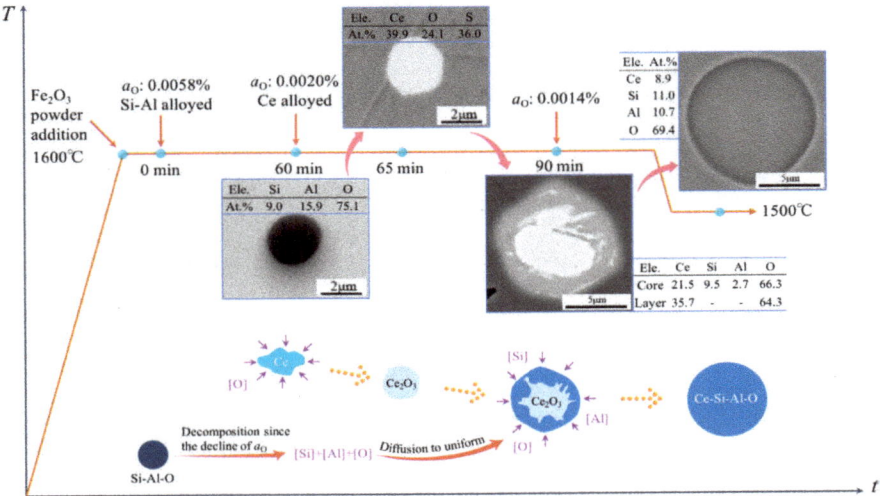

Figure 6. Inclusions evolution in 253MA steel after Al addition.

4. Conclusions

The liquid region is near the SiO_2 corner in the phase diagram of Al_2O_3-SiO_2-Ce_2O_3 systems optimized by the CALPHAD technique, implying the possibility of modifying Ce_2O_3/Ce_2O_2S to liquid phases. Then the thermodynamic model coded in the user defined package computes the appropriate aluminum addition, about 0.01%, to control the inclusion compositions' inner boundaries of liquid region. The spherical liquid inclusions are found after 30 min of cerium addition when Si-Al alloys are chosen as the deoxidant instead of pure silicon. The thermodynamic and experimental results can support the theory and data groundwork to remit the nozzle clogging of 253MA steel in the future.

Author Contributions: Conceptualization, methodology Y.L. and C.L.; Validation, T.Z.; Writing—review and editing, Y.L.; Supervision, C.L.; Funding acquisition, T.Z. and M.J.

Funding: This research was funded by Natural Science Foundation of Chongqing (cstc2018jcyjAX0792), Open Project from Key Laboratory of Ecological Utilization of Multi-metallic Mineral of Education Ministry (NEMM2018003), Research Project from Chongqing Committee of Education (KJQN201801408, KJZD-M201801401) and the Introduce Talents Research Start-up Fund in Central South University of China.

Conflicts of Interest: The authors declare no conflict of interest.

References

1. Qi, J.; Liu, C.J.; Zhang, C.; Jiang, M. Effect of Ce_2O_3 on structure, viscosity, and crystalline phase of CaO-Al_2O_3-Li_2O-Ce_2O_3 slags. *Metall. Mater. Trans. B* **2017**, *48*, 11–16. [CrossRef]
2. Wang, X.; Ou, D.R.; Shang, L.; Zhao, Z.; Cheng, M. Sealing performance and chemical compatibility of SrO-La_2O_3-Al_2O_3-SiO_2 glasses with bare and coated ferritic alloy. *Ceram. Int.* **2016**, *42*, 14168–14174. [CrossRef]
3. Iftekhar, S.; Pahari, B.; Okhotnikov, K.; Jaworski, A.; Stevensson, B.; Grins, J.; Edén, M. Properties and structures of RE_2O_3-Al_2O_3-SiO_2 (RE = Y, Lu) glasses probed by molecular dynamics simulations and solid-state NMR: The roles of aluminum and rare-earth Ions for dictating the microhardness. *J. Phys. Chem. C* **2012**, *116*, 18394–18406. [CrossRef]
4. Yang, X.H.; Long, H.; Cheng, G.G.; Wu, C.C.; Wu, B. Effect of refining slag containing Ce_2O_3 on steel cleanliness. *J. Rare Earths* **2011**, *29*, 1079–1083. [CrossRef]
5. Chen, L.; Ma, X.; Wang, L.; Ye, X. Effect of rare earth element yttrium addition on microstructures and properties of a 21Cr-11Ni austenitic heat-resistant stainless steel. *Mater. Des.* **2011**, *32*, 2206–2212. [CrossRef]
6. Matway, R.J.; McGuire, M.F.; Mehta, J. Steel Alloy Having Improved Creep Strength. U.S. Patent 5,393,487, 28 February 1995.
7. Liu, C.J.; Qiu, J.Y.; Sun, L.F. Liquidus and phase equilibrium in CaO-SiO_2-Nb_2O_5-10%La_2O_3 system. *ISIJ Int.* **2018**, *58*, 612–619. [CrossRef]
8. Qiu, J.Y.; Liu, C.J. Subsolidus phase relations in the CaO-SiO_2-Nb_2O_5-La_2O_3 quarternary system at 1273K. *ISIJ Int.* **2017**, *57*, 2107–2114. [CrossRef]
9. Liu, C.J.; Qiu, J.Y. Phase equilibrium relations in the specific region of CaO-SiO_2-La_2O_3 system. *J. Eur. Ceram. Soc.* **2018**, *39*, 2090–2097. [CrossRef]
10. Kojola, N.; Ekerot, S.; Jönsson, P. Pilot plant study of clogging rates in low carbon and stainless steel grades. *Ironmak. Steelmak.* **2011**, *38*, 81–89. [CrossRef]
11. Zhou, S.C. Study of the clogging of the submersible nozzle in the continuous casting of stainless steel RE-253MA. *Metallurgist* **2013**, *57*, 510–515.
12. Memarpour, A. *An Experimental Study of Submerged Entry Nozzles (SEN) Focusing on Decarburization and Clogging*; Dalarna University: Falun, Sweden, 2011.
13. Memarpour, A.; Brabie, V.; Jönsson, P.G. Studies of effect of glass/silicon powder coatings on clogging behaviour of submerged entry nozzles when using REM alloyed stainless steels. *Ironmak. Steelmak.* **2011**, *38*, 229–239. [CrossRef]
14. Tuttle, R.B.; Smith, J.D.; Peaslee, K.D. Casting simulation of calcium Titanate and calcium Zirconate nozzles for continuous casting of aluminum-killed steels. *Metall. Mater. Trans. B* **2007**, *38*, 101–108. [CrossRef]
15. McPherson, N.A.; McLean, A. *Continuous Casting Volume 6—Tundish Tomold Transfer Operations*; Iron and Steel Society: Warrendale, PA, USA, 1992.

16. Redlich, O.; Kister, A.T. Algebraic representation of thermodynamic properties and the classification of solutions. *Ind. Eng. Chem.* **1948**, *40*, 345–348. [CrossRef]
17. Pelton, A.D. A general "geometric" thermodynamic model for multicomponent solutions. *Calphad* **2001**, *25*, 319–328. [CrossRef]
18. Li, Y.D.; Liu, C.J.; Zhang, T.S.; Jiang, M.F.; Peng, C. Liquid inclusions in heat-resistant steel containing rare earth elements. *Metall. Mater. Trans. B* **2017**, *48*, 956–965. [CrossRef]
19. Bondar, I.A.; Galakhov, F.Y. Phase equilibria in the system Y_2O_3-SiO_2-Al_2O_3. *Bull. Acad. Sci. USSR Div. Chem. Sci.* **1964**, *13*, 1231–1232. [CrossRef]
20. Kolitsch, U.; Seifert, H.J.; Ludwig, T.; Aldinger, F. Phase equilibria and crystal chemistry in the Y_2O_3-Al_2O_3-SiO_2 system. *J. Mater. Res.* **1999**, *14*, 447–455. [CrossRef]
21. Kolitsch, U.; Seifert, H.J.; Aldinger, F. Phase relationships in the system Gd_2O_3-Al_2O_3-SiO_2. *J. Alloys Compd.* **1997**, *257*, 104–114. [CrossRef]

© 2019 by the authors. Licensee MDPI, Basel, Switzerland. This article is an open access article distributed under the terms and conditions of the Creative Commons Attribution (CC BY) license (http://creativecommons.org/licenses/by/4.0/).

Article

Investigation on the Slag-Steel Reaction of Mold Fluxes Used for Casting Al-TRIP Steel

Kaitian Zhang [1,2], Jianhua Liu [2] and Heng Cui [1,*]

1. Collaborative Innovation Center of Steel Technology, University of Science and Technology Beijing, Beijing 100083, China; zhangkaitianbk@163.com
2. Engineering Research Institute, University of Science and Technology Beijing, Beijing 100083, China; liujianhua@metall.ustb.edu.cn
* Correspondence: cuiheng@ustb.edu.cn; Tel.: +86-136-7123-9796

Received: 14 March 2019; Accepted: 29 March 2019; Published: 1 April 2019

Abstract: The reaction between [Al] in molten steel and (SiO_2) in the liquid slag layer was one of the restrictive factors in the quality control for high Al-TRIP steel continuous casting. In this work, the composition and property variations of two slags during a slag-steel reaction were analyzed. Accordingly, the crystalline morphologies of slag were discussed and the solid layer lubrication performance was evaluated by Jackson α factors. In addition, a simple kinetics equilibrium model was established to analyze the factors which affected SiO_2 consumption. The results reflected that slag-steel reacted rapidly in the first 20 minutes, resulting in the variation of viscosity and the melting temperature of slags. The slag-steel reaction also affected the crystal morphology significantly. Slag was precipitated as crystals with a higher melting temperature, a higher Jackson α factor, and a rougher boundary with the consumption of SiO_2 and the generation of Al_2O_3. In other words, although generated Al_2O_3 acted as a network modifier to decrease the viscosity of the liquid slag layer adjacent slab shell, the consumption of SiO_2 led to the deterioration of the lubrication performance in the solid slag layer adjacent copper, which was detrimental to the quality control for high Al-TRIP steel. Finally, a kinetics equilibrium model indicated that it is possible to reduce a slag-steel reaction by adjusting factors, such as the diffusion coefficient k, c_{SiO_2}, ρ_f and L_f, during the actual continuous casting process.

Keywords: Al-TRIP steel; slag; slag-steel reaction; crystalline morphology; equilibrium model

1. Introduction

While the "environmentally friendly, safety, long-life and low-cost" have become an international consensus of the automobile industry, attention for developing advanced automobile strength steel have been paid to high Al-TRIP (Transformation Induced Plasticity), which had a good combination of high strength and high toughness [1–4]. However, 1.35% Al-TRIP steel, currently in industrial production, is a kind of typical peritectic steel with volume shrinkage during peritectic reaction L + δ → γ [5]. As a result, the inhomogeneous growth of the solidification shell increases the incidence risk of the slab surface defects [6]. During the continuous casting process, slag plays an important role in counteracting this phenomenon. It controls the quality of slab through the properties of non-metallic inclusions absorption, lubrication, thermal transmission, etc. Therefore, a system of slag with suitable functions, such as the chemical composition, viscosity, and crystallization morphology, was greatly befitting for a continuous casting process [7,8]. Study on the behavior of slag, especially for the physicochemical characteristics of liquid slag film cling to the slab shell and the crystallization properties of solid slag film attaching to copper, will provide profound guidance for enhancing the slab surface quality of high Al-TRIP steel.

Many investigations on slag used for continuous casting of high Al content steel have been carried out. Conventional slags were mainly based on the CaO-SiO_2-CaF_2 system. However, because of Equation (1), which would take place during the continuous casting of high Al steel, the chemical composition of slag would change rapidly into an obvious increase in Al_2O_3 and decrease in SiO_2, leading to the deterioration of slag physicochemical properties and casting [9].

$$4[Al] + 3(SiO_2) \rightarrow 2(Al_2O_3) + 3[Si] \tag{1}$$

Yu et al. [10] tried to use a high SiO_2 with a low basicity flux to counteract the increase in basicity caused by the (SiO_2)-$[Al]$ reaction, and the casting experiments showed that the slab had a good surface quality. Zhang et al. [11] also studied the variation in viscosity and crystallization properties by adjusting Al_2O_3/SiO_2, however, the improvement effect on continuous casting was not significant. Some researchers have tried to suppress the deterioration in the performance of slag by adding fluxing agents [12,13]. Park et al. [13] studied the addition of CaF_2 that was beneficial to the reduction of the viscosity of high Al steel slag, but because of the high basicity of Na_2O, when the content of CaF_2 exceeds 8%, the viscosity-reducing effect was weakened significantly.

Primarily, it was the drastic slag-steel reaction during the continuous casting process that led to slag performance deterioration. Many attempts based on conventional slag still find it difficult to avoid the slag-steel reaction effectively. Therefore, some researchers began to explore the study of non-reactive or weakly reactive slag [6,14–21]. Among them, Wang et al. [16,17] studied the effect of various oxide additions on the crystallization behavior and heat transfer properties of the CaO-Al_2O_3 system continuous casting slag and provided some guidelines for the design of a reasonable CaO-Al_2O_3 system slag. Cho et al. [18] designed the CaO-Al_2O_3 based slag for continuous casting of 1.45% Al steel and found that the reactivity of Equation (1) was reduced significantly, resulting in a certain improvement of the slab quality. Seo et al. [19,20] focused on the lubrication of a solid layer of slag used for high Al-AHSS (Advanced High Strength Steel) steel and proposed to control the crystalline morphology by Jackson α factors. It was found that the crystallization at the slag film was too strong to affect the lubrication, which ultimately led to defects in the slab.

In summary, the current design schemes of slag still could not continuously cast high Al-TRIP steel perfectly caused by exorbitant Al_2O_3 and free of SiO_2 content in slag after the reaction. And the additions of fluxing agents, such as Li_2O and Na_2O, would also lead to cost problems for industrial production. As a result, Cho et al. [21] proposed a slag feeding technology to improve the castability and surface quality of a continuous cast TWIP (Twinning-induced plasticity) steel. Therefore, low-reactive slag, rather than non-reactive slag, might be a suitable and economic method for the continuous casting of high Al-TRIP steel. The influence of consuming SiO_2 during continuous cast Al-TRIP steel slab still requires further study.

Based on the investigation of the composition variation of CaO-SiO_2 system slag in an actual Al-TRIP steel continuous casting process, a kind of low-reactive slag based on CaO-SiO_2-Al_2O_3 system was adopted in this work. The initial low-reactive slag and corresponding slag samples after 10 min, 20 min, and 120 min of slag-steel high-temperature reaction were analyzed to determine their compositions and properties variation. Next, the crystalline morphologies were investigated by BSE (Back Scattering Electron) and Jackson α factors were adopted to evaluate solid layer lubrication performance. Finally, a kinetics equilibrium model of (SiO_2) in liquid slag was established to analyze the factors that affected SiO_2 consumption.

2. Experiments

2.1. Materials Preparation

The chemical composition of high Al-TRIP steel is listed in Table 1, samples were cut with a weight of 300 g from an industrial slab. The initial compositions of two types of industrial slag were tested by XRF (X-Ray Fluorescence) shown in Table 2. Among them, *S/A* was the content ratio of SiO_2 and

Al$_2$O$_3$, which represented the reactivity of slag samples; R was the basicity of slag samples, which was represented by the content ratio of CaO and SiO$_2$; η represented the viscosity of slag samples at 1300 °C; and T_m represented the melting temperature of slag samples. It is worth mentioning that sample A was used as the control group in the actual Al-TRIP steel continuous casting production. And sample B was used to high-temperature static balance experiment.

Table 1. Main chemical composition of Al-TRIP (Transformation Induced Plasticity) steel, wt.%.

Fe	C	Si	Mn	P	S	Alt	N
bal.	0.16	0.16	1.49	0.008	0.001	1.35	0.0016

Table 2. Main chemical compositions and properties of initial slag samples, wt.%.

Samples	CaO	SiO$_2$	Al$_2$O$_3$	CaF$_2$	BaO	Na$_2$O	S/A	R(C/S)	η [1] (Pa·S)	T_m [2] (°C)
A	30.57	27.31	3.03	17.26	5.65	16.17	9.02	0.147	0.087	841.88
B	29.50	20.90	16.16	18.72	10.93	3.79	1.29	0.147	0.154	1109.35

[1,2] Calculated by Factsage software 7.2, and η were calculated at 1300 °C.

2.2. Experimental Method

Figure 1 shows the schematic diagram of the high-temperature static balance experimental apparatus. 300 g steel was put into Si-Mo high-temperature furnace with zirconia crucible. The steel was heated to 1550 °C by 10 °C/min and then stayed the temperature with argon at the rate of 5 L/min to prevent the oxidation of steel. To simulate the situation of a continuous addition of slag in the industry casting process, excess 60 g pre-melted slag was added on the surface of molten steel and taken out of the furnace after continuing heat preservation for 10 min, 20 min, and 120 min. Finally, after crushing and grinding, the slag chemical compositions were measured by XRF. The viscosities of 1300 °C and melting temperature for samples were calculated by Factsage software 7.2 (GTT-Technologies, Aachen, Germany). In addition, part of the slag sample was embedded in the epoxy resin and sprayed with gold, and then, the microstructure and crystal composition of slag sample B during the high-temperature static experiment were observed by Scanning Electron microscopy (Zeiss, Heidenheim, Germany) in Back Scattering Electron mode.

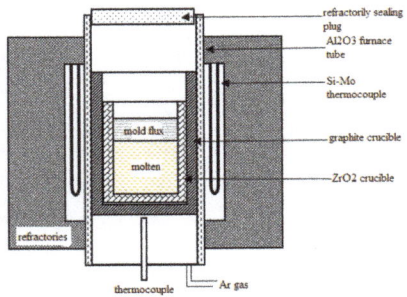

Figure 1. Schematic of the experimental apparatus.

3. Results

3.1. Compositions Variation of Slag during the Slag-Steel Reaction

The main compositions of slag sample A and B in different reaction time (10 min, 20 min, and 120 min) with 1.35% Al-TRIP steel are shown in Figure 2. In the first 10 min, the content of SiO$_2$ decreased sharply while the content of Al$_2$O$_3$ increased sharply. This means that because of the strong reaction driving force, which resulted in a low content of Al$_2$O$_3$ and high content of SiO$_2$ in initial slag samples, both sample A and B reacted rapidly with steel. In the following 10 min, the reaction (1) continued, however, the reactive rates of both samples gradually calmed down by the reason of the actual content ratio of Al_2O_3/SiO_2 in slag being increased. In other words, the reaction driving force was weakened. After 20 min, there was no obvious change in the tendency of compositions for sample A and sample B, indicating that the reaction had reached a kinetics equilibrium, and the equilibrium compositions were nearly 20 min after the beginning of the slag-steel reaction. Eventually, there was only 5.69% SiO_2 content in slag sample B and its composition had moved from the $CaO-SiO_2-Al_2O_3$ system to $CaO-Al_2O_3$ system. On the other hand, the final content of SiO_2 was 19.84% in sample A, which kept sample A in the $CaO-SiO_2-Al_2O_3$ system. In addition, because of the evaporation, no Na_2O was detected in sample B after the reaction.

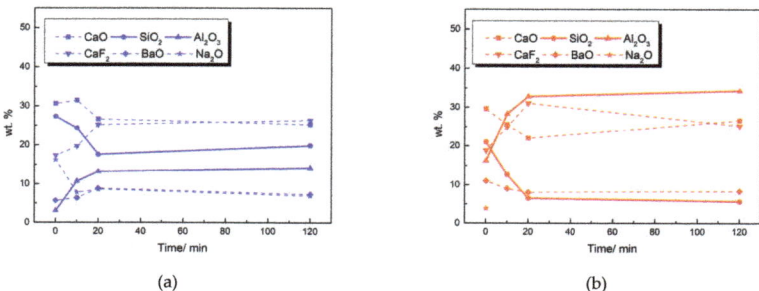

Figure 2. Variation of main compositions for slag sample: (**a**) Sample A; (**b**) sample B.

3.2. Properties Variation of Slag during the Slag-Steel Reaction

The viscosity at 1300 °C and melting temperature of slag samples before and after the reaction were calculated by Factsage software, shown in Figure 3. It was apparent that the melting temperature increased significantly and then gradually stabilized during the reaction. This was almost consistent with the composition of Al_2O_3 with a high melting temperature. The viscosity increased rapidly in the first 10 min of the reaction, and then decreased and gradually stabilized. In general, the variation of the composition and properties of sample A were consistent with that of sample B, indicating that the slag-steel reaction during the actual continuous casting process can be simulated through the high-temperature static balance experiment, so as to analyze the influence of the change of $(SiO_2)/(Al_2O_3)$ content on the properties of slag.

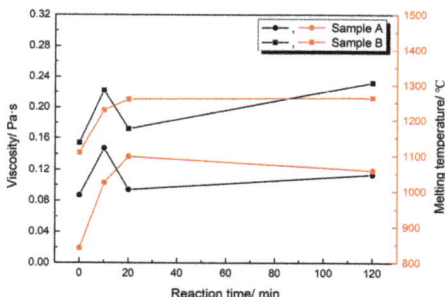

Figure 3. Variation of viscosity and melting temperature for slag sample A and B.

3.3. Revolution of Crystalline Morphology for Slag during the Slag-Steel Reaction

Figure 4 was the morphology of the original slag sample B. The distribution of slag components for industrial slag was mainly composed of three parts unevenly. Part I was only blocked Al_2O_3 with a high melting temperature. It precipitated prematurely during the continuous casting process, and it performed as amphoteric oxide depending on the basicity of slag. Part II was mainly composed of C, which was a matrix of slag and played a role in assisting melting. Part III was more complex, the main compositions were P1 ($Al_2O_3 \cdot 4SiO_2$), P2 ($2CaSiO_3 \cdot CaAl_2O_4 \cdot BaO$) and a small amount of P3 (CaF_2). Among them, P1 existed alone, while P3 existed around P2.

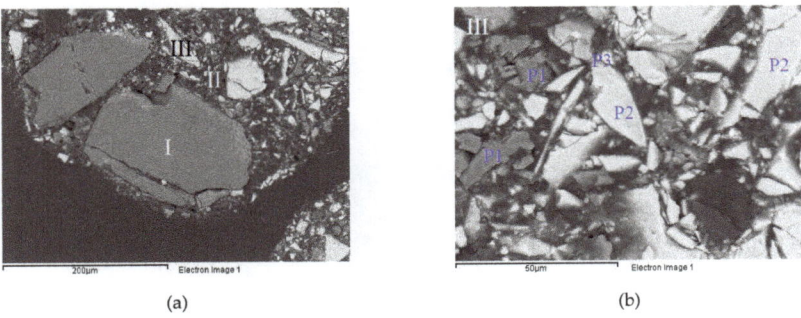

Figure 4. Morphology of original slag: (**a**) full view; (**b**) details of Part III.

The crystal morphology for slag sample B after the slag-steel reaction experiment (10 min, 20 min and 120 min respectively) are shown in Figure 5. The significant difference in the microstructure of the crystal morphology of the slag with different extent of slag-steel can be clearly observed. As shown in Figure 5a, after the slag-steel reaction lasted for 10 min, the slag matrix was smooth, with a dendritic crystalline phase on the surface but without large crystalline phase. After the slag-steel reacted for 20 min, as shown in Figure 5b, the slag became rough, and there were few large-size planar granular crystalline phases with smooth boundaries on the surface. After the reaction lasted for 120 min, as shown in Figure 5c, the crystalline phase size further increased and the boundary was rougher.

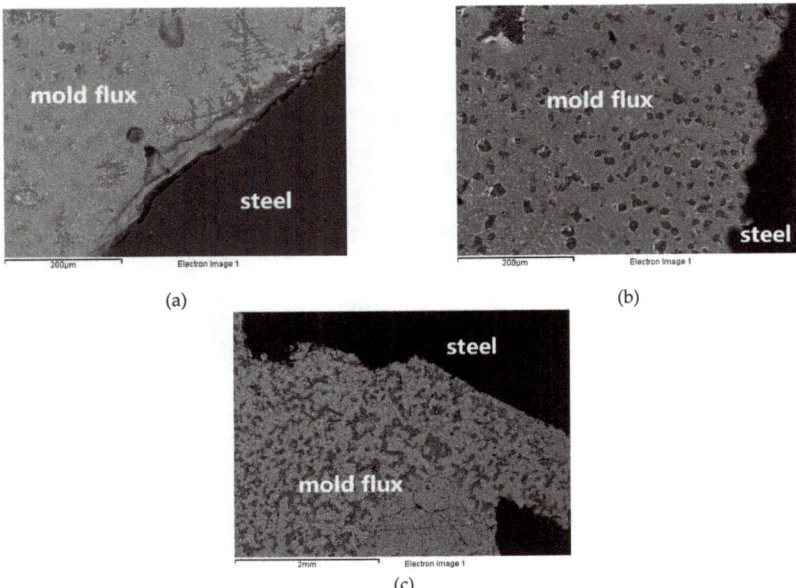

Figure 5. Crystals morphology of slag after reaction: (**a**) 10 min; (**b**) 20 min; (**c**) 120 min.

Specifically, Figure 6 shows the details of the slag morphology after the slag-steel reaction for 10 min. The dendritic crystalline phase was CaF_2, which was a benefit to lubrication. The matrix of slag crystallized uniformly along the CaF_2 dendritic, and the composition was $3CaO \cdot 3SiO_2 \cdot 2Al_2O_3 \cdot BaO$. No cuspidine ($3CaO \cdot 2SiO_2 \cdot CaF_2$) with high melting temperature was observed, indicating the slag had good performance on the uniform thermal transmission and melting characteristic so far.

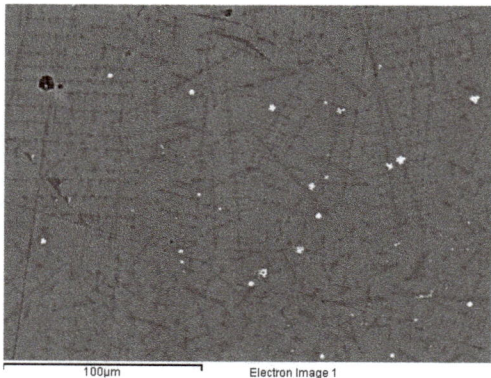

Figure 6. Details of crystals morphology after reaction lasted for 10 min.

After the reaction lasted for 20 min, as shown in Figure 7, the morphology of CaF_2 (P1) was a granular structure with a smooth boundary, which was favorable for the lubrication performance. The slag crystallized with P1 as the core, and is surrounded by P2 ($CaAl_2O_4$) and few CaF_2 dendrites. The main component of a boundary (P3) was aluminosilicates of Ca and Ba, and Ba had a tendency to make up for the formation of fluoride by Ca. The matrix of the slag was $CaAl_2O_4$ and a small amount of $CaSiO_3$, resulting from the large consumption of SiO_2.

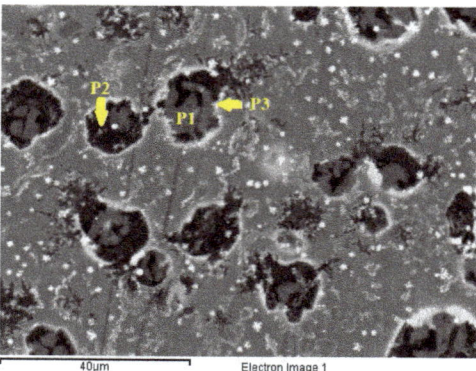

Figure 7. Details of crystals morphology after reaction lasted for 20 min.

As the reaction progressed to 120 min (Figure 8), the morphology was similar to that in Figure 6. However, the distribution was dense and uneven with the larger size and sharp boundary, leading to an unfavorable lubrication performance. With the growth of the CaF_2 boundary, the surrounding Ba was constantly absorbed into the core, forming BaF_2 and Ba-containing calcium aluminate (P2) eventually. The dendritic CaF_2 (P3) still existed between the CaF_2 core and the boundary. Compared with Figures 6 and 7, the crystal composition in Figure 8 was more complicated. There were Ba-containing $CaAl_2O_4$ and $CaAl_4O_7$ with high melting temperature.

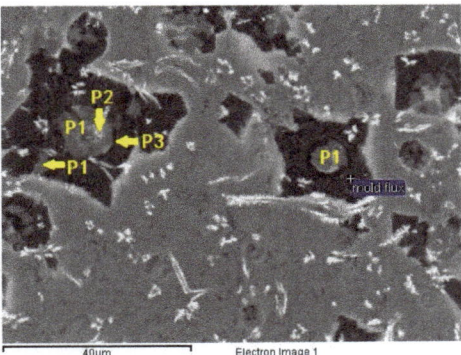

Figure 8. Details of crystals morphology after reaction lasted for 120 min.

In addition, Figure 9 was the crystalline morphology at the slag-steel interface after reaction for 120 min. The crystal was dominated by gehlenite and alumina with less fluoride. There were P1 ((Ba, Ca)F_2), P2 (Ca_2SiO_4), P3 ($3CaAl_4O_7 \cdot BaO$), and P4 (Al_2O_3). Since the Ba^{2+} electrostatic potential was smaller than Ca^{2+}, P1 was mainly composed of BaF_2. P2 was $2CaO \cdot SiO_2$, which was converted from wollastonite ($CaSiO_3$) with a large consumption of SiO_2. In addition, the duration of the slag-steel reaction was 120 min, which was close to the reaction equilibrium. Considering the kinetic conditions in the experimental furnace, the Al_2O_3 generated by the reaction was concentrated at the slag-steel interface and difficult to diffuse. Therefore, amounts of large-size crystals P3 and P4 could be observed, which had a significant effect on the heat transfer performance of slag-steel interface, and further affected the melting and consumption of slag.

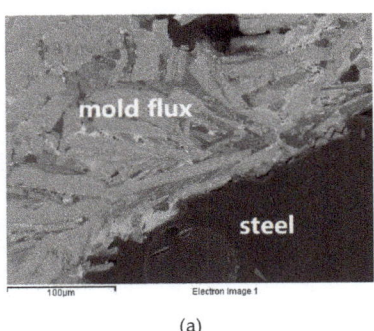
(a)

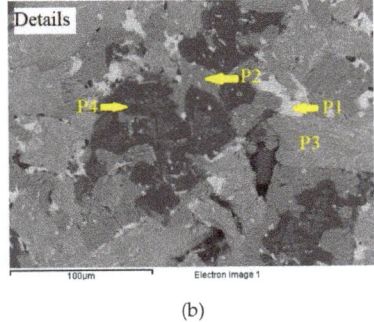
(b)

Figure 9. Morphology of original slag: (**a**) full view; (**b**) details of Part III.

4. Discussion

4.1. The Relationship between the Properties and Compositions Variation of Slag

In an alkalescent melting system, according to previous researchers [16,22], the existential form of Al_2O_3 was $[AlO_4]^{5-}$ tetrahedron, which could incorporate into $[SiO_4]^{4-}$ tetrahedral units to act as the network former, resulting in the increase of viscosity. With the content of $(CaO)/(Al_2O_3)$ having decreased, Al_2O_3 existed in the form of $[AlO_6]^{9-}$ octahedral, which could act as a network modifier, leading to a decrease of viscosity. Therefore, the viscosity of both slags was not decreased monotonically, it fluctuated slightly in the later period of reaction.

In summary, there were obvious differences in the crystalline phase in each part of the slag after the slag-steel reaction, mainly reflected in the content of SiO_2. Different crystalline phase performed different lubrication effect. The Jackson α factor was adopted in this work to evaluate the roughness of the crystalline phase. The crystals tended to be more faceted and anisotropic when their Jackson α factor increased [20]. The main crystalline phases and their corresponding melting temperature and Jackson α factors were shown in Table 3. As the reaction progressed, the content of SiO_2 was gradually reduced while Al_2O_3 increased significantly, resulting in a gradual replacement of $CaSiO_3$ and $CaAl_2O_4$ (with lower Jackson α factors) by $CaAl_4O_7$ (with higher Jackson α factor). From the perspective of crystallography, crystals with higher Jackson α factor consisted of crystallographic planes with different orientations, which were distributed at different angles to the heat dissipation direction. This was favorable for the formation of crystals with regular geometrical shapes, and thus the anisotropy of crystals was increased, which resulted in the decrease of lubrication performance in solid slag layer adjacent to the copper. In general, although generated Al_2O_3 acted as network modifier to decrease the viscosity of liquid slag layer adjacent slab shell, the consumption of SiO_2 led to the deterioration of lubrication performance in solid slag layer adjacent copper, which was detrimental to the quality control for high Al-TRIP steel.

Table 3. Typical crystals of slag sample B and corresponding Jackson α factors, wt.%.

Time	Crystal Phase	Melting Temperature	Jackson α Factors [1]
10 min	CaF_2 (dendritic)	1418 °C	2.11
	$CaSiO_3$	1540 °C	3.7
20 min	CaF_2 (faceted)	1418 °C	2.11
	$CaSiO_3$ (less)	1540 °C	3.7
	$CaAl_2O_4$	1604 °C	3.51
120 min	CaF_2 (faceted)	1418 °C	2.11
	$CaAl_2O_4$	1604 °C	3.51
	$CaAl_4O_7$	1765 °C	7.57
120 min (slag-steel interface)	CaF_2 (faceted)	1418 °C	2.11
	$CaAl_4O_7$	1765 °C	7.57
	Al_2O_3	2054 °C	6.12

[1] Based on reference [20].

4.2. Kinetics Equilibrium Model of Slag during the Reaction

Non-linear fitting of SiO_2 content in both slags through Origin software is shown in Figure 10. These content variations had good agreement with function $c_t - c_0 = a \times [1 - \exp(b \times t)]$. From the perspective of mathematics, coefficient a influenced the limit of the function, and b influenced the inflection time of the function.

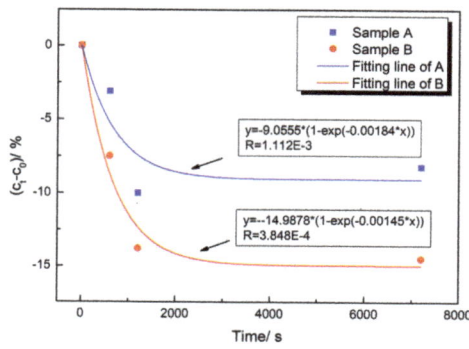

Figure 10. Details of crystals morphology after reaction lasted for 120 min.

Therefore, the reactivity of the slag was closely related to coefficient a and b. Theoretically, during the casting of 1.35% Al-TRIP steel, a certain content of SiO_2 in the powder slag was consumed because of reaction (1), and the rest of SiO_2, accompanied by molten slag, was flowed into the gap between solidification shell and copper. Therefore, the kinetics equilibrium balance of SiO_2 content in the liquid slag layer could be expressed as Figure 11.

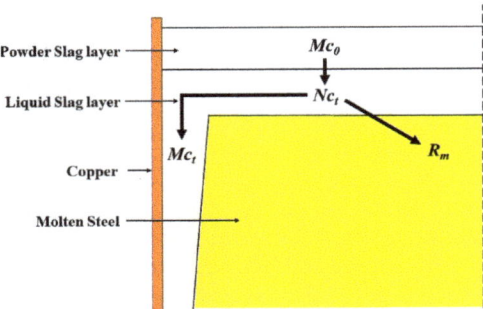

Figure 11. Kinetics equilibrium balance of SiO_2 content in liquid slag layer.

The c_0 was the content of SiO_2 in the initial powder slag. At time t, the SiO_2 content in the liquid slag decreased to c_t. Thus, the SiO_2 content flow into the gap between copper and slab shell at this time was also c_t. The R_m was the consumption of SiO_2 caused by the slag-steel interface reaction. Assuming that the supplementary of liquid slag from powder slag layer was equal to consumed slag from liquid slag layer during the continuous casting, and the SiO_2 content in the liquid slag layer was uniform at time t. Thus, the mass balance of SiO_2 in the liquid slag layer could be expressed as follows:

$$\frac{d[SiO_2]_t}{dt} = \frac{M[SiO_2]_0 - 100R_m - M[SiO_2]_t}{N} \quad (2)$$

where, $[SiO_2]_t$ was SiO_2 content in the liquid slag layer at time t, %; $[SiO_2]_0$ was SiO_2 content in initial powder slag, %; M was the consumed rate of liquid slag, kg/s; N was the mass of liquid slag layer, kg; R_m was the SiO_2 consumption rate from slag-steel interface reaction, kg/s; and t was time, s.

The slag-steel reaction consumed SiO_2 and produced Al_2O_3. In a constant temperature condition, the SiO_2 consumption rate R_m was associated with [Al] content in steel and the initial SiO_2 content in powder slag. If the reaction was adequate during the experiments, R_m could be expressed as follows:

$$R_m = Akc_{Al}^m c_0^n \quad (3)$$

where, A was reaction area, cm^2; k was diffusion coefficient related to flow velocity and viscosity. Thus, the Equation (1) could be rewritten as Equation (4) by an integral method.

$$c_t - c_0 = -\frac{100}{\rho_s v_s Q} \times Akc_{Al}^m c_0^n \times \left[1 - \exp\left(\frac{\rho_s v_s Q}{\rho_f L_s}\right)\right] \quad (4)$$

where, c_{Al} was the [Al] content in steel, %; ρ_s was molten steel density, g/cm^3; ρ_f was slag density, g/cm^3; Q was slag consumption for steel, g/g; V_s was equivalent casting speed, cm/s; L_f was thickness of liquid slag layer, cm. Combined with Figure 10, the coefficients a and b could be written as follows:

$$a = -\frac{100}{\rho_s V_s Q} \times (k(V_s c_{Al})^m c_{SiO2}^n \quad (5)$$

$$b = -\frac{\rho_s V_s Q}{\rho_f L_f} \qquad (6)$$

In this work, the temperature, c_{Al}, ρ_s, Q, L_f and V_s could be considered as constant, while ρ_f, L_f, c_{SiO_2} and k were different. Therefore, from the perspective of the slag-steel reactivity, k and c_{SiO_2} influenced the limit SiO_2 consumption of the reaction and ρ_f and L_f influenced the SiO_2 consumption rate of the reaction.

5. Conclusions

This work investigated the variations of slag components, properties, and crystalline phases during a slag-steel reaction. The following conclusions can be made:

(1) The components and properties variated rapidly in the first 20 min of the slag-steel reaction and then stabilized gradually. Specifically, slag moved from $CaO-SiO_2-Al_2O_3$ system to $CaO-Al_2O_3$ system, and the basicity of the slag was increased significantly because of the large consumption of SiO_2. However, the generated Al_2O_3 also acted as network modifier in the slag, resulting in a limited decrease in viscosity of liquid slag layer adjacent to the slab. In addition, a small amount of SiO_2 still existed in the slag at the reaction equilibrium point.

(2) The components variation during the slag-steel reaction also affected the crystal morphology significantly. Accompanied by the consumption of SiO_2 and the generation of Al_2O_3, the crystallization phase of the slag tended to be dendritic $CaF_2 \rightarrow$ faceted CaF_2 and $CaSiO_3 \rightarrow CaAl_2O_4 \rightarrow CaAl_4O_7$. These phenomena indicated that as the reaction progressed, the slag used for 1.35% Al-TRIP steel was precipitated as crystals with a higher melting temperature, higher Jackson α factor, and rougher boundary. As a result, the anisotropy of crystals was increased and the lubrication performance in the solid slag layer adjacent the copper was deteriorated.

(3) According to the nonlinear fitting of the slag composition variation during the reaction, it was found that the slag-steel reaction existed the consumption limit of SiO_2. A kinetics equilibrium mold of slag was also derived, which indicated that the perspective of slag, the diffusion coefficient k and c_{SiO_2} affected the limit SiO_2 consumption of the reaction, and ρ_f and L_f affected the SiO_2 consumption rate. Therefore, it is possible to reduce the slag-steel reaction by adjusting these parameters during the actual continuous casting process, and it can provide a new idea for the research of slag used for high Al-TRIP steel.

Author Contributions: Investigation, K.Z. and H.C.; project administration, H.C. and J.L.; writing—original draft, K.Z.; writing—review & editing, K.Z. and H.C.

Funding: This research was funded by the National Natural Science Foundation of China (No. U1860106).

Conflicts of Interest: The authors declare no competing interests.

References

1. Frommeyer, G.; Brux, U.; Neumann, P. Supra-ductile and high-strength manganese-TRIP/TWIP steels for high energy absorption purposes. *ISIJ Int.* **2003**, *43*, 438–446. [CrossRef]
2. Li, Y.; Lin, Z.; Jiang, A.; Chen, G. Use of high strength steel sheet for lightweight and crashworthy car body. *Mater. Design* **2003**, *24*, 177–182. [CrossRef]
3. Blazek, K.; Yin, H.B.; Skoczylas, G.; McClymonds, M.; Frazee, M. Development and evaluation of lime-alumina-based mold powders for casting high-aluminum TRIP steel grades. *Iron Steel Tech.* **2011**, *8*, 232–240.
4. Yan, W.; Chen, W.Q.; Yang, Y.D.; Alex, M. Mold fluxes for continuous casting of steel with high aluminum steel. In Proceedings of the Materials Science and Technology (MS&T), Montreal, QC, Canada, 27–31 October 2013.
5. Takashi, M.; Jin, Y.; Tsuyoshi, M. Surface quality improvement of Al-TRIP steel slab. In Proceedings of the 13th China-Japan Symposium on Science and Technology of Iron and Steel, Beijing, China, 17–19 November 2013.
6. Wang, W.L.; Lu, B.X.; Xiao, D.A. Review of mold flux development for the casting of high-Al steels. *Metall. Mater. Trans. B* **2016**, *47B*, 384–389. [CrossRef]

7. Shin, S.H.; Yoon, D.W.; Cho, J.W.; Kim, S.H. Controlling shear thinning property of lime silica-based mold flux system with borate additive at 1623K. *J. Non-Cry. Solids* **2015**, *425*, 83–90. [CrossRef]
8. Zhang, S.F.; Zhang, X.; Liu, W.; Lv, X.; Bai, C.; Wang, L. Relationship between structure and viscosity of $CaO-SiO_2-Al_2O_3-MgO-TiO_2$ slag. *J. Non-Cry. Solids* **2014**, *402*, 214–222. [CrossRef]
9. Wang, W.L.; Blazek, K.; Cramb, A. A study of the crystallization behavior of a new mold flux used in casting of Transformation-Induced-Plasticity steels. *Metall. Mater. Trans. B* **2008**, *39B*, 66–74. [CrossRef]
10. Yu, X.; Wen, G.H.; Tang, P.; Ma, F.J.; Wang, H. Behavior of mold slag used for 20Mn23Al nonmagnetic steel during casting. *J. Iron Steel Res. Int.* **2011**, *18*, 20–25. [CrossRef]
11. Zhang, Z.T.; Wen, G.H.; Liao, J.L.; Sridhar, S. Observations of crystallization in mold slags with varying Al_2O_3/SiO_2 ratio. *Steel Res. Int.* **2010**, *81*, 516–528. [CrossRef]
12. Yu, X.; Wen, G.H.; Tang, P.; Wang, H. Investigation on viscosity of mould fluxes during continuous casting of aluminum containing TRIP steels. *Ironmak. Steelmak.* **2009**, *36*, 623–630. [CrossRef]
13. Park, H.S.; Kim, H.K.; Sohn, I. Influence of CaF_2 and Li_2O on the viscous behavior of calcium silicate melts containing 12wt pct Na_2O. *Metall. Mater. Trans. B* **2011**, *42B*, 324–330. [CrossRef]
14. Li, J.L.; Shu, Q.F.; Cho, K. Effect of TiO_2 addition on crystallization characteristics of $CaO-Al_2O_3$-based mould fluxes for high Al steel casting. *ISIJ Int.* **2015**, *55*, 830–836. [CrossRef]
15. Shi, C.B.; Seo, M.D.; Cho, J.W.; Kim, S.H. Crystallization characteristics of $CaO-Al_2O_3$-based mold flux and their effects on in-mold performance during high-aluminum TRIP steels continuous casting. *Metall. Mater. Trans. B* **2014**, *45B*, 1081–1097. [CrossRef]
16. Xiao, D.; Wang, W.L.; Lu, B.X. Effects of B_2O_3 and BaO on the crystallization behavior of $CaO-Al_2O_3$-Based mold flux for casting high-Al steels. *Metall. Mater. Trans. B* **2015**, *46B*, 873–881. [CrossRef]
17. Lu, B.X.; Chen, K.; Wang, W.L.; Jiang, B.B. Effects of Li_2O and Na_2O on the Crystallization Behavior of Lime-Alumina-Based Mold Flux for Casting High-Al Steels. *Metall. Mater. Trans. B* **2014**, *45B*, 1496–1509. [CrossRef]
18. Cho, J.W.; Blazek, K.; Frazee, M.; Yin, H.; Park, J.H.; Moon, S.W. Assessment of $CaO-Al_2O_3$ based mold flux system for high aluminum TRIP casting. *ISIJ Int.* **2013**, *53*, 62–70. [CrossRef]
19. Seo, M.D.; Shi, C.B.; Baek, J.Y.; Cho, J.W.; Kim, S.H. Kinetics of isothermal melt crystallization in $CaO-SiO_2-CaF_2$-based mold fluxes. *Metall. Mater. Trans. B* **2015**, *46B*, 2374–2383. [CrossRef]
20. Guo, J.; Seo, M.D.; Shi, C.B.; Cho, J.W.; Kim, S.H. Control of crystal morphology for mold flux during high-aluminum AHSS continuous casting process. *Metall. Mater. Trans. B* **2016**, *47B*, 2211–2221. [CrossRef]
21. Cho, J.W.; Yoo, S.; Park, M.S.; Park, J.K.; Moon, K.H. Improvement of castability and surface quality of continuously cast TWIP slabs by molten mold flux feeding technology. *Metall. Mater. Trans. B* **2017**, *48B*, 187–196. [CrossRef]
22. Park, J.H.; Min, D.J.; Song, H.S. Amphoteric behavior of alumina in viscous flow and structure of $CaO-SiO_2$ (-MgO)-Al_2O_3 slag. *Metall. Mater. Trans. B* **2004**, *35B*, 269–275. [CrossRef]

© 2019 by the authors. Licensee MDPI, Basel, Switzerland. This article is an open access article distributed under the terms and conditions of the Creative Commons Attribution (CC BY) license (http://creativecommons.org/licenses/by/4.0/).

Article

High Quality Steel Casting by Using Advanced Mathematical Methods

Tomas Mauder * and Josef Stetina

Energy Institute, Brno University of Technology, 616 69 Brno, Technicka 2, Czech Republic; stetina@fme.vutbr.cz
* Correspondence: mauder@fme.vutbr.cz; Tel.: +420-541-14-3252

Received: 15 November 2018; Accepted: 30 November 2018; Published: 4 December 2018

Abstract: The main concept of this paper is to utilize advanced numerical modelling techniques with self-regulation algorithm in order to reach optimal casting conditions for real-time casting control. Fully 3-D macro-solidification model for the continuous casting (CC) process and an original fuzzy logic regulator are combined. The fuzzy logic (FL) regulator reacts on signals from two data inputs, the temperature field and the historical steel quality database. FL adjust the cooling intensity as a function of casting speed and pouring temperature. This approach was originally designed for the special high-quality high-additive steel grades such as higher strength grades, steel for acidic environments, steel for the offshore technology and so forth. However, mentioned approach can be also used for any arbitrary low-carbon steel grades. The usability and results of this approach are demonstrated for steel grade S355, were the real historical data from quality database contains approximately 2000 heats. The presented original solution together with the large steel quality databases can be used as an independent CC prediction control system.

Keywords: continuous casting; fuzzy logic; optimal cooling; steel quality prediction

1. Introduction

The continuous casting technology is a well-known predominant process how the steel is produced in the world. CC is already fully-grown technology and successfully casting million tonnes of classic low-carbon and low-alloy steel grades per year. In spite of this fact, casting of special high-strength grades, steels for acidic environment, steels for the offshore technology, high alloyed tool steels, can be still challenging in order to ensure a constant steel quality through the whole casting process. Flick and Stoiber [1] described present issues and future trends in the CC technology. The quality of the steel is still a discussed topic. In CC the solid shell is permanently subjected to thermal and mechanical stresses and it can give rise to crack formation, see Birat et al. [2]. Figure 1 shows CC installation and mechanical tensile/compression stresses in specific locations. The rejected slabs, by the quality control system, which need to be scraped is very uneconomic. The breakdown situation caused by a low quality of these steels could have a catastrophic effect on material and human losses. This quality issue can be handled by using the advanced numerical modelling methods, optimization-regulation techniques and statistical evaluations of the real casting data.

Historically, there are many papers which combine the numerical modelling and optimization-regulation approaches, such as Santos et al. [3] who applied the genetic algorithm, Zhemping et al. [4] verified the usage of the ant colony algorithm, Zheng et al. [5] attempted to use the swarm optimization on 2-D solidification model, Mauder and Novotny [6] showed the possibility of using classical mathematical programming method and compared the results with simple Nelder-Mead algorithm, Ivanova [7] constructed basic predictive control algorithm for 2-D solidification model and so forth. Rao et al. [8] publish a comprehensive review dealing with parameters optimization of selected casting processes. Unfortunately, these works are far from the use on the real casting process, because they often calculate very simplified solidification models (simple geometry, 2-D mesh, simple boundary conditions, contain a small steel database, etc.) and it is proven that the numerical results of these models in compare with the 3-D fine-mesh validated solidification models are poor, see Mosayebidorcheh and Bandpy [9]. Moreover, the optimization algorithms are often based on black-box approach that generally needs a large number of optimization iterations before the optimal solution is found. This is not a problem in the case of simple solidification models, which calculates the temperature field very fast. However, in the case of the complex 3D numerical simulations it is not possible to use these optimization algorithms for the real time control.

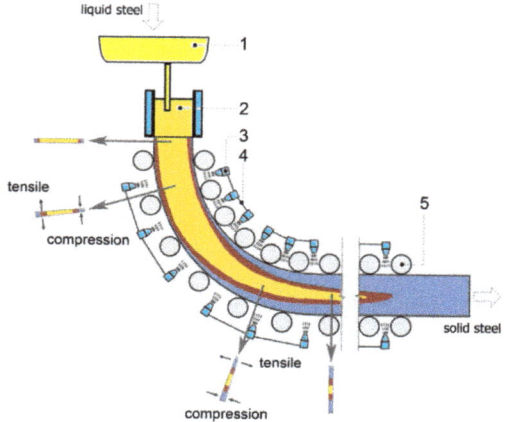

Figure 1. Scheme of the continuous casting and mechanical stresses during bending and straightening. 1—tundish; 2—mould; 3—nozzle; 4—cooling circuit; 5—roller.

This paper describes both, the original 3-D solidification model, the so-called Brno Dynamic Solidification Model® (BrDSM) and advanced optimal control algorithm based on the fuzzy logic (FL-BrDSM). This unique combination represents a tool for achievement of high steel quality products. Besides of the numerical and fuzzy regulation models, special attention must be concentrated to proper setting of thermophysical parameters of investigated steel grade, experimental measurement of boundary conditions and statistical evaluations of the real casting data.

Presented approach is demonstrated on the real radial slab caster with twelve cooling loops in the secondary cooling for the special grade of steel S355. This steel was selected expediently because it is mainly used for shipbuilding projects, marine mechanical systems and deep-water ocean offshore structural projects were the quality of steel S355 plates is essential. The mechanical properties for these grades of steel are specified by the European Standards.

2. Solidification Model—BrDSM

The core of presented approach is the 3-D transient solidification model. The computing precision, accuracy, robustness and fast computational times are essential. In order to reach these specifications a suitable numerical scheme should be selected. Numerical mesh independence tests as well as tests of the calculation speed on different CPU/GPU and mainly a properly validation on the real casting data were made.

The transient 3-D solidification model is based on Fourier-Kirchhoff Equation [2] which can be expressed as:

$$\frac{\partial (h\rho)}{\partial \tau} + v_{cast}\frac{\partial (h\rho)}{\partial z} = \nabla \cdot \left(k_{eff}\nabla T\right), \tag{1}$$

with boundary conditions:

$$T(x,y,z)|_{z=0} = T_{pouring} \qquad \text{the meniscus,} \qquad (2)$$

$$-\lambda \frac{\partial T}{\partial n} = 0 \qquad \text{the exit area,} \qquad (3)$$

$$-\lambda \frac{\partial T}{\partial n} = \frac{\dot{m}_{water} c_{water}(T_{out} - T_{in})}{S_{mold}} \qquad \text{in the mould,} \qquad (4)$$

$$-\lambda \frac{\partial T}{\partial n} = \frac{\pi(1/2)d_{rol}}{S_{rol}} htc_{rol}(T_{rol} - T_{amb}) + \sigma\varepsilon\left(T_{rol}^4 - T_{amb}^4\right) \qquad \text{beneath the rollers,} \qquad (5)$$

$$-\lambda \frac{\partial T}{\partial n} = htc(T - T_{water}) + \sigma\varepsilon\left(T^4 - T_{water}^4\right) \qquad \text{beneath the spray cooling,} \qquad (6)$$

$$-\lambda \frac{\partial T}{\partial n} = 0.84(T - T_{amb})^{4/3} + \sigma\varepsilon\left(T^4 - T_{amb}^4\right) \qquad \text{within the solidified shell free surface.} \qquad (7)$$

k_{eff} is the effective thermal conductivity (W/mK); T is the temperature (K); $T_{pouring}$ is the pouring temperature (K); h is the specific enthalpy (J/kg); ρ is the density (kg/m^3); τ is the time (s); v_{cast} is the casting speed (m/min); z is the direction of casting (m), T_{rol} is the roller temperature; T_{water} is the cooling water temperature; T_{amb} is the ambient temperature; $\dot{m}_{water}$ is the mass water flow in the mould (kg/s); c_{water} is the specific heat capacity of water (J/kgK); htc is the heat transfer coefficient beneath spraying surface (W/m^{2}K); σ is the Stefan-Boltzman constant (W/m^{2}K^4) and ε is the emissivity of the slab surface (-).

The latent heat released during the solidification in Equation (1) is substituted by the enthalpy h and can be expressed as:

$$h = \int_0^T \left(c(\zeta) - \Delta H \frac{\partial f_s}{\partial \zeta}\right) d\zeta, \qquad (8)$$

where ΔH is the latent heat (J/kg) and f_s is the solid fraction (-). The Enthalpy method was used for modelling the solidification process, see Mauder et al. [10]. The Enthalpy method is robust method because it ensures energy conservation and there is no discontinuity at either the liquidus or the solidus temperatures because the solidification/melting path is characterized strictly by decreasing/increasing enthalpy. In the Equation (1) the enthalpy is calculated in the first step as the primary variable and the temperature is calculated from a defined enthalpy-temperature relationship in the second step Equation (8).

In the presented solidification model, Equation (8) is substituted by an enthalpy-temperature function obtained by the InterDendritic Solidification (IDS) package created by the Miettinen [11]. The IDS allows calculation of enthalpy, thermal conductivity, density and other thermophysical parameters as a function of temperature from 1600 °C–0 °C.

Fluid flow of the liquid steel can be calculated by the computational fluid dynamics (CFD) techniques. However, computational times are even with the use of new modern modelling approaches far from the real duration of the simulated process. The influence of heat transfer by the fluid flow is account in the effective thermal conductivity term k_{eff}. The thermal conductivity is increased by the flow of liquid steel at different distances from the meniscus, where k_{eff} is represented by Zhang et al. [12] as follows:

$$k_{eff} = \begin{cases} k & T \leq T_{sol} \\ 4k + \frac{3k(T-T_{liq})}{(T_{pouring}-T_{liq})} & T \geq T_{liq} \wedge 0 \leq z \leq 1\,\text{m} \\ k + \frac{3k(T-T_{sol})}{(T_{liq}-T_{sol})} & T_{sol} \geq T \geq T_{liq} \wedge 0 \leq z \leq 1\,\text{m} \\ k + \frac{k(T-T_{liq})}{(T_{pouring}-T_{liq})} & T \geq T_{liq} \wedge 1\,\text{m} \leq z \leq 3\,\text{m} \\ k & z \geq 3\,\text{m,} \end{cases} \qquad (9)$$

where k is the thermal conductivity (W/mK); T_{sol} is the temperature of solidus (K); T_{liq} is the temperature of liquidus (K); and z is the distance from the meniscus (m).

The heat transfer in the mould region described by Equation (4) is calculated from the heat balance between heat removal from the mould walls and internal water cooling in-out temperature difference. The real data comes directly from the measurement of cooling water temperatures and flow rates, which are affected by the casting speed and steel carbon content.

The situation beneath the rollers Equation (5) also comes from the heat balance assumption. The removed heat from the slab surface is equal to the heat, which is radiated to the surroundings by the rollers. In the case of internal cooled rollers, Javurek et al. [13] published a paper on how to determine boundary conditions in the case of dry casting.

In the spraying area, the strand is cooled by the water or water–air mixture sprays. Nozzle parameters like air and water flow, nozzle position and impact angles have an effect on the cooling efficiency, the htc from the Equation (6) respectively. Totten et al. [14] showed several empirical formulas describing how to deal with the heat transfer coefficient beneath the nozzle. However, these empirical formulas include many constants and parameters and their correct determination for a particular cooling setup is difficult. Proper determination of boundary conditions is crucial in case to obtain a real results, see Lopez et al. [15].

The advantage of the model discussed in this paper is that it obtains its heat transfer coefficients from measurements of the spraying characteristics of all nozzles used by the caster on a so-called hot plate in the experimental laboratory, according to Raudensky et al. [16]. Thus, the model takes htc as a function of water flow, casting speed, surface temperature, air pressure and can be expressed mathematically as:

$$htc = f\left(\dot{m}_{water}, v_{cast}, T_{surface}, p_{air}\right). \tag{10}$$

3. Numerical Formulation and Massive Parallelization

There are several discretization schemes how to solve the Equations (1)–(7). After large-scale numerical studies two numerical schemes proved to be good candidates for solving the case of 3-D transient heat transfer problem with solidification phenomena, Simple Explicit (SE) and Alternating Direction Implicit (ADI), see Mauder et al. [10]. The ADI scheme proposed by Douglas and Gunn is unconditionally stable and retains the second-order accuracy when applied to 3-D problems. However, ADI scheme has very limited possibility of parallelization on multiple CPU and massive parallelization on GPU is out of question. In the case of complex 3-D solidification-optimization models, the requirement for the real time control can be reached only by the use of the massive parallelization techniques. In Table 1 there are computation times for ADI and SE schemes tested on three different types of meshes (coarse-mesh with 10 000 nodes, fine-mesh with 125,000 nodes and the very-fine mesh with 1,000,000 nodes) on Intel(R) Core(TM) i7-3770 CPU @ 3.4GHz 16GB RAM. More information about efficiency, robustness and accuracy of the examined numerical schemes can be found in Mauder et al. [10].

Table 1. Computation time for non-parallel and parallel solution (s).

Numerical Scheme	Computational Time (s)		
	Coarse mesh	Fine-mesh	Very-fine mesh
SE—1 CPU	11.54	887.14	54,362.12
SE—12 CPU	135.28	1122.13	35,124.54
ADI—1 CPU	41.81	714.73	36,210.41
ADI—12 CPU	58.61	824.32	32,416.32
SE—GPU	18.67	97.72	972.49

This is the reason why the numerical core in presented solidification model is created by SE scheme. On the one hand, SE has its limitation in the form of stability condition, thus for fine meshes the time step has to be very small. However, due to the use of massive parallelization on GPU this issue is compensated. The calculation on GPU is more than 50 times faster for very-fine meshes. The SE scheme applied to Equation (1) can be in the form:

$$h_{i,j,k}^{n+1}\rho_{i,j,k}^{n+1} = h_{i,j,k}^n \rho_{i,j,k}^n + \Delta\tau \left[\dot{Q}x + \dot{Q}y + \dot{Q}z\right] - \Delta\tau\, v_{cast} \frac{h_{i,j,k+1}^n \rho_{i,j,k+1}^n - h_{i,j,k}^n \rho_{i,j,k}^n}{\Delta z_{k-1}}, \quad (11)$$

where:

$$\dot{Q}_\Psi = 2 \frac{\left(\frac{0.5}{k_{\zeta+1}^n} + \frac{0.5}{k_\zeta^n}\right)^{-1} \frac{T_{\zeta+1}^n - T_\zeta^n}{\Delta\Psi_\zeta} - \left(\frac{0.5}{k_\zeta^n} + \frac{0.5}{k_{\zeta-1}^n}\right)^{-1} \frac{T_\zeta^n - T_{\zeta-1}^n}{\Delta\Psi_{\zeta-1}}}{\Delta\Psi_\zeta + \Delta\Psi_{\zeta-1}} \quad \Psi \in \{x,y,z\}\ \zeta \in \{i,j,k\}. \quad (12)$$

The index $n+1$ is associated to the future time; n is the index corresponding to the actual time; $\Delta\tau$ is the increment of the time; x, y and z are the directions; $\Delta\Psi$ are increments in the directions and ζ are the positions. The time step $\Delta\tau$ has to fulfil a well-known stability condition:

$$\Delta\tau \leq \frac{1}{\left(\frac{2k(T)}{\rho(T)c(T)}\right)\left(\sum_\Psi \frac{1}{\Delta\Psi^2}\right) + \frac{v_{cast}}{\Delta z}}. \quad \Psi \in \{x,y,z\} \quad (13)$$

The numerical model has a non-equidistant mesh in all directions. The reason for this is that the largest temperature gradients are near the surface. In axis z (direction of casting) the nodes are adapted to the real rollers and nozzles positions to get more accurate determination of boundary conditions in secondary cooling zone. Mesh independence test in Figure 2 indicates when the convergence is achieved. In x axis is the number of elements, in y-left axis is the temperature in several points on the top surface (distance from meniscus 1, 2, 4, 10, 15, 20 m) and in the y-right axis is the value of the metallurgical length. This study shows that the number of elements should be at least 2 500,000. Finer mesh do not affected the results significantly. These results with respect the results obtained in Table 1 show again the needs of SE GPU parallel scheme in order to calculate complex 3-D solidification model in the real time.

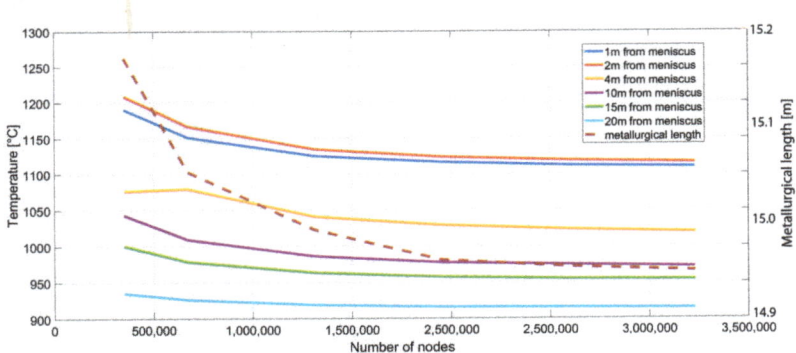

Figure 2. Temperatures and metallurgical length for different mesh densities.

The mesh decomposition is sketched out in Figure 3. The primary mesh is created on CPU, then is divided into many sub-meshes (regions). The boundary conditions and initial condition are set on CPU. Further, numerical regions are separately send to the GPU units where Equations (10) and (11) are calculated. After the temperatures in future time step are calculated, the values in separated regions are send back to CPU where boundary conditions are recalculated and again send to the GPU. This loop is repeated until final temperature field is reached.

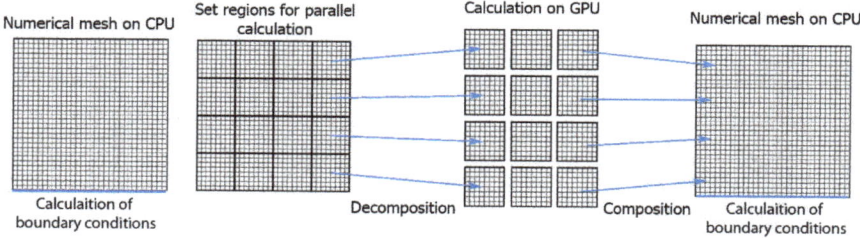

Figure 3. Numerical mesh CPU-GPU decomposition.

4. Fuzzy Logic Regulator—FL-BrDSM

The BrDSM can calculate the temperature distribution of the strand in real time for given casting parameters such as the initial temperature distribution, intensity of cooling, speed of casting and so forth. The problem is how to set input parameters in the optimal way to get a required (optimal) temperature distribution. This is referred to as an inverse problem. For this reason, supervision-system/optimal-regulator based on fuzzy logic has been created. The FL regulator treats the solidification model as a black-box and when the FL regulator changes input parameters such as casting speed and intensity of cooling, the black-box returns the final temperature field for new input parameters. Based on the temperature field the FL regulator adjusts input parameters and repeats the process in a closed-loop until the optimal solution is found.

Several regulation approaches have been tested in the past, such as PID regulation, FL regulation, Model Predictive Control (MPC) regulation and their combinations. The most promising regulation outcome for dynamic changes in process parameters was reached by using a combination of MPC and FL regulation, see Stetina et al. [17]. This approach generally requires a large number of iterations because several future scenarios are calculated. For the real time simulation, this is possible only with use of GPU solidification model.

The most important parameters for the MPC/FL regulator are so called optimal surface temperature intervals. These temperature intervals should guarantee smooth temperature (cooling) profiles and high quality of final steel with zero or minimum surface defects. The optimal cooling strategy is to keep the surface temperatures in these intervals, see Figure 4. These intervals are set by the user and they are distinct for different grades of steel. In Section 5 is discussed how to set these intervals.

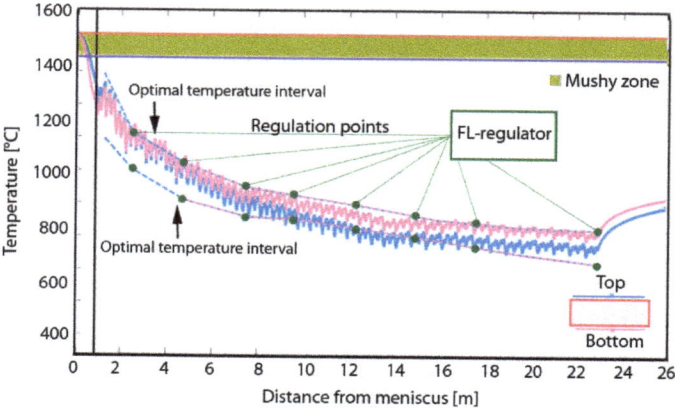

Figure 4. Optimal temperature intervals.

FL regulator extracts temperatures in the regulation points from the computed result (output from BrDSM) and by compare them with the prescribed temperature values to determine their errors. With all the temperature information, the FL regulator infers modifications for each cooling circuits. The temperatures in the regulation points are influenced by all previous cooling circuits, which should be taken account. The closer cooling circuit has a bigger influence on the regulation point, thus artificial parameter so call impact parameter in the range from 0 to 10 is created. The regulator evaluates the errors and impacts by the fuzzy rules. These rules give the value of modification for each cooling circuit. One circuit can get several different modifications from following control points proportional to the impact. Some modifications can require increase of cooling, the other decrease of cooling. The algorithm uses the final value for modification of one cooling loop as a sum of these values.

Membership functions for all fuzzy statements are trapezoidal-shaped uniformly distributed through the corresponding interval. These intervals are professionally/expertly set based on caster geometry and another caster parameters. As a defuzzification method, the standard centre of gravity was chosen.

The combination of the FL regulator with the GPU solidification model can predict future temperature states and it works like the MPC system. The detailed description of the FL regulator including linguistic variables and linguistic rules, input and output data and so forth, is described in Mauder et al. [18].

5. Steel S355 and Real Casting Data

A frequent occurrence of surface defects in slabs of the cast steel grade S355 was the reason for statistical and numerical investigation. The importance of high quality of the steel S355 was mentioned in the introduction. The chemical composition is listed in Table 2.

Table 2. Chemical composition of steel S355.

Weight Fraction	Ni	Mn	Mo	Si	Nb	Ti	Cu
wt%	max 0.300	1.400–1.550	max 0.080	0.5	max 0.060	max 0.020	max 0.200
Weight Fraction	V	Al	P	C	Cr	S	Ca
wt%	max 0.020	0.020–0.060	0.030	0.160–0.180	max 0.200	0.020	0.002

Three typical defects were found: transverse facial cracks, star cracks and longitudinal facial cracks. The most common defects were the star cracks. These defects primarily appear at the top surface of the strand and possible causes for these defects are hard cooling and tensile stresses at the straightening area [2]. Therefore, the top surface of the strand and the temperatures at the straightening area were investigated.

The real casting data were statistically evaluated from approximately 2000 heats cast in 2011 and 2012 in EVRAZ VITKOVICE STEEL machinery in the Czech Republic. The evaluation of the statistical hypothesis shows that the surface temperature in the unbending point significantly influences the surface quality of slabs. The heats were divided into two groups. Heats where surface defects were found and heats without surface defects. The surface temperature was measured by a pyrometer at the straightening area. The results in a form of histogram are shown in Figure 5. The heats without surface defects were fitted by the Gauss curve to obtain optimal temperature intervals at unbending point.

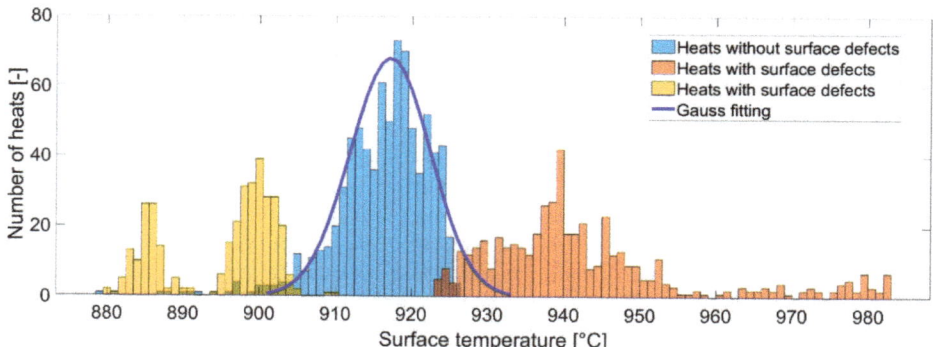

Figure 5. Surface temperatures in unbending point with and without defects.

Hypothesis tests show that we cannot reject the hypothesis on the 5% level of significance stating that the surface temperature at the straightening area influences the presence of surface defects. This leads back to the following idea: if we will keep the surface temperature in some range, the occurrence of surface defects would be minimized. The optimal temperature intervals at unbending point were set according to the statistical results from heats without defects 916.03 ± 6.89 °C.

There are three pyrometers distributed on the examined caster. First one at end of the mould, the second one at unbending point (Figure 5) and the last one at the end of tertiary cooling zone. The most significant area is at unbending point from the perspective of cracks. However, smooth decreasing of surface temperatures through the casting process has positive influence on the final steel quality. Thus, the optimal temperature intervals for the fuzzy regulator were obtained by the curve fitting of these three points using following function:

$$y = xa^b + c, \qquad (14)$$

where y is the temperature, x is distance from the meniscus and constant a, b and c were found by using curve fitting tool. These constants have to be found for each grade of steel separately. Optimal temperature intervals obtained this way are used as the input parameters for FL-BrDSM. Thermophysical properties for the steel grade S355 as functions of the temperature were calculated using the InterDendritic Solidification (IDS) thermodynamic-kinetic package (version 1.3.1, Helsinki University of Technology, Helsinki, Finland). The results are presented in Figure 6a. The solidification model of the IDS is a so-called "grey box", that is, it combines empirical or semi-emprical submodels with physically conceived submodels. The IDS model has been created at Aalto University in Helsinki [11] and is further developed, see Louhenkilpi et al. [19]. The IDS model consists of two main submodels for the simulation of interdendritic solidification (solves solidification from liquidus temperature to 1000 °C, i.e., the formation of, for example, ferrite or austenite) and simulation of solid state austenability (solves solidification from 1000 °C to temperature, 25 °C, that is, the formation of proeutectoid ferrite, cementite, perlite, bainite and martensite. The model also supports basic calculations of properties that influence the strength behaviour of steel and can serve as a basis for predictive crack criteria, see Figure 6b).

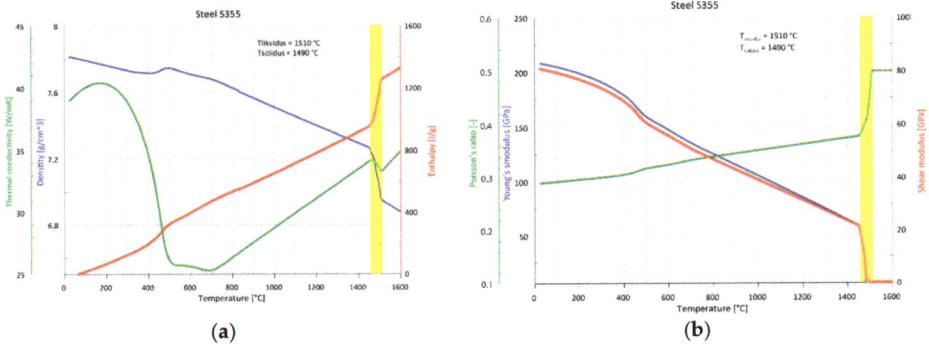

Figure 6. Temperature dependent physical properties for the steel grade S355, (**a**) thermophysical properties; (**b**) Mechanical properties.

6. Results and Discussion

The input parameters for the FL-BrDSM were: the thermophysical properties of steel grade S355 (from IDS); optimal surface temperature intervals (from statistical evaluation); speed of casting (set by the user); meniscus temperature (from historical/actual casting data in case of offline/online simulation); heat fluxes at the mould (from historical/actual casting data as a function of casting speed); and maximal and minimal water flows at the secondary cooling (from the caster specification). The output parameters for the FL-BrDSM were the temperature field (from BrDSM) and the optimal cooling intensity at the secondary cooling zone (from the FL regulator).

The geometry of the investigated CC machine is as follows: a mould length of 900 mm; a radius of 8000 mm; a length of secondary cooling after an unbending point of 8500 mm; length of tertiary cooling of 2000 mm. The secondary cooling zone is divided into 12 independent cooling loops according to Figure 7. For more detailed caster description see Mauder et al. [18].

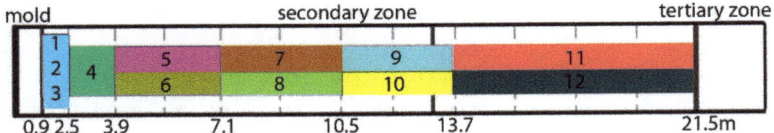

Figure 7. Position of cooling loops.

Firstly, the FL regulator was tested for static process parameters. The optimal temperature field for the average casting speed 1.7 m/min is shown in Figure 8. The cross-section size of the slab was 1530 mm × 250 mm and the numerical mesh was created from over 1.5 million nodes. Because the CC is a dynamic process and the casting speed varies in time, it is necessary to keep surface temperatures at constant values in order to achieve a high steel quality. Thus, optimization was carried out for two different casting speeds in order to obtain the optimal relationship between cooling intensities and the casting speed, see Table 3. These results can be used for the real CC process in case of steel grade S355 casting.

The cooling intensity in loops 9 and 11 (last two loops at top surface) reaches its minimal allowed values for casting speed of 1.5 m/min. This means that for lower values of casting speed is even a better solution possibly exists. The last two cooling loops at top surface should therefore be replaced by smaller nozzles, which can operate smaller water flows (soft cooling). Moreover, significant savings of water consumption can be reached. In the case of casting speed 1.9 m/min, the maximal allowed cooling intensity values (physical limitations of pumps) were reached. This casting speed should not be exceeded, otherwise the surface temperature increases over the optimal temperature interval and the number of defects may rise.

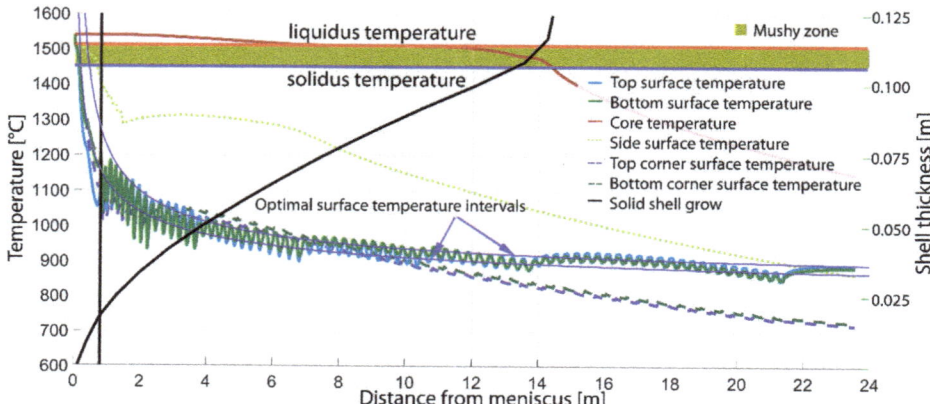

Figure 8. Optimal temperature field after fuzzy regulation.

Table 3. Optimal cooling intensities for cooling loops and different casting speeds.

Casting Speed (m/min)	Loop 1 (L/min)	Loop 2 (L/min)	Loop 3 (L/min)	Loop 4 (L/min)	Loop 5 (L/min)	Loop 6 (L/min)
1.5	96.6	125.4	103.9	124.6	65.3	98.5
1.9	98.1	132.6	109.5	146.3	98.3	103.3
Casting Speed (m/min)	Loop 7 (L/min)	Loop 8 (L/min)	Loop 9 (L/min)	Loop 10 (L/min)	Loop 11 (L/min)	Loop 12 (L/min)
1.5	32.0	65.7	22.0	34.7	31.2	43.7
1.9	46.9	85.7	26.7	58.3	99.5	106.2

Another set of simulations were carried out for dynamic changes in process parameters such as casting speed or casting temperature. The following results show 1 h casting process where a drop of the casting speed from 1.9 to 1.2 m/min at time 30 min and the increase of the casting speed from 1.2 to 1.9 m/min occurs at time 40 min. The Figure 9 shows the response in cooling intensities for all 12 cooling loops on casting speed drop calculated by FL regulator. Because MPC were used, several future scenarios were calculated and optimal dynamic changes were found. The value of temperature errors after each regulation points (surface temperatures after each cooling loop) are shown in Figure 10. The maximal temperature error was approximately 20 °C but average temperature error was less than 10 °C. This can be consider as a very good regulation results.

Static and dynamic simulations proved regulation possibilities of the presented FL-BrDSM solution. The computation time without GPU was 27.25 min for static and 47.20 min for dynamic simulation. With GPU approach, the computation time was less than 1 min and 2 min in case of dynamic simulation respectively. This declares that there is no problem to solve complex fine-mesh (more than 3 million nodes) 3D transient solidification models and their optimal regulation in the real time on the real CC machine.

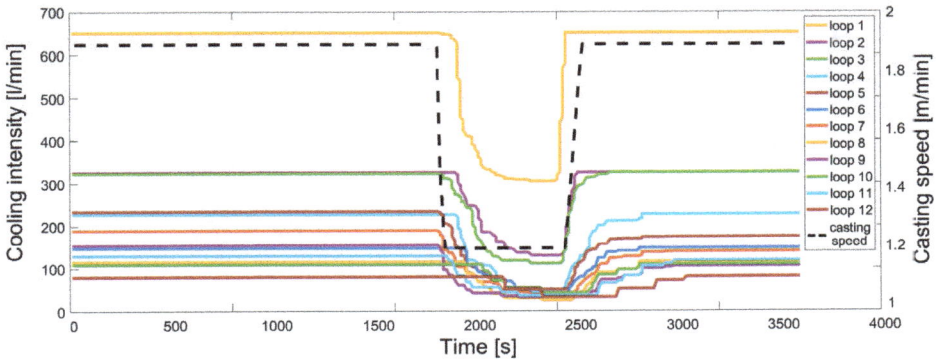

Figure 9. Dynamic response of FL regulator to casting speed drop.

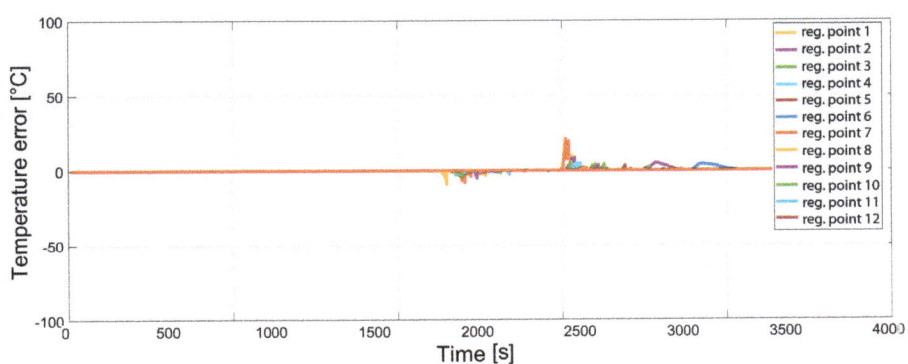

Figure 10. Surface temperature errors after each cooling loop.

7. Conclusions

The FL-BrDSM was tested for many different fuzzy parameters, for different casting temperatures and casting speeds constraints, for different caster and slab geometries and for different steel grades. This paper is focused on the quality improvement of the steel grade S355 and demonstration of the FL-BrDSM. The first part of the work was focused on detail description of solidification model and its GPU version. Than FL regulator was described. Last part of the work was statistical evaluation of real historical casting data for steel grade S355. The influence of the surface temperature at the straightening area to the occurrence of surface defects was statistically evaluated as significant. From the statistic results the optimal (recommended) temperature field was obtained and used as the input to the FL-BrDSM. For different casting speeds the optimal cooling intensities were found. The control of the CC process using recommended cooling curves can decrease the number of surface defects. The dynamic simulation shows real time regulation possibilities in the case of casting speed drop. The same approach can be applied for any grade of steel. The main advantage of the presented approach is a small number of evaluations before the optimal solution is reached and its overall versatility. This also allows for the on-line real-time regulation of a real CC process.

Author Contributions: Conceptualization, T.M. and J.S.; Methodology, J.S.; Software, T.M.; Validation, T.M. and J.S.; Formal Analysis, J.S.; Investigation, T.M.; Resources, T.M.; Data Curation, J.S.; Writing-Original Draft Preparation, T.M.; Writing-Review & Editing, T.M.; Visualization, T.M.; Supervision, J.S.; Project Administration, J.S.; Funding Acquisition, J.S.

Funding: This research was funded by the project NETME+, LO1202, with the financial support from the Ministry of Education, Youth and Sports of the Czech Republic under the "National Sustainability Programme I and APC was funded by the Open Access fond Brno University of Technology.

Acknowledgments: The authors gratefully acknowledge funding from the Open Access fond Brno University of Technology and Specific research on BUT FSI-S-17-4444.

Conflicts of Interest: The authors declare no conflict of interest. The funders had no role in the design of the study; in the collection, analyses, or interpretation of data; in the writing of the manuscript and in the decision to publish the results.

References

1. Flick, A.; Stoiber, C. Trends in continuous casting of steel: Yesterday, today and tomorrow. In Proceedings of the METEC InSteelCon, Düsseldorf, Germany, 27 June–1 July 2011; pp. 80–91.
2. Birat, P.; Chow, C.; Emi, T.; Emling, W.H.; Fastert, H.P.; Fitzel, H.; Flemings, M.C.; Gaye, H.R.; Gilles, H.L.; Glaws, P.C.; et al. *The Making, Shaping and Treating of Steel: Casting Volume*, 11th ed.; The AISE Steel Foundation: Pittsburgh, PA, USA, 2003; p. 1000, ISBN 978-0-930767-04-4.
3. Santos, C.A.; Spim, J.A.; Garcia, A. Mathematical modeling and optimization strategies (genetic algorithm and knowledge base) applied to the continuous casting of steel. *Eng. Appl. Artif. Intell.* **2003**, *16*, 511–527. [CrossRef]
4. Zhemping, J.; Wang, B.; Xie, Z.; Lai, Z. Ant Colony Optimization Based Heat Transfer Coefficient Identification for Secondary Cooling Zone of Continuous Caster. In Proceedings of the IEEE International Conference Industrial Technology, Shenzhen, China, 20–24 March 2007; pp. 558–562. [CrossRef]
5. Zheng, P.; Guo, J.; Hao, X.-J. Hybrid Strategies for Optimizing Continuous Casting Process of Steel. In Proceedings of the IEEE International Conference Industrial Technology, Hammamet, Tunisia, 8–10 December 2004; pp. 1156–1161. [CrossRef]
6. Mauder, T.; Novotny, J. Two mathematical approaches for optimal control of the continuous slab casting process. In Proceedings of the Mendel 2010—16th International Conference on Soft Computing, Brno, Czech Republic, 26–28 June 2010; pp. 395–400, ISBN 978-80-214-4120-0.
7. Ivanova, A.A. Predictive Control of Water Discharge in the Secondary Cooling Zone of a Continuous Caster. *Metallurgist* **2013**, *57*, 592–599. [CrossRef]
8. Rao, R.V.; Kalyankar, V.D.; Waghmare, G. Parameters optimization of selected casting processes using teaching-learning-based optimization algorithm. *Appl. Math. Modell.* **2014**, *38*, 5592–5608. [CrossRef]
9. Mosayebidorcheh, S.; Bandpy, M.G. Local and averaged-area analysis of steel slab heat transfer and phase. change in continuous casting process. *Appl. Thermal Eng.* **2017**, *118*, 724–733. [CrossRef]
10. Mauder, T.; Charvat, P.; Stetina, J.; Klimes, L. Assessment of Basic Approaches to Numerical Modeling of Phase Change Problems-Accuracy, Efficiency, and Parallel Decomposition. *J. Heat Transf.* **2017**, *139*, 5. [CrossRef]
11. Miettinen, J. *IDS Solidification Analysis Package for Steels: User Manual of DOS Version 2.0.0*; Helsinki University of Technology: Helsinki, Finland, 1999; p. 22, ISBN 9512246600.
12. Zhang, J.; Chen, D.F.; Zhang, C.Q.; Wang, S.G.; Hwang, W.S.; Han, M.R. Effects of an even secondary cooling mode on the temperature and stress fields of round billet continuous casting steel. *J. Mater. Process. Technol.* **2015**, *222*, 315–326. [CrossRef]
13. Javurek, M.; Ladner, P.; Watzinger, J.; Wimmer, P. Secondary cooling: Roll heat transfer during dry casting. In Proceedings of the METEC ESTAD, Düsseldorf, Germany, 15–19 June 2015; p. 8.
14. Totten, G.E.; Bates, C.E.; Clinton, N.A. *Handbook of Quenchants and Quenching Technology*; Haddad, M.T., Ed.; ASM International: Materials Park, OH, USA, 1993; p. 507, ISBN 0-87170-448-X.
15. Ramírez-López, A.; Muñoz-Negrón, D.; Palomar-Pardavé, M.; Romero-Romo, M.A.; Gonzalez-Trejo, J. Heat removal analysis on steel billets and slabs produced by continuous casting using numerical simulation. *Int. J. Adv. Manuf. Technol.* **2017**, *93*, 1545–1565. [CrossRef]
16. Raudensky, M.; Hnizdil, M.; Hwang, J.Y.; Lee, S.H.; Kim, S.Y. Influence of Water Temperature on The Cooling Intensity of Mist Nozzles in Continuous Casting. *Mater. Tehnol.* **2012**, *46*, 311–315; ISSN 1580-2949
17. Stetina, J.; Mauder, T.; Klimeš, L. Utilization of Nonlinear Model Predictive Control to Secondary Cooling during Dynamic Variations. In Proceedings of the AISTech, Pittsburgh, PA, USA, 16–19 May 2016; p. 14.

18. Mauder, T.; Sandera, C.; Stetina, J. Optimal control algorithm for continuous casting process by using fuzzy logic. *Steel Res. Int.* **2015**, *86*, 785–798. [CrossRef]
19. Louhenkilpi, S.; Laine, J.; Miettinen, J.; Vesanen, R. New Continuous Casting and Slab Tracking Simulators for Steel Industry. *Mater. Sci. Forum* **2013**, *762*, 691–698. [CrossRef]

© 2018 by the authors. Licensee MDPI, Basel, Switzerland. This article is an open access article distributed under the terms and conditions of the Creative Commons Attribution (CC BY) license (http://creativecommons.org/licenses/by/4.0/).

Article

Effect of Cooling Rate on the Formation of Nonmetallic Inclusions in X80 Pipeline Steel

Xianguang Zhang, Wen Yang, Haikun Xu and Lifeng Zhang *

School of Metallurgical and Ecological Engineering, University of Science and Technology Beijing, Beijing 100083, China; xgzhang@ustb.edu.cn (X.Z.); wenyang@ustb.edu.cn (W.Y.); sdm_xhk@163.com (H.X)
* Correspondence: zhanglifeng@ustb.edu.cn; Tel.: +86-10-62332267

Received: 5 March 2019; Accepted: 25 March 2019; Published: 29 March 2019

Abstract: Nonmetallic inclusions have a strong influence on the hydrogen-induced cracking (HIC) and sulfide stress cracking (SSC) in pipeline steels, which should be well controlled to improve the steel resistance to HIC and SSC. The effects of cooling rate on the formation of nonmetallic inclusions have been studied both experimentally and thermodynamically. It was found that the increasing cooling rate increased the number density and decreased the size of the inclusions, while the inverse results were obtained by decreasing the cooling rate. Furthermore, as the cooling rate decreased from 10 to 0.035 K/s, the inclusions were changed from Al_2O_3-CaO to Al_2O_3-CaO-MgO-CaS. At a high cooling rate, the reaction time is short and the inclusions cannot be completely transformed which should be mainly formed at high temperatures. While, at low cooling rate, the inclusions can be gradually transformed and tend to follow the equilibrium compositions.

Keywords: pipeline steel; inclusions; cooling rate; composition

1. Introduction

High-grade pipeline steels have been widely used in the construction of long-distance oil and gas transportation systems, which requires good combination of high strength, toughness, corrosion resistance and weldability. Hydrogen-induced cracking (HIC) and sulfide stress cracking (SSC) have been recognized as the vital threats to the safety of the pipeline operation, and caused significant economic losses throughout the world [1–8]. It has been acknowledged [5,9] that nonmetallic inclusions are one of the dominant factors affecting the HIC and SSC in pipeline steels, especially the large inclusions with string shape. Therefore, it is crucial to control the inclusions in pipeline steels to improve the steel resistance to HIC and SSC.

There are several kinds of inclusions in pipeline steel due to its complex alloy elements. The inclusions are mainly identified as aluminum oxides, calcium oxides or calcium sulfide, magnesium oxides and manganese compounds [10]. It was reported that the calcium aluminate with low melting temperature can be elongated into string shape after rolling, which is harmful to the performance of pipeline steels [11–13]. Researchers proposed that the inclusions should be controlled to CaO-Al_2O_3-CaS, CaO-CaS type [11] or Al_2O_3-CaS type [14,15] to reduce their detrimental effects. Hence, the type and composition of the inclusions are important to pipeline steels which should be well controlled. Inclusions are usually controlled by adjusting compositions of slag or/and steel during steelmaking process [16–21]. In addition, the nonmetallic inclusions can transform even during and after solidification. It was reported that in stainless steel the inclusion of MnO-SiO_2 can transform to MnO-Cr_2O_3 spinel during heat treatment and the transformation rate depended on both temperature and size of the inclusions [22]. Therefore, the evolution of the inclusions during the cooling process is also very important for the formation of inclusions in the final product.

Cooling rate is an important parameter during the continuous casting, which largely changes the temperature and time of the steels experienced during the solidification and subsequent solid-state cooling process. Hence, the cooling rate may have a strong influence on the precipitation and transformation behaviors of the inclusions in pipeline steels. However, the previous works were mainly focused on the influences of compositions of the slag or steels on the formation of the inclusions. Little attention has been paid to the evolution of inclusions at different cooling rates. Goto et al. [23,24] investigated the effects of cooling rate during solidification on the formation of oxides in plain low carbon steels. It was found that [23,24] the amount, size and compositions of the oxides are strongly influenced by the cooling rate. However, according to the best of the authors' knowledge, no systematic work has been reported regarding the effects of cooling rate on the formation of inclusions in high-grade pipeline steels, especially the type and composition of the inclusions, which requires a stricter control in inclusions. In the current work, the number, size distributions and compositions of inclusions in X80 pipeline steel formed at different cooling rates were studied systematically to clarify the effects of cooling rate on the precipitation, growth and transformation of the inclusions in pipeline steel.

2. Materials and Methods

X80 pipeline steel was used in this study, the composition of which is shown in Table 1. The production procedure of X80 pipeline steel is "basic oxygen furnace (BOF)→ladle furnace (LF) refining→RH refining→Ca treatment→soft-blowing→continuous casting". Molten steel specimens were taken from the ladle furnace after the Ca treatment and soft-blowing for 600 s, and quenched into water immediately. Figure 1a shows a schematic of the thermal history for re-melting and cooling treatments of the specimens used in this study. The specimens were first re-melted at 1873 K for 600 s in a Si-Mo resistance furnace under an argon atmosphere in MgO crucibles, as schematically shown in Figure 1b. After that, the crucibles were taken out and cooled at three different cooling rates by employing water, air or furnace, respectively, as schematically shown in Figure 1a.

The cooling rate in air was evaluated from Equation (1) [25] which relates the secondary dendrite arm spacing with the cooling rate and it is valid for a carbon content below 0.53 mass%.

$$\lambda = 148 C_R^{-0.38}, \tag{1}$$

where, λ represents the secondary dendrite arm spacing (μm), and C_R is the cooling rate (K/s). The average secondary dendrite arm spacing was measured to be around 191 μm, and thus the average cooling rate in air was estimated to be 0.51 K/s. The dendritic structure for the samples cooled in furnace and in water did hardly produce clear images. The cooling rates in furnace and in water were estimated to be 0.035 K/s [26] using a pyrometer and 10 K/s using thermocouples [27], respectively, from the temperature of 1873 to 983 K.

After cooling to room temperature, the samples were cut and polished for analyzing the distribution of the inclusions. The morphology, number, area and chemical composition of inclusions in the samples were detected by using an automated Scanning Electron Microscope-Energy Dispersive Spectrometer (SEM-EDS, FEI, Pittsburgh, PA, USA) inclusion analysis system (Aspex Explorer) operated at 15 kV. The minimum detectable inclusion was set as approximately 0.5 μm, and the maximum diameter was chosen as the size of inclusion. The area of around 17 mm^2 were scanned for the of inclusions analyses for each sample.

Table 1. Chemical composition of the alloy used in this study (mass%).

C	Si	Mn	P	S	Al	Ti	Ca	O
0.06	0.20	1.72	0.014	0.0015	0.03	0.015	0.0016	0.0012

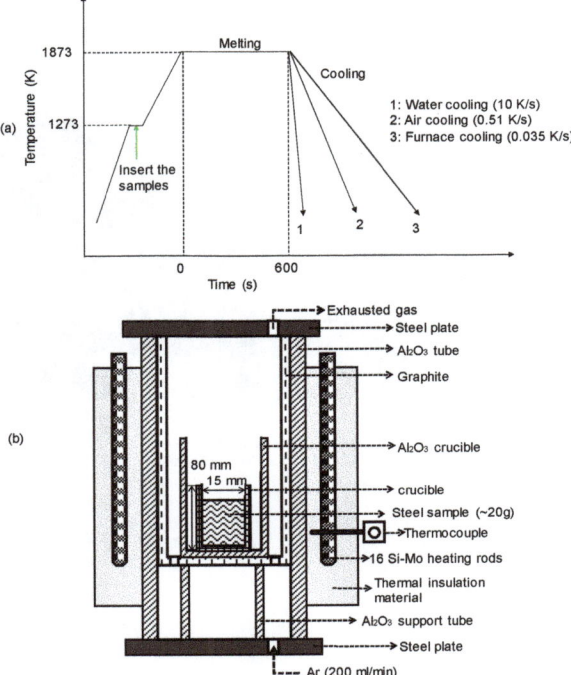

Figure 1. Schematic illustrations of (**a**) the thermal history for re-melting and cooling treatments of the specimens and (**b**) the Si-Mo resistance furnace employed in this study.

Two statistical parameters, area fraction and number density, are defined as the following two equations to characterize nonmetallic inclusions.

$$AF = \frac{A_{inclusion}}{A_{total}}, \tag{2}$$

where, AF is the area fraction of inclusions, ppm; $A_{inclusion}$ is the total area of the detected inclusions, μm^2; and A_{total} is the area scanned, mm^2.

$$ND = \frac{n}{A_{total}}, \tag{3}$$

where, ND is the number density of inclusions, per mm^2; and n is number of detected inclusions on the area of A_{total}.

3. Results and Discussion

3.1. Characterization of the Inclusions Formed at Different Cooling Rates

The element mapping of a typical inclusion formed at the cooling rate of 10 K/s (water cooling) is shown in Figure 2a. The morphology of the inclusion is spherical and it mainly contains the elements of Al, Ca, O, and a small amount of Mg. The composition and size distributions of inclusions were plotted in ternary phase diagrams, as shown in Figure 2b–c, where the scanning area and number of inclusions are presented and the black star indicates the average composition. Most of the inclusions are the Al$_2$O$_3$-CaO, with a very small amount of MgO and CaS. The diameters of the inclusions are mostly smaller than 3.5 µm. Figure 2d shows the variation in composition of the inclusions against its equivalent diameter. The contents of Al_2O_3 and CaO are almost stable regardless the size of the inclusions, and the contents of MgO and CaS are very low.

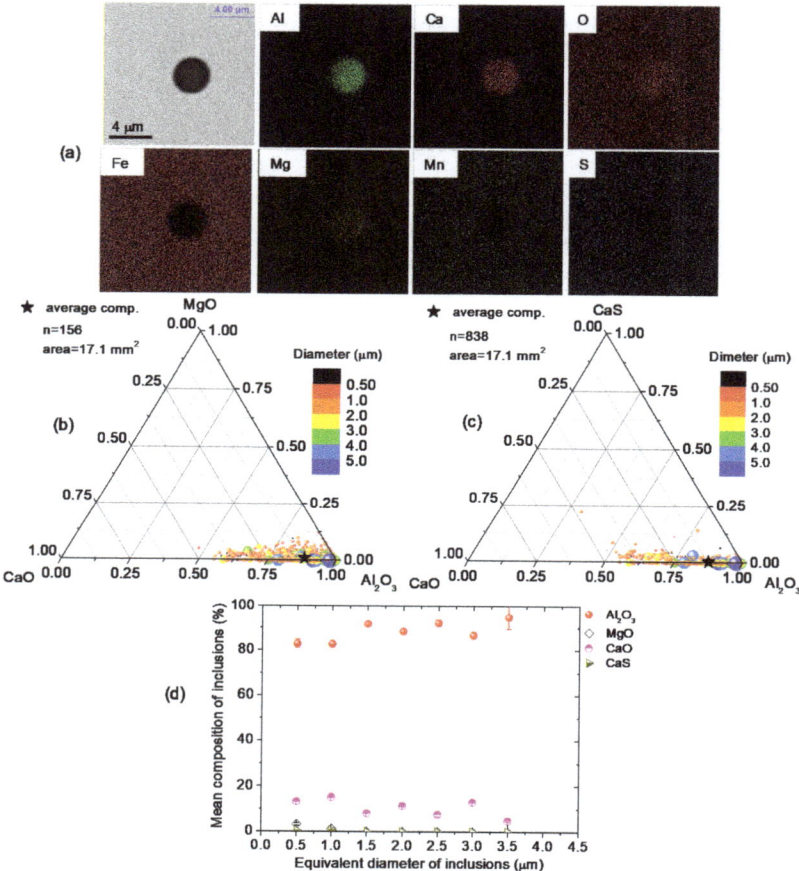

Figure 2. (a) Element mapping of a typical inclusions formed at the cooling rate of 10 K/s. Summary of the (b–c) composition and size distributions of inclusions and (d) variations in composition of inclusions against its diameter formed at the cooling rate of 10 K/s.

The element mapping of a typical inclusion formed at the cooling rate of 0.51 K/s (air cooling) is shown in Figure 3a. The morphology of the inclusion is spherical and it mainly contains the elements of Al, Ca, O. The content of Mg seems slightly higher than that of the water-cooled specimen shown in Figure 2a. The composition and size distributions of inclusions are shown in Figure 3b–c. Most of the inclusions are the Al_2O_3-MgO-CaO and Al_2O_3-CaO-CaS. In addition, the diameters of the inclusions are mostly smaller than 3 µm. The average composition of the inclusions is about 62% Al_2O_3-3% MgO-26% CaO-9% CaS. Figure 3d shows the variation in composition of the inclusions as a function of its diameter. With the increase in diameter of inclusions, the Al_2O_3 and CaO contents are increased, while the CaS content is decreased. Around 10% CaS was contained in inclusions with the diameter of 0.5 µm.

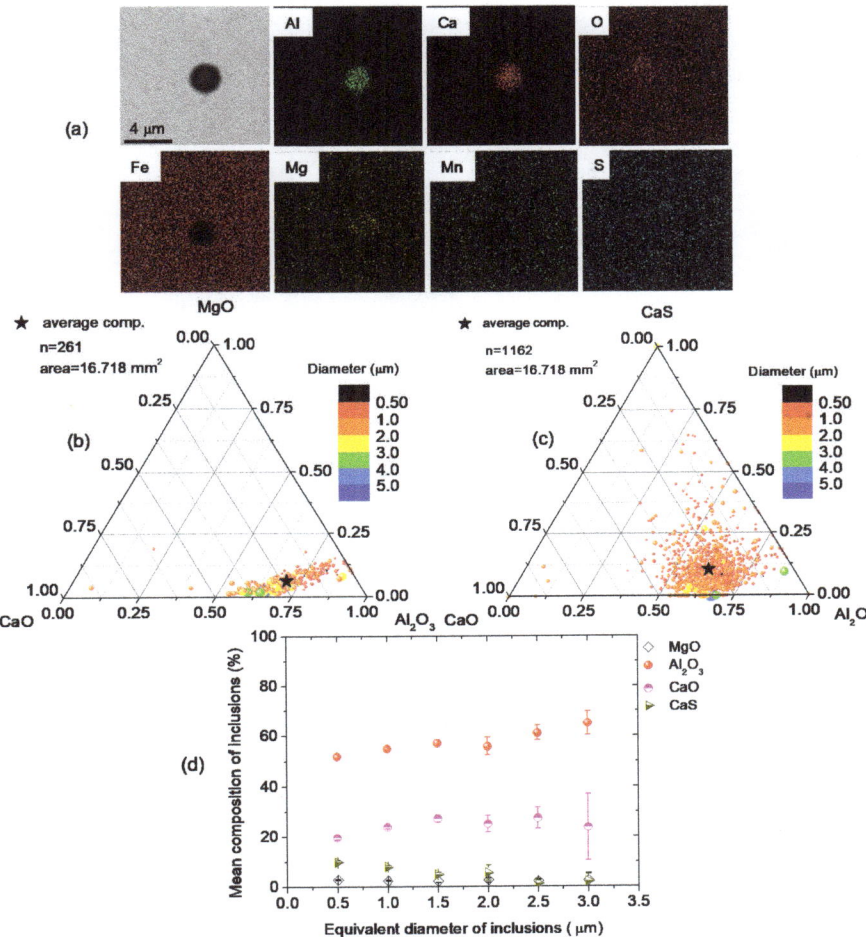

Figure 3. (a) Element mapping of a typical inclusions formed at the cooling rate of 0.51 K/s. Summary of the (b–c) composition and size distributions of inclusions and (d) variations in composition of inclusions against its diameter formed at the cooling rate of 0.51 K/s.

The element mapping of a typical inclusion at the cooling rate of 0.035 K/s (furnace cooling) is shown in Figure 4a. The morphology of the inclusion is spherical and it mainly contains the elements of Al, Ca, O, Mg and S. The contents of Mg and S are much higher than that of the high cooling rates as shown in Figures 2a and 3a. More importantly, it can be noticed that the inclusion is not as uniform as that of the inclusions formed at a high cooling rates (Figures 2a and 3a). According to the element distribution, the out-layer of the inclusion is CaS, while the core is composed by Al_2O_3-MgO-CaO. The composition and size distributions of the inclusions formed at the cooling rate of 0.035 K/s (furnace cooling) are shown in Figure 4b,c. Most of the inclusions were the Al_2O_3-MgO-CaO and Al_2O_3-CaO-CaS. The diameters of inclusions are mostly larger than 3 μm and the number of inclusions is low. The average composition of inclusions is about 76% Al_2O_3-7% MgO-3% CaO-14% CaS.

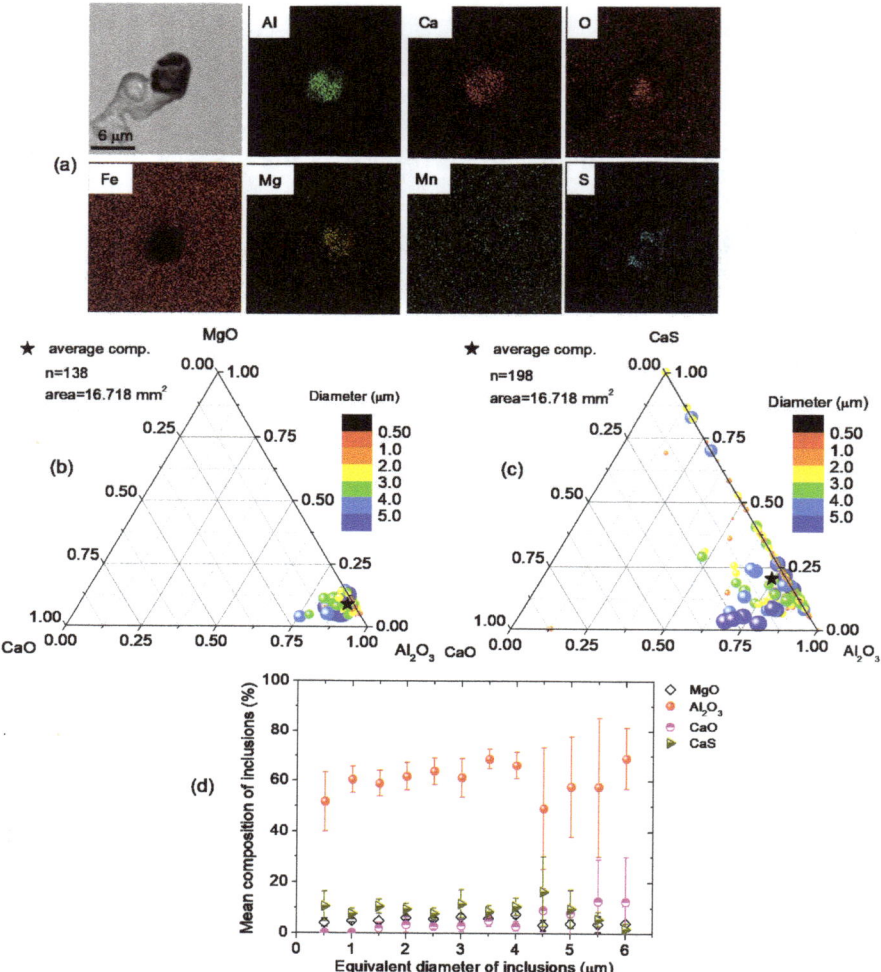

Figure 4. (a) Element mapping of a typical inclusions formed at the cooling rate of 0.035 K/s. Summary of the (b–c) composition and size distributions of inclusions and (d) variations in composition of inclusions against its diameter formed at the cooling rate of 0.035 K/s.

Figure 4d shows the variation in composition of inclusions against its diameter. The content of Al_2O_3 increases with the increase in diameter of the inclusions up to 4 μm, then it has a large fluctuation with the further increase in diameter. The content of CaO is slightly increased in the inclusions larger than 4 μm. The contents of CaS and MgO do not show strong dependence on the size of the inclusions.

Table 2 summarized the mean composition of the inclusions that were formed at different cooling rates. At a cooling rate of 10 K/s, the inclusions are mostly Al_2O_3-CaO with a very small amount of CaS and MgO. When the cooling rate is decreased to 0.51 K/s, the contents of CaO, CaS and MgO are increased. As the cooling rate is further decreased to 0.035 K/s, the contents of CaS and MgO are further increased, while the CaO content is decreased due to the transition of the CaO to CaS.

Table 2. Average composition of the inclusions formed at various cooling rates in X80 pipeline steel (mass%).

Cooling Rate, K/s	Al_2O_3	CaO	CaS	MgO
10	86.23	12.38	0.34	1.05
0.51	61.72	26.3	9.16	2.82
0.035	75.91	3.02	14.24	6.83

3.2. Effects of Cooling Rate on the Formation of the Inclusions

Figure 5a,b shows the variation in number density and area fraction of inclusions as a function of its diameters formed at different cooling rates, respectively. The peak position of the number density appears at the diameter of around 1 μm for the cases of various cooling rates, then it gradually decreases with the increase in diameter of the inclusions. The low cooling rate of 0.035 K/s results in relatively uniform distributions in size and area fractions of the inclusions. The high cooling rates of 10 and 0.51 K/s are similar in number density and area fraction distributions. The total number density of the inclusions formed at various cooling rates is summarized in Figure 5c. The number density increases dramatically with the increase in cooling rate from 0.035 to 0.51 K/s, then it becomes almost constant with the further increase in heating rate. The area fraction and average diameter of the inclusions formed at different cooling rates are shown in Figure 5d,e, respectively. With the increase in cooling rate from 0.035 to 0.51 K/s, the area fraction and average diameter are decreased, and then they become almost constant.

The number density of the inclusions should be related with the nucleation of the inclusions. According to the classical homogenous nucleation theory, as for a homogeneous nucleation of an inclusion in liquid steel, the activation energy for nucleation can be written as [28]

$$\Delta G^* = \frac{16\pi\sigma^3}{3\Delta G_v^2} = \frac{16\pi}{3}\sigma^3 \frac{V_m^2}{(RT\ln \eta)^2} \qquad (4)$$

where σ is the interfacial tension between the inclusion and liquid steel, ΔG_v the driving force for the inclusion nucleation, V_m the molar volume of the inclusion and η is the supersaturation, viz., the ratio of activities of precipitating elements to the solubility product.

The critical radius, r*, for nucleation is given by

$$r^* = -2\sigma/\Delta G_v \qquad (5)$$

The nucleation rate is written as [28]

$$I = K_Z \Gamma_A N_A \exp\left(\frac{-\Delta G^*}{kT}\right)\exp(-\tau/t), \qquad (6)$$

where K_Z is the Zeldovitch factor, the value for K_Z is about 10^{-2}. Γ_A is the jump frequency of atom A on the nucleus surface and is proportional to its diffusion coefficient in the liquid phase. N_A is the atomic concentration of atom A per unit volume. k is the Boltzmann's constant and T is the absolute temperature. τ is the incubation time, when the time t is larger than τ, the nucleation rate is stationary. Indeed, the pre-exponent factor is usually assumed to be constant and in the range of 10^{35}–10^{45} (m^{-3}·s^{-1}). Therefore, according to Equation (6), the nucleation rate is inversely proportional to the activation energy ΔG^*.

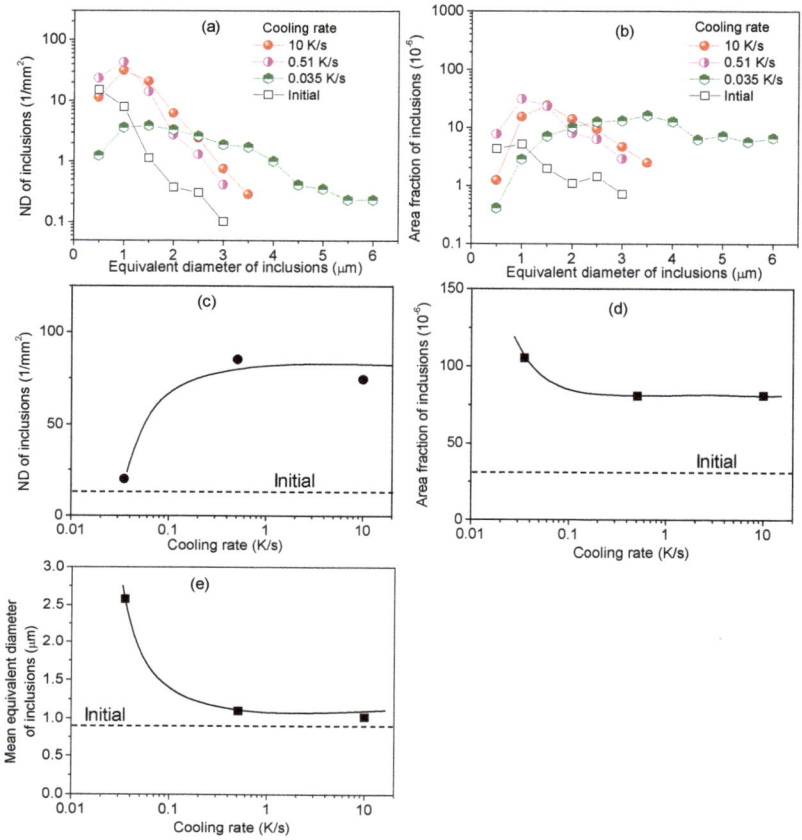

Figure 5. Variation in (**a**) number density and (**b**) area fraction of inclusions against its diameter under different cooling rates. The effects of cooling rate on the (**c**) number density, (**d**) area fraction and (**e**) mean equivalent diameter of the inclusions.

At a high cooling rate, the supersaturation (driving force) for inclusion nucleation is larger, due to the increase in elements segregation [24], which decreased the activation energy (ΔG^*) and the critical radius (r^*) for nucleation, according to Equations (4) and (5). This indicates that the nucleation become easier at a high driving force. On the other hand, the increase in driving force promotes the nucleation rate according to Equation (6). This should result in the high number density of the inclusions at a high cooling rate. It needs to point out that the supersaturation may be not always increase with the increase in cooling rate, and it may have an upper limit above which the supersaturation may not change obviously. This can be the reason for the almost constant number density of inclusions above 0.51 K/s. Furthermore, because the driving force for nucleation is changing in the present work, which is difficult to be evaluated, qualitative analysis is carried out in this study.

The low cooling rate has a low driving force for nucleation results in low number density of inclusions. On the other hand, at a low cooling rate, the inclusions have longer time to grow, the Ostwald-ripening effect is pronounced which results in coarse and low number density of inclusions.

In order to understand the effects of cooling rate on the composition of the inclusions, the contents of the various inclusions against the temperature in the pipeline steel was calculated by Fact Sage, which is shown in Figure 6a. The steel was melted at 1873 K as indicated in by a broken line in the phase diagram, at which all the inclusions were in liquid phase. With the decrease in temperature, the amount of liquid inclusions including CaO(l) and Al_2O_3(l) is increased at first and then it is decreased. During the solidification, the liquid inclusions gradually disappeared, and the solid inclusions including CaS, MgO-Al_2O_3 spinels and CaO-2Al_2O_3 are formed. The calculated average composition of the inclusions against the temperature obtained by combining the contents of each phase is shown in Figure 6b.

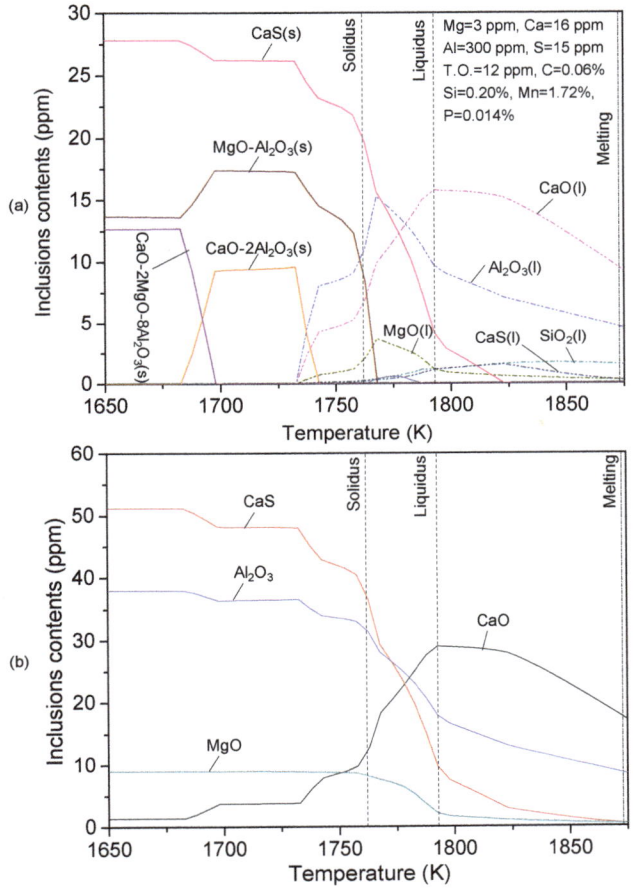

Figure 6. (a) Phase transformation and (b) average composition of inclusions during cooling of X80 pipeline steel. The CaO(l), Al_2O_3(l), SiO_2(l), CaS(l) and MgO(l) are all belong to one slag phase in (a).

At the cooling rate of 10 K/s, the reaction time is very short, inclusions could not be completely transformed. The inclusions should be mostly formed in the liquid steel. Therefore, the Al_2O_3-CaO type inclusion was dominantly formed. However, at the slow cooling rate, there is longer time for inclusion to transform. The solid inclusions including CaS, MgO-Al_2O_3, CaO-2Al_2O_3 and CaO-2MgO-8Al_2O_3 can be formed. Therefore, the contents of CaS and MgO are increased with the decrease in cooling rate. Consequently, the CaO content is decreased due to the transformation of CaO into CaS.

It should be noticed that the thermodynamic prediction is based on the concept of equilibrium, which is useful to help understand the precipitation sequence of the inclusions during cooling, but it cannot be used directly to discuss the amount of inclusions formed. For example, the measured CaS content at the heating rate of 0.035 K/s is much lower than that of the thermodynamic prediction in Figure 6b. This should be attributed to the slower precipitation kinetics of CaS, which can be apparently seen in Figure 4a, that only the out-layer of the inclusions was composed by CaS.

From the above discussion it can be concluded that, at the rapid cooling rate, the inclusions are mostly formed at high temperatures, which is retained to room temperature. While, at the slow cooling rate, the inclusions can be gradually transformed and tends to follow the equilibrium compositions at low temperature.

4. Conclusions

In the current study, the effects of cooling rate on the formation of inclusions in X80 pipeline steel were investigated. The main results obtained were summarized as follows:

(1) High cooling rate resulted in large number density and fine inclusions, while, the inverse results were obtained at low cooling rate. At a high cooling rate the driving force for nucleation is larger, which increased the nucleation frequency of inclusions, thus resulted in high number density of inclusions. On the contrary, at low cooling rate, the nucleation frequency is lower and the Ostwald-ripening effect is pronounced, which caused the low number density and coarse inclusions.

(2) As the cooling rate decreased from 10 to 0.035 K/s, the inclusions were changed from Al_2O_3-CaO to Al_2O_3-CaO-MgO-CaS. At a high cooling rate, the reaction time is short and the inclusions cannot be completely formed, which should be mainly formed at high temperatures. At the low cooling rate, the inclusions can be gradually transformed and tends to follow the equilibrium compositions at low temperature.

Author Contributions: X.Z., W.Y. and L.Z., conceived and designed the experiments; X.Z., W.Y. and H.X. performed the experiments and collected the data; X.Z., and W.Y. analyzed the data; X.Z. and H.X. writing—original draft preparation; W.Y. and L.Z. discussed with the results and revised the manuscript. L.Z. supervised and project administration.

Funding: This research was financially supported by the National Key R&D Program of China (2017YFB0304001, 2016YFB0300102), National Science Foundation China (Grant No. 51874031, No. 51725402 and No. U1860206), the Fundamental Research Funds for the Central Universities (Grant No. FRF-TP-15-067A1), Beijing Key Laboratory of Green Recycling and Extraction of Metals (GREM) and the High Quality Steel Consortium (HQSC) and Green Process Metallurgy and Modeling (GPM2) at the School of Metallurgical and Ecological Engineering at University of Science and Technology Beijing (USTB), China.

Conflicts of Interest: The authors declare no conflict of interest.

References

1. National Energy Board. *Report of Public Inquiry Concerning Stress Corrosion Cracking on Canadian Oil and Gas Pipelines*; National Energy Board: Calgary, AB, Canada, 1996.
2. Baker, M., Jr. Stress Corrosion Cracking Studies. Integrity Management Program DTRS56-02-D-70036; TTO 8; bkmuduli, Department of Transportation, Office and Pipeline Safety, 2004. Available online: https://www.scribd.com/document/81927365/Scc-Report-Full-Text (accessed on 1 January 2019).
3. Parkins, R.N. A review of stress corrosion cracking of high pressure gas pipelines. In Proceedings of the CORROSION 2000, Orlando, FL, USA, 26–31 March 2000; NACE International: Huston, TX, USA, 2000.
4. Torres-Isla, A.; Salinas-Bravo, V.M.; Albarran, J.L.; Gonzalez-Rodriguez, J.G. Effect of hydrogen on the mechanical properties of X-70 pipeline steel in diluted $NaHCO_3$ solutions at different heat treatments. *Int. J. Hydrogen Energy* **2005**, *30*, 1317–1322. [CrossRef]
5. Hara, T.; Asahi, H.; Ogawa, H. Conditions of hydrogen-induced corrosion occurrence of X65 grade line pipe steels in sour environments. *Corrosion* **2004**, *60*, 1113–1121. [CrossRef]
6. Kane, R.D.; Cayard, M.S. NACE committee report 8 × 294: Review of published literature on wet H_2S cracking. In Proceedings of the CORROSION 1999, San Antonio, TX, USA, 25–30 April 1999; NACE International: San Antonio, TX, USA, 1999.
7. Stephen, S.N. Corrosion of carbon steel by H_2S in CO_2 containing oilfield environments. In Proceedings of the CORROSION 2006, San Diego, CA, USA, 12–16 March 2006; NACE International: Houston, TX, USA, 2006.

8. Kittel, J.; Smanio, V.; Fregonese, M.; Garnier, L.; Lefebvre, X. Hydrogen induced cracking (HIC) testing of low alloy steel in sour environment: Impact of time of exposure on the extent of damage. *Corros. Sci.* **2010**, *52*, 1386–1392. [CrossRef]
9. Huang, F.; Liu, J.; Deng, Z.J.; Cheng, J.H.; Lu, Z.H.; Li, X.G. Effect of microstructure and inclusions on hydrogen induced cracking susceptibility and hydrogen trapping efficiency of X120 pipeline steel. *Mater. Sci. Eng. A* **2010**, *527*, 6997–7001. [CrossRef]
10. Atkinson, H.V.; Shi, G. Characterization of inclusions in clean steels: A review including the statistics of extremes methods. *Prog. Mater. Sci.* **2003**, *48*, 457–520. [CrossRef]
11. Wang, X.; Li, X.; Li, Q.; Huang, F.; Li, H.; Yang, J. Control of stringer shaped nonmetallic inclusions of CaO-Al$_2$O$_3$ System in API X80 linepipe steel plates. *Steel Res. Int.* **2014**, *85*, 155–163. [CrossRef]
12. Liu, D.; Zhai, W.; Liu, Y.; Meng, D. Nonmetallic inclusions removal process of pipeline steel in NISCO. *Met. World* **2015**, *4*, 69–71.
13. Ma, Z.; Huang, Z.; Hu, H. Improvement of controlling techniques of inclusions in pipeline steel. *Technol. Bao Steel* **2014**, *5*, 14–17.
14. Zhao, D.; Li, H.; Bao, C.; Yang, J. Inclusion Evolution during Modification of Alumina Inclusions by Calcium in Liquid Steel and Deformation during Hot Rolling Process. *ISIJ Int.* **2015**, *55*, 2115–2124. [CrossRef]
15. Xu, J.; Huang, F.; Wang, X. Formation Mechanism of CaS-Al$_2$O$_3$ Inclusions in Low Sulfur Al-Killed Steel after Calcium Treatment. *Metall. Mater. Trans. B* **2016**, *47*, 1217–1227. [CrossRef]
16. Miao, K.; Haas, A.; Sharma, M.; Mu, W.; Dogan, N. In-situ observation of calcium aluminate inclusions dissolution into steelmaking slag. *Metall. Mater. Trans. B* **2018**, *49*, 1612–1623. [CrossRef]
17. Reis, B.H.; Bielefeldt, W.V.; Vilela, A.C.F. Efficiency of inclusion absorption by slags during secondary refining of steel. *ISIJ Int.* **2014**, *54*, 1584–1591. [CrossRef]
18. Tripathi, N.N.; Beskow, K.; Nzotta, M.; Sandberg, A.; Du, S. Impact of slag refractory lining reactions on the formation of inclusions in steel. *Ironmak. Steelmak.* **2004**, *31*, 514–518.
19. Yan, P.; Huang, S.; Pandelaers, L.; Dyck, J.V.; Guo, M.; Blanpain, B. Effect of the CaO-Al$_2$O$_3$-Based top slag on the cleanliness of stainless steel during secondary metallurgy. *Metall. Mater. Trans. B* **2013**, *44*, 1105–1119. [CrossRef]
20. Dong, W.; Ni, H.; Zhang, H.; Lü, Z. Effect of slag composition on the cleanliness of 28MnCr5 gear steel in the refining processes. *Int. J. Miner. Metall. Mater.* **2016**, *23*, 269–275. [CrossRef]
21. Ren, Y.; Zhang, L. Thermodynamic model for prediction of slag-steel-inclusion reactions of 304 stainless steels. *ISIJ Int.* **2017**, *57*, 68–75. [CrossRef]
22. Takahashi, I.; Sakae, T.; Yoshida, T. Changes of the nonmetallic inclusion by heating. *TetsuHagane* **1967**, *53*, 350–352.
23. Goto, H.; Miyazawa, K.I.; Yanmada, W.; Tanaka, K. Effect of cooling rate on composition of oxides precipitated during solidification of steels. *ISIJ Int.* **1995**, *35*, 708–714. [CrossRef]
24. Goto, H.; Miyazawa, K.; Yamaguchi, K.; Ogibayashi, S.; Tanaka, K. Effect of cooling rate on oxide precipitation during solidification of low carbon steels. *ISIJ Int.* **1994**, *34*, 414–419. [CrossRef]
25. El-Bealy, M.; Thomas, B.G. Prediction of dendrite arm spacing for low alloy steel casting processes. *Metall. Mater. Trans. B* **1996**, *27*, 689–693. [CrossRef]
26. Luo, Y.; Conejo, A.N.; Zhang, L.; Chen, L.; Cheng, L. Effect of superheat, cooling rate, and refractory composition on the formation of nonmetallic inclusions in non-oriented electrical steels. *Metall. Mater. Trans. B* **2015**, *46*, 2348–2360. [CrossRef]
27. Yang, C.W.; Lv, N.B.; Zhuo, X.J.; Wang, X.; Wang, W. Study of MnS precipitation on Ti-Al complex de-oxidation inclusions. *Iron Steel* **2010**, *45*, 34–36. [CrossRef]
28. Rocabois, P.; Lehmann, J.; Gaye, H.; Wintz, M. Kinetics of precipitation of nonmetallic inclusions during solidification of steel. *J. Cryst. Growth* **1999**, *198*, 838–843. [CrossRef]

 © 2019 by the authors. Licensee MDPI, Basel, Switzerland. This article is an open access article distributed under the terms and conditions of the Creative Commons Attribution (CC BY) license (http://creativecommons.org/licenses/by/4.0/).

Article

Comparison of Transverse Uniform and Non-Uniform Secondary Cooling Strategies on Heat Transfer and Solidification Structure of Continuous-Casting Billet

Yanshen Han [1], Xingyu Wang [1], Jiangshan Zhang [1], Fanzheng Zeng [2], Jun Chen [2], Min Guan [3] and Qing Liu [1,*]

1. State Key Laboratory of Advanced Metallurgy, University of Science and Technology Beijing, Beijing 100083, China; yansh@xs.ustb.edu.cn (Y.H.); 18503430960@163.com (X.W.); zjsustb@163.com (J.Z.)
2. Xiangtan Iron & Steel Co., Ltd of Hunan Valin., Xiangtan 411101, China; fanzhengz111@163.com (F.Z.); chenjun0621@126.com (J.C.)
3. Jiangsu Boji Spraying Systems Co., Ltd., Yangzhou 225267, China; guanmin@boji.sg
* Correspondence: qliu@ustb.edu.cn; Tel.: +86-133-3111-6466

Received: 9 April 2019; Accepted: 8 May 2019; Published: 10 May 2019

Abstract: Water flux distribution largely influences the heat transfer and solidification of continuously-cast steel billets. In this paper, a secondary cooling strategy of transverse non-uniform water flux (i.e., higher flux density on billet center), was established and compared with the uniform cooling strategy using mathematical modeling. Specifically, a heat transfer model and a cellular automaton finite element coupling model were established to simulate the continuous casting of C80D steel billet. The water flux was measured using different nozzle configurations to assist the modeling. The mathematical results were validated by comparing the surface temperature and the solidification structure. It is shown that the non-uniform cooling strategy enables the increase of corner temperature and reduction in surface temperature difference, while a higher reheating rate is found on the surface center of the billet. Moreover, the non-uniform cooling strategy can enhance the cooling effect and refine the solidification structure. Accordingly, the liquid pool length is shortened, and the equiaxed crystal density is increased along with the decreased equiaxed crystal ratio. The uniform cooling strategy contributes to reducing internal cracks of billet, and the non-uniform one is beneficial for surface quality and central segregation. For C80D steel, the non-uniform cooling strategy outperforms the uniform one.

Keywords: numerical simulation; C80D steel; transverse water flux distribution; heat transfer; solidification structure

1. Introduction

Continuous casting is the main method of steel production at present, and its cooling and solidification process largely influences the quality of blanks. During casting, the cooling of the blanks is realized in the mold, secondary cooling zone, and air-cooling zone. Since the cooling conditions in the mold and the air-cooling zone are relatively steady, the control of solidification mainly focuses on the secondary cooling zone. Secondary cooling is particularly important, as it is closely related to the heat transfer and solidification structure of the blank, and therefore influences the formation of defects, such as cracks [1–3], central pipe [4], and segregation [5–7]. To minimize the defects, water flux and its distribution were frequently resorted to as a means to improve the product quality through both numerical simulation and industrial trials [8–10].

Secondary cooling is achieved by spraying water on the blank using arranged nozzles and is directly influenced by the water flux distribution. The nozzle arrangement varies for different cross-sectional casting blanks, such as slab, billet, and bloom. For a continuous casting slab, there is a wide range of water flux on the slab surface owing to its large section size, and the nozzle collocations are usually complex in the secondary cooling zone. Therefore, the water flux distribution needs to be taken into consideration along both the transverse direction and casting direction [10–13]. Moreover, the transverse water flux distribution can be used as a method to improve the quality of slab by the collocations of nozzles. Wang et al. [14] optimized the water flux distribution along slab width direction through the arrangement of spraying nozzles, and the centerline macro-segregation and transverse cracking were improved significantly. Long et al. [15] proposed that a uniform solidified shell in the transverse direction was beneficial to mitigating slab central macro-segregation, thanks to the optimized transverse water flux distribution.

As for small sectional blanks, typically including square billet and rectangular billet, there is usually only one spraying nozzle along the transverse direction on one side of the billet. The transverse water flux distribution is not as complicated as slab and normally assumed to be uniform in the literature [6–8,16–18]. Zeng et al. [6] investigated the solidification structure and macro-segregation of a rectangular billet under different secondary cooling conditions. Ma et al. [18] optimized secondary cooling water distribution to improve billet quality. Their works mainly focus on the total water flux and its distribution along the casting direction, and is not related to the uniformity of transverse water flux distribution, which depends on the nozzle type, installation parameter, and inevitable aging and clogging. The change of transverse water flux distribution obviously affects the solidification behavior of the billet. What is worth mentioning is that Assuncao et al. [19] compared uniform and non-uniform water flux distribution on the thermal behavior of a round billet in a recent work. They found that there is a big difference between them, and the thermal behavior of the billet is more accurate using the measured water flux distribution. However, round billet and square billet still differs greatly, especially at the corner of square billet, which is largely associated with corner defects. Until recently, the transverse water flux distributions of square billet under different nozzle collocations have drawn little attention, and the heat transfer behavior and solidification structure of the billet are still unclear under different transverse water flux distributions.

The present work shed lights on the transverse water flux distribution in the secondary cooling zone of square billet. Two similar continuous casters with different types and arrangements of spraying nozzles were studied. Transverse water flux distribution was measured using an apparatus, and two cooling strategies were obtained, including transverse uniform and non-uniform distribution of secondary cooling water. A heat transfer model and a cellular automaton finite element (CAFE) coupling model were used to analyze the heat transfer and solidification structure of the billet, respectively. The main differences between the two cooling strategies were compared on the solidification behavior of the billet, and the advantages of each cooling strategy were discussed.

2. Brief Description of the Studied Continuous Caster

Two similar continuous casters are involved in the present study. Both casters are four-strand circular-arc casters with a curved mold that mainly produce billets with a small section size (150 mm × 150 mm). The two casters have their own features and produce different steel grades. At the mold, electromagnetic stirrer (EMS, Produced by Hunan Kemeida Electric Co., Ltd., Yueyang, China) is applied on each caster and the stirrer can generate a rotating magnetic field. The schematic diagrams of the casters and nozzle arrangements are shown in Figure 1. Caster 1 and Caster 2 have similar cooling zones, however, the length of each cooling zone and nozzle arrangement at the secondary cooling zone are different. For Caster 1, a full cone jet nozzle and flat jet nozzle (Produced by Jiangsu Boji Spraying Systems Co., Ltd., Yangzhou, China) are used in the foot-roller zone and the secondary cooling zone, respectively. Moreover, all the nozzles overspray on the billet surface and the flat jet nozzles deflect at an angle of 30°. With regard to Caster 2, two full cone jet nozzles (Produced by Beijing Zhongye Metallurgical Equipment Manufacturing Co., Ltd., Beijing, China) are installed on the width direction in the foot-roller zone, and one full cone jet nozzle is installed on the width direction that just fully covers the billet surface in the secondary cooling zone.

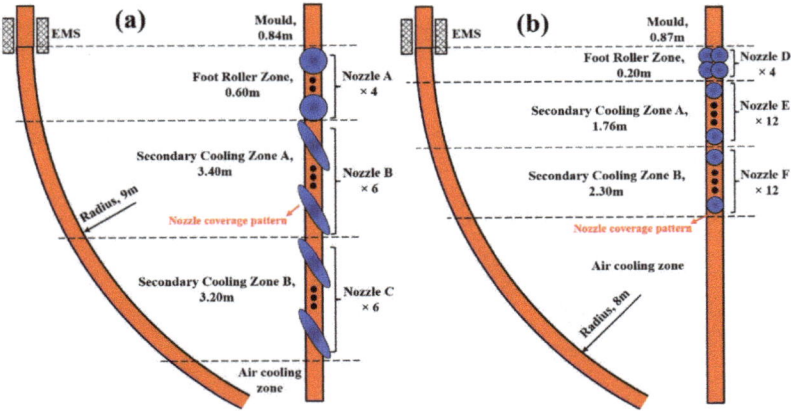

Figure 1. Schematic diagrams of continuous casters and nozzle arrangements: (**a**) Caster 1; (**b**) Caster 2.

3. Spraying Nozzle Measurement of Water Flux Distribution

The water flux distribution largely influences the local thermal behavior of billet during continuous casting [11,19]. It is essential to clarify the effect of water flux distribution on the secondary cooling. Under the practical installation conditions, transverse water flux distributions were measured in the secondary cooling zone of Caster 1 and Caster 2. The water flux distribution along the casting direction was not taken into consideration due to the limitation of the experimental set-up.

3.1. Experimental Apparatus

Measurements were carried out using an experimental apparatus that can quantify the water flow rate, spraying angle, and water flux distribution of the spraying nozzle. Figure 2 shows the schematic diagram of the experimental apparatus. The apparatus includes water tank, gas tank, water-gas regulator, control system, and water collector. Water tank and gas tank supply sufficient water pressure and gas pressure, and the water-gas regulator is used to adjust the water pressure and the gas pressure to a desired value. The water collector consists of a row of grooves and one pressure sensor is installed under each groove. The width of each groove is 25 mm.

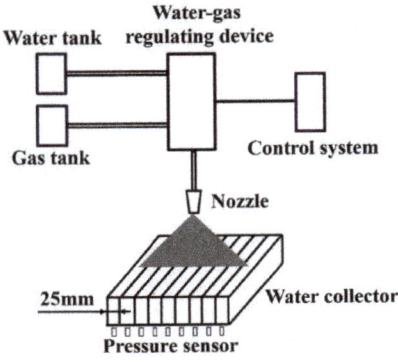

Figure 2. Schematic diagram of experimental apparatus.

3.2. Water Flux Distribution Measurement

Six actual water spraying nozzles were adopted in the experiment. Among the nozzles, Nozzle A, Nozzle B, and Nozzle C were used in the foot-roller zone, secondary cooling zone A, and secondary cooling zone B of Caster 1, respectively. Nozzle D, Nozzle E, and Nozzle F were used in the foot-roller zone, secondary cooling zone A, and secondary cooling zone B of Caster 2, respectively. The type number and jet type of the nozzles were different. Nozzle A, Nozzle D, Nozzle E, and Nozzle F were full cone jet nozzles, while Nozzle B and Nozzle C were flat jet nozzles. In practical casting process, the water pressure and jet distance of nozzles are listed in Table 1. The spraying angle and water flux distribution were measured, as shown in Table 1 and Figure 3.

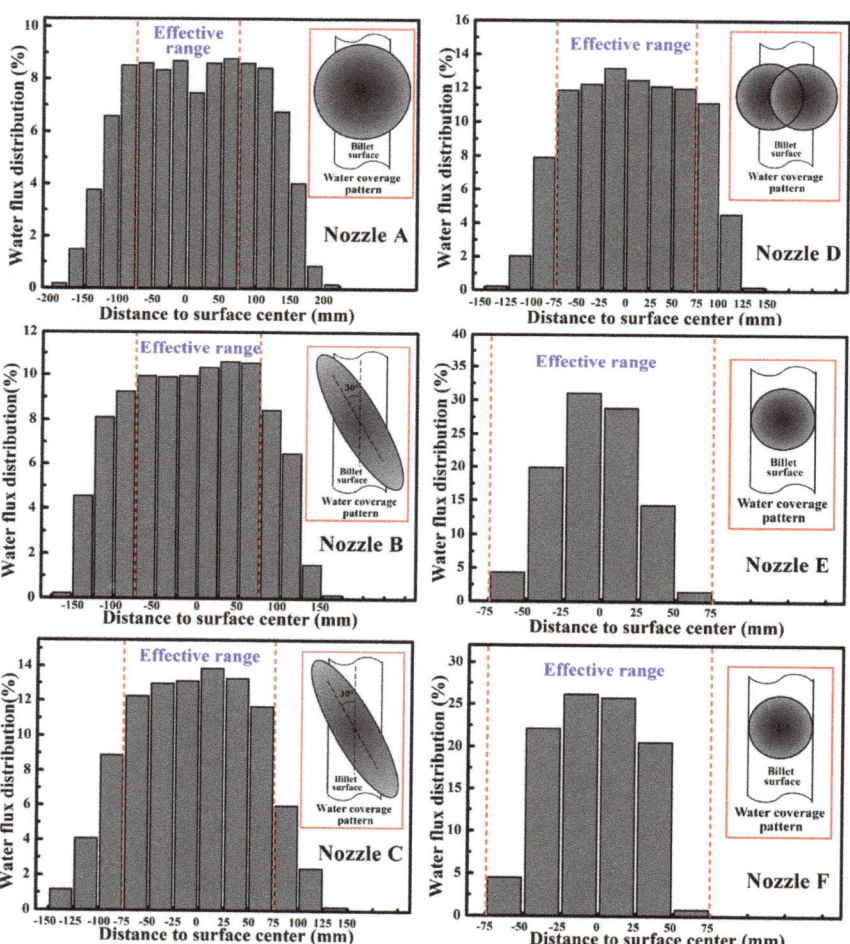

Figure 3. Water flux distributions of spraying nozzles under specific collocations.

In Figure 3, it can be seen that the water flux distribution is non-uniform and generally symmetric for each spraying nozzle, and there is more water on the spraying center and less water on the spraying edge. As the billet width is 150 mm, an effective range on the billet surface is counted from −75 mm to 75 mm. In the effective range, the water flux distributions of Nozzle A, Nozzle B, Nozzle C, and Nozzle D are generally uniform, while the water flux of Nozzle E and Nozzle F is unevenly distributed with a higher water flow rate on the center and a lower one around the corner. Corresponding to different cooling zones of the continuous casters, it can be found that the transverse water flux distributions are relatively uniform in the foot-roller zones of Caster 1 and Caster 2. In the secondary cooling zone, the transverse water flux distributions of Caster 1 and Caster 2 are uniform and non-uniform, respectively. It indicates that Caster 1 and Caster 2 have two different cooling strategies, namely transverse uniform and non-uniform secondary cooling strategies.

Table 1. Main parameters of spraying nozzles.

Nozzle	Type number	Jet type	Pressure/MPa	Jet Distance/mm	Spraying Angle
Nozzle A	CONCAST 7565L	Full cone	0.52	300	60.4°
Nozzle B	CONCAST RE2-8-80/28	Flat	0.52	300	87.6°/26.4°
Nozzle C	CONCAST RE1-4-80/28	Flat	0.52	300	82.4°/28.4°
Nozzle D	ZY3/8PZ72667QZ1	Full cone	0.40	125	61.6°
Nozzle E	3665	Full cone	0.80	125	53.1°
Nozzle F	2265	Full cone	0.60	125	53.5°

4. Numerical Modeling and Validation

Caster 1 was chosen in the numerical model to study the solidification behaviors of the billet under different secondary cooling strategies. The water flux distribution in the secondary cooling zone of Caster 2 was applied on Caster 1 to compare with the original one, and two cooling modes were determined as follows.

Mode 1: Original water flux distribution in the secondary cooling zone was adopted on Caster 1, and the transverse water flux distribution was regarded as uniform on the billet surface, as shown in Figure 4a.

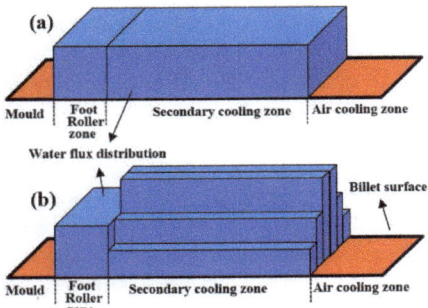

Figure 4. Schematic diagram of water flux distribution in the foot-roller zone and secondary cooling zone during continuous casting: (a) Mode 1; (b) Mode 2.

Mode 2: Water flux distribution in the secondary cooling zone of Caster 2 was applied on Caster 1. The transverse distribution of secondary cooling water is non-uniform, and there is more water on the surface center and less water on the surface edge of the billet, as shown in Figure 4b.

In the simulation, all the other parameters of Mode 1 and Mode 2 were the same except for the different transverse water cooling in the secondary cooling zone.

4.1. Model Development

4.1.1. Heat Transfer Model

Some assumptions were made to simplify the mathematical model.

1. The heat transfer in casting direction and meniscus was neglected.

2. The convective heat transfer was equivalent to the conductive heat transfer by increasing the thermal conductivity of liquid and mushy zone.
3. The heat transfer of radiation, contact with the supporting roller, and cooling water in the secondary cooling zone were included by an integrated heat transfer coefficient.
4. The water flux distribution along the casting direction of each segment was uniform in the secondary cooling zone.
5. The density, solid fraction, and thermal conductivity of the steel were temperature dependent.
6. The dimensional change caused by solidification shrinkage was neglected on the analysis of temperature and solidification structure of the billet.

Based on the assumptions, a two-dimensional unsteady state heat transfer equation was expressed as follows:

$$\rho C_{P,eff} \frac{\partial T}{\partial \tau} = \frac{\partial}{\partial x}(\lambda_{eff} \frac{\partial T}{\partial x}) + \frac{\partial}{\partial y}(\lambda_{eff} \frac{\partial T}{\partial y}) \tag{1}$$

where ρ is the density of steel in kg/m^3; $C_{p,eff}$ is the effective specific heat in J/(kg·K); T is the temperature of steel in K; τ is the calculation time in s; λ_{eff} is the effective thermal conductivity in W/(m^2·K); x is the distance from the billet center along transverse direction in m, and y is the distance from billet center on the inner radius face in thickness direction in m.

The effect of latent heat on the solidification of steel at the mushy zone was incorporated into the effective specific heat, and represented by enthalpy, as shown in the Equations (2) and (3).

$$C_{p,eff} = C_P - L \cdot \frac{\partial f_s}{\partial T} \tag{2}$$

$$H = \int_0^T C_P dT + L(1 - f_s) \tag{3}$$

where C_P is the actual specific heat in J/(kg·K); L is the latent heat in J/kg; H is the enthalpy in J/kg and f_s is the solid fraction.

A slice moving method was applied in the simulation. To simplify the calculation, the geometric model of the slice adopted a quarter of the cross section of the billet with a thickness of 10 mm and the rounded corners were neglected, namely with a dimension of 75 mm × 75 mm × 10 mm. It was assumed the slice moved along the casting direction from the mold to the foot-roller zone, the secondary cooling zone, and finally the air-cooling zone. In the mold, the heat flux was obtained from a heat balance of the cooling water, and equalized to the empirical equation, as shown in Table 2. To consider the effect of the gap on the heat transfer, a decreasing heat flux was used from the billet center to the corner, and the heat flux at the corner decreased with the increase of distance from the meniscus [10]. In the foot-roller zone and the secondary cooling zone, the surface temperature of the billet is mostly above the Leidenfrost point. The heat flux was characterized by an integrated heat transfer coefficient, which was a function of water flux density. In the air-cooling zone, the radiation was the main heat transfer pattern and the heat extraction was obtained according to the Stefan–Boltzmann law.

Table 2. Water flow rates, boundary conditions, and calculated formulae of different cooling zones.

Zone	Water Flow Rate, m^3/h	Boundary Condition	Calculated Formula
Mold	113	q_m	$q_m = 2680000 - \beta \sqrt{t}$
Foot-roller zone	8.06	$q_f = h_f(T - T_f)$	$h_f = 480(W/60)^{0.351}$
Secondary cooling zone A	12.96	$q_s = h_s(T - T_s)$	$h_s = 1570W^{0.55}(1 - 0.0075T_s)/2.16$
Secondary cooling zone B	5.76	$q_s = h_s(T - T_s)$	$h_s = 1570W^{0.55}(1 - 0.0075T_s)/2.16$
Air-cooling zone	-	$q_a = \sigma\varepsilon(T^4 - T_a^4)$	$\sigma = 5.67 \times 10^{-8}, \varepsilon = 0.8$

The water flow rates, boundary conditions, and calculated formulae [20–22] of different cooling zones are listed in Table 2. In Table 2, q_m, q_f, q_s, and q_a are the heat flux of the mold, the foot-roller zone, the secondary cooling zone, and the air-cooling zone, respectively, W/m^2; h_f and h_s are the heat transfer coefficient of the foot-roller zone and the secondary cooling zone, respectively, W/(m^{2}K); T, T_f, T_s, and T_a are the temperature of the billet surface, the water temperature of the foot-roller zone, the water temperature of the secondary cooling zone, and the environment temperature of the air-cooling zone, respectively, K; β is a coefficient between the shape of the mold and parameter of casting, W/(m^2·s$^{1/2}$); t is the holding time in the mold, s; σ is Stefan–Boltzmann constant, W/(m^2·K^4); ε is the radiation coefficient. The numerical values of q_m, β, h_f, and h_s can be obtained by a calculation using casting parameters.

In the simulation, the boundary conditions of Mode 1 and Mode 2 were treated as the same in the mold, the foot-roller zone, and the air-cooling zone. As for the secondary cooling zone, the boundary condition of Mode 1 was uniform. For Mode 2, the billet surface was divided into three sections along the transverse direction, and the boundary condition of each section was separated according to local water flux density, as shown in Figure 4. Whether Mode 1 or Mode 2, the boundary condition along the casting direction was constant at each secondary cooling zone.

4.1.2. Nucleation Model

Procast software was used to establish the nucleation model and dendritic tip growth model. A continuous nucleation distribution function, $dn/d(\Delta T)$ was used to describe the grain density change [23], and dn is induced by the increase of undercooling, $d(\Delta T)$. The distribution function is expressed by Equation (4):

$$\frac{dn}{d(\Delta T)} = \frac{n_{max}}{\sqrt{2\pi}\Delta T_\sigma} \exp\left[-\frac{1}{2}\left(\frac{\Delta T - \Delta T_{max}}{\Delta T_\sigma}\right)^2\right] \quad (4)$$

where ΔT is the calculated local undercooling in K; ΔT_{max} is the mean undercooling in K; ΔT_σ is the standard deviation in K, and n_{max} is the maximum nucleation density in m^{-3}.

4.1.3. Dendritic Tip Growth Model

The KGT (Kurz, Giovanola, Trivedi) model [24,25] was used as the model of growth kinetics of a dendrite tip in the C80D steel. Based on the marginal stability criterion, Equation (5) is obtained.

$$V^2 A + VB + C = 0 \quad (5)$$

where V is the growth velocity of a dendrite tip, and A, B, and C are expressed by Equations (6) to (8).

$$A = \frac{\pi^2 \Gamma}{p^2 D^2} \quad (6)$$

$$B = \frac{mC_0(1-k_0)\xi_C}{D[1-(1-k_0)Iv(P)]} \quad (7)$$

$$C = G \quad (8)$$

where Γ is the Gibbs–Thomson coefficient; P is the Peclet number for solute diffusion; D is the diffusion coefficient in the liquid; m is the liquidus slope; C_0 is the initial concentration; k_0 is the partition coefficient; $Iv(P)$ is the Ivantsov function; $\xi_c = \pi^2/(k_0 P)$ closes to unity at a low temperature gradient, and G is the temperature gradient. For the dendrite growth regime, G has little effect on the growth velocity V and can be regarded as zero.

Moreover, the undercooling at dendrite tip, ΔT, is expressed as Equation (9). The relationship between the undercooling ΔT and growth velocity V can be calculated by substituting an arbitral value of the Peclet number into Equations (5) and (9). The material properties [26,27] of C80D steel that were used in the simulation are given in Table 3. The partition coefficients and liquidus slopes were assumed to be constants from liquidus to solidus.

$$\Delta T = mC_0\left[1 - \frac{1}{1-(1-k_0)Iv(P)}\right] + \frac{2\Gamma}{r} \quad (9)$$

where r is the dendrite tip radius, $r = (2D \cdot P)/V$.

Table 3. Chemical composition of C80D steel and related parameters.

Composition	C	Si	Mn	P	S
Mass fraction, %	0.82	0.20	0.73	0.017	0.004
Partition coefficient, k_0	0.35	0.52	0.75	0.06	0.025
Liquidus slope, m	−60	−8	−5	−34	−40
Diffusivity in liquid, D, m²/s	2.0×10^{-8}	2.4×10^{-9}	2.0×10^{-8}	4.7×10^{-9}	4.5×10^{-9}

To accelerate the computational speed, a set of values for the undercooling and growth velocity of the dendrite tip are calculated in Equations (5) and (9). Then the following Equation (10) is obtained.

$$V = a_2 \Delta T^2 + a_3 \Delta T^3 \quad (10)$$

where a_2 and a_3 are the fitting coefficients of the multinomial of dendrite tip growth velocity. The values of a_2 and a_3 were calculated using the material properties shown in Table 3, and the specific values are 3.041×10^{-6} m/(s·K²) and 2.146×10^{-5} m/(s·K³), respectively.

4.2. Material Properties

C80D steel was studied and its main chemical composition has been given in Table 3. Variations of solid fraction, enthalpy, thermal conductivity, and density with temperature were calculated using a thermodynamic database from JMatPro software according to the steel composition, as shown in Figure 5. The flow of liquid steel, which results from the initial flow from the submerged nozzle, natural convection, and external force of mold electromagnetic stirring (MEMS), can enhance the heat transfer during solidification. The thermal conductivities of the liquid and the mushy zone were enlarged to consider the effect of fluid flow [17,28]. The thermal conductivity of solid adopted calculated values, five times the thermal conductivity of solidus was adopted for the liquid phase, and the thermal conductivity of the mushy zone was assumed to change with temperature linearly.

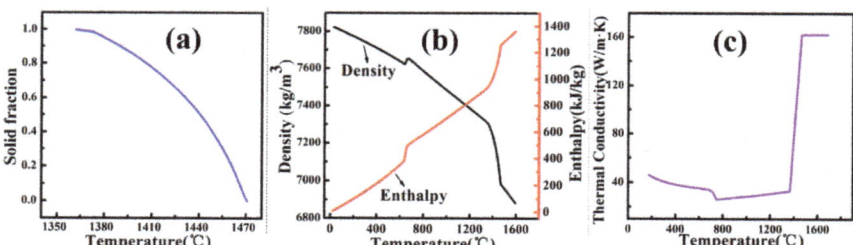

Figure 5. Calculated thermo-physical properties of C80D steel, (a) solid fraction, (b) density and enthalpy, and (c) thermal conductivity.

4.3. Model Validation

4.3.1. Casting Parameters

Industrial tests were conducted on Caster 1 to calibrate the heat transfer model and the CAFE coupling model. The main casting parameters of the tests are listed in Table 4.

Table 4. Casting parameters used for C80D steel billet.

Item	Value
Casting speed	2.4 m/min
Pouring temperature	1485 °C
Superheat	15 °C
Water flux of mold cooling	113 m^3/h
Temperature difference between inlet and outlet of mold water	7.9 °C
Ambient temperature	25 °C

4.3.2. Heat Transfer Model Validation

The surface temperature of the billet was measured using an infrared radiation pyrometer, the error range of which is ±1.5%. During the measurement, the pyrometer was perpendicular to the surface center of the side arc and peak values were adopted as the local temperature. The temperature of the billet was also calculated by the heat transfer model under practical casting parameters. Figure 6 shows the comparison between the calculated temperature and the measured temperature. A good agreement was obtained between the mathematical and the actual results within the error range. It can be concluded that the heat transfer model is generally reliable.

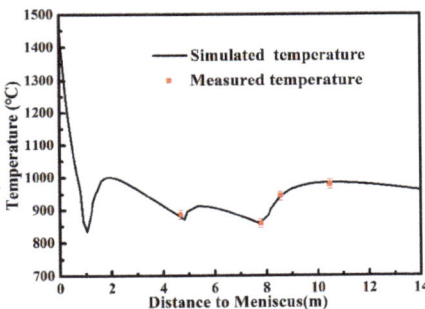

Figure 6. Temperature profiles of surface center.

4.3.3. CAFE Coupling Model Validation

The transverse billet sample was sliced, polished and etched by hydrochloric acid, and then the actual solidification structure was revealed. From the billet surface to the center, the solidification structure features with an outer chilled layer, a columnar crystal zone, a mixed crystal zone, and an equiaxed crystal zone. Moreover, the columnar crystal deflects on account of MEMS. The deflection was neglected in the simulation. MEMS can also enlarge the equiaxed crystal zone, which is considered by adjusting the maximum nucleation density. The solidification structure was simulated using the CAFE coupling model. The maximum nucleation density was adjusted to match the simulated solidification structure and the actual one, and finally was set as 2.5×10^9 m^{-3}. Figure 7 shows the comparison of the actual solidification structure and the simulated one. Each zone of the actual solidification structure and the simulated one shares basically the equivalent area. This result indicates the selected nucleation parameters are reasonable and the CAFE coupling model is reliable to simulate the solidification structure of C80D steel.

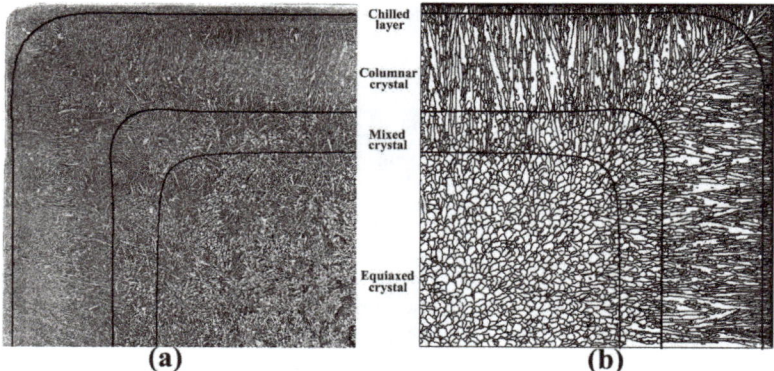

Figure 7. Comparison of (**a**) actual solidification structure and (**b**) simulated solidification structure.

5. Results and Discussion

Following the casting parameters shown in Table 4, comparisons were made between transverse uniform and non-uniform distribution of secondary cooling water on the heat transfer behavior and solidification structure.

5.1. Comparison of Surface Temperature

Figure 8 comparatively presents the 3D topography of the surface temperature variation under the two cooling strategies. Their temperature variations generally show a similar and typical tendency. The surface temperature starts at around 1500 °C and rapidly drops to less than 800 °C caused by the intense cooling in the mold and foot-roller zone. Thereafter, the temperature fluctuates in different sections of the secondary cooling zone owing to the change of cooling intensity, and finally decreases slowly in the air-cooling zone. As the cooling condition in the mold and the foot-roller zone are the same, the surface temperature distributions of Mode 1 and Mode 2 exhibit no difference. In the secondary cooling zone of Mode 1, a wide range low-temperature zone (less than 700 °C) is observed near the billet corner, and the temperature difference between the corner and the center is large. For Mode 2, the cooling is strengthened at the center and weakened at the corner owing to more water on the surface center and less water on the surface edge. As a result, the low-temperature zone is narrowed and the temperature difference along the width direction becomes small. When billet enters into the air-cooling zone, the billet surface reheats and the temperature distribution tends to become uniform gradually.

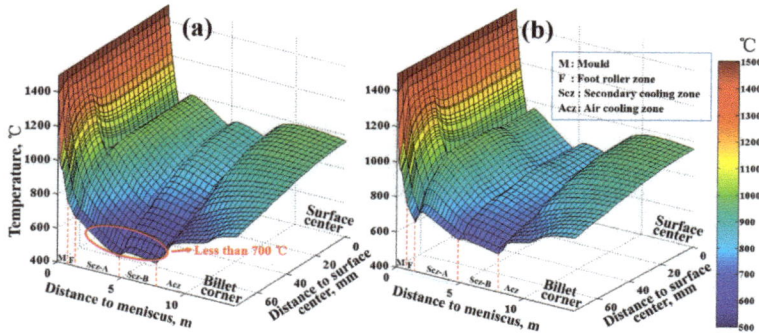

Figure 8. Temperature variations of the billet surface during continuous casting: (**a**) Mode 1; (**b**) Mode 2.

To analyze the temperature variation quantitatively, the temperature variations of billet center, surface center, and billet corner are highlighted in Figure 9. It shows that the temperature of the billet center keeps nearly constant until a fast drop at the final stage of solidification, and there exists little difference between Mode 1 and Mode 2. The corner temperature in the secondary cooling zone of Mode 1 is below 600 °C mostly, while that of Mode 2 is all above 600 °C. For Mode 2, the temperature difference of the billet corner and surface center is 174 °C, much lower than that of Mode 1 with a value of 356 °C, which means a more uniform transverse temperature distribution of Mode 2, beneficial for surface quality [29].

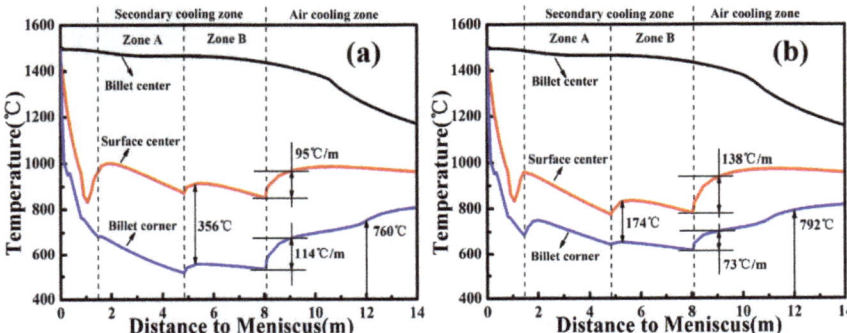

Figure 9. Temperature variations of the billet center, surface center, and billet corner: (**a**) Mode 1; (**b**) Mode 2.

When the billet enters into the air-cooling zone, the reheating rates of the billet corner and surface center for Mode 1 are 114 °C/m and 95 °C/m, respectively. As for Mode 2, the reheating rate of the billet corner is decreased to 73 °C/m and that of surface center is increased to 138 °C/m. These effects may result in reheat-type internal cracks as the reheating rate exceeds 100 °C/m [30]. At the straightening point, the corner temperature of Mode 2 is 792 °C, which is 32 °C higher than that of Mode 1. The corner of the billet is a crack-prone area on account of the low temperature. The transverse non-uniform secondary cooling strategy is able to increase the corner temperature and contribute to avoiding the third temperature zone of low ductile for C80D steel, thus reducing the possibility of generating corner cracks [29].

5.2. Comparison of Internal Temperature

The temperature distributions of a one-fourth cross section at various positions are exhibited in Figure 10. Except for the temperature near the billet corner, the difference of internal temperature between Mode 1 and Mode 2 is not as significant as the surface temperature. It indicates water flux distribution mainly affects the surface and subsurface temperature. At the end of the foot-roller zone, the liquid core shape is nearly square and no difference exists between Mode 1 and Mode 2. When billet reaches the end of the secondary cooling zone A, the liquid core shape becomes arc-shaped and the shell thickness of Mode 1 is slightly smaller than that of Mode 2. Moreover, the difference becomes more obvious at the end of secondary cooling zone B. When the billet arrives at the straightening point, the temperature distributions of Mode 1 and Mode 2 are nearly the same except for a low-temperature zone at the billet corner of Mode 1.

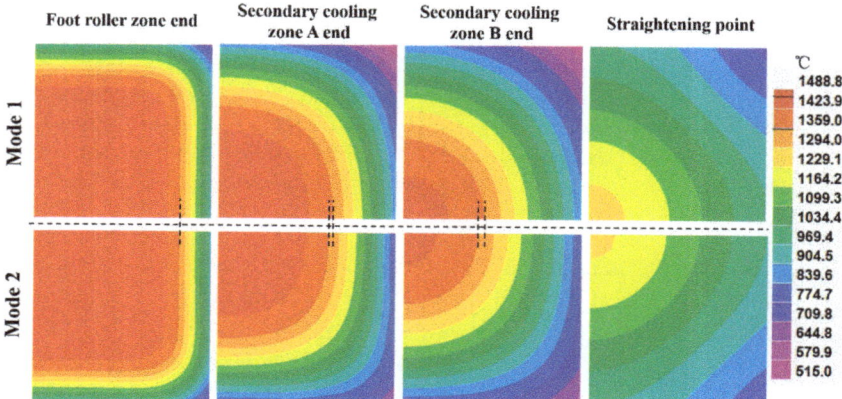

Figure 10. Temperature distributions of one-fourth cross section at various positions.

Figure 11 presents the internal temperature distribution from the meniscus to the air-cooling zone. The temperature near the billet surface between the two cooling modes has a big difference. It indicates that the surface and subsurface areas are more sensitive to water flux distribution. As the solidus temperature of C80D steel is 1359 °C, the shell thicknesses of billet are evaluated at the end of the secondary cooling zone for Mode 1 and Mode 2, which are 41.9 mm and 43.8 mm, respectively. Meanwhile, the liquid pool lengths of Mode 1 and Mode 2 can also be obtained, which are 10.60 m and 10.32 m, respectively. Compared with Mode 1, the shell thickness at the end of the secondary cooling zone is increased by 1.9 mm and the liquid pool length is decreased by 0.28 m for Mode 2. These effects mean that the transverse non-uniform secondary cooling strategy increases the cooling intensity on billet.

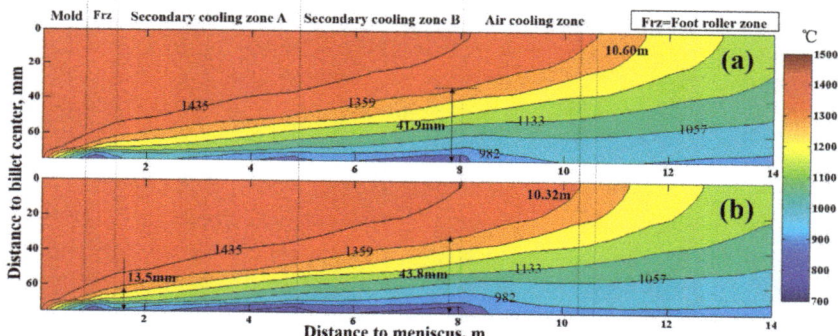

Figure 11. Internal temperature distribution from meniscus to air-cooling zone: (**a**) Mode 1; (**b**) Mode 2.

5.3. Comparison of Solidification Structure

The solidification structures under the two cooling strategies were simulated using the CAFE coupling model and presented in Figure 12. Different kinds of structures are demarcated by solid lines. As seen from Figure 12, the columnar crystal zone and the mixed crystal zone of Mode 2 are longer than that of Mode 1, and the equiaxed crystal zone shows the opposite result. The columnar crystal is closely related to temperature gradient during solidification [31], and liable to grow at a big temperature gradient.

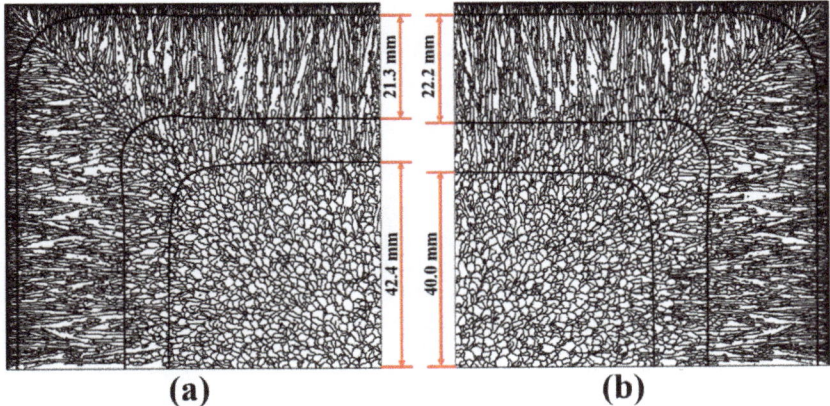

Figure 12. Simulated solidification structure of one-fourth cross section: (**a**) Mode 1; (**b**) Mode 2.

At the end of secondary cooling zone A, the temperature distribution and gradient from the billet center to surface center were calculated using the heat transfer model, as shown in Figure 13. It can be seen that there are differences between Mode 1 and Mode 2 on the temperature distribution and gradient, and the differences become unclear at the position of 30 mm from the billet center. Within the range of 30 mm to 50 mm from the billet center, where the solidification proceeds, the temperature distribution and gradient are highlighted. It can be seen that the temperature gradient of Mode 2 is bigger than that of Mode 1. Therefore, the growth of columnar crystal for Mode 2 is promoted. Consequently, the solidification structure of Mode 2 exhibits a relatively wider columnar crystal zone.

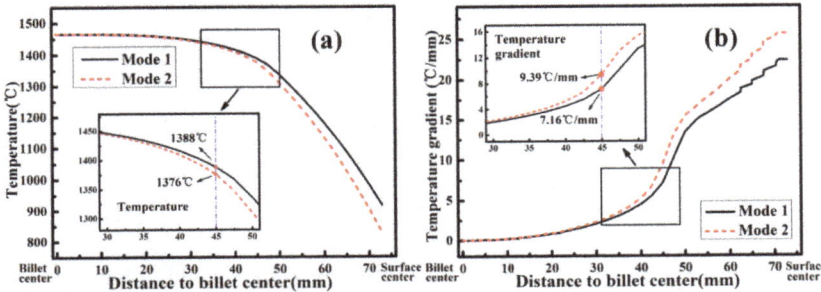

Figure 13. Calculated temperature (**a**) variation and (**b**) gradient from the billet center to surface center at the end of secondary cooling zone A.

According to the solidification structures shown in Figure 12, the equiaxed crystal ratio was calculated in light of areal percentage. Meanwhile, the number of equiaxed crystal was counted to calculate the equiaxed crystal density. Table 5 shows the equiaxed crystal ratio and density of different cooling strategies. The equiaxed crystal density of Mode 2 is increased and the equiaxed crystal ratio is decreased compared with Mode 1. It indicates that the transverse non-uniform secondary cooling strategy is able to increase the compactness of central equiaxed crystal and decrease the equiaxed crystal ratio. It has been documented that the equivalent effect can be acquired by increasing the secondary cooling intensity [6,7], and that intense secondary cooling is beneficial to reducing the central segregation of high carbon steel billet with a small section size [32]. Transverse non-uniform secondary cooling strategy enhances the cooling effect using the same amount of water, and therefore shows potential to minimize the central segregation of the billet.

Table 5. Equiaxed crystal ratio and density of different cooling strategies.

Cooling Strategy	Equiaxed Crystal Density	Equiaxed Crystal Ratio
Mode 1	0.83 /mm^2	32.0%
Mode 2	0.89 /mm^2	28.4%

5.4. Summary

Table 6 summarizes the main comparative factors between transverse uniform and non-uniform secondary cooling strategies according to the above findings. Compared with transverse uniform secondary cooling strategy, the non-uniform strategy features a higher corner temperature, smaller transverse temperature difference, shorter liquid pool length, narrower central equiaxed crystal zone, and finer central equiaxed crystal of billet. These effects are beneficial to improving the surface quality and central segregation of billet. Disadvantageously, the reheating rate of the surface center is increased and may lead to reheat-type internal cracks in billet when the transverse distribution of secondary cooling water is non-uniform. In a word, the uniform cooling strategy contributes to reducing internal cracks of billet, and the non-uniform one is beneficial for surface quality and central segregation. According to the demands of different steel grades, the continuous caster can adopt a proper cooling strategy, which is associated with nozzle collocations. C80D steel has a high demand for central segregation and surface quality, but a low demand for internal cracks as the cracks can be eliminated in the rolling process. Therefore, transverse non-uniform secondary cooling strategy is a better choice for the production of C80D steel.

Table 6. Main comparative factors between transverse uniform and non-uniform secondary cooling strategies.

Item	Uniform	Non-Uniform
Transverse temperature difference in the secondary cooling zone, °C	356	174
Corner temperature at straightening point, °C	760	792
Maximum reheating rate of surface center, °C/m	95	138
Maximum reheating rate of billet corner, °C/m	114	73
Liquid pool length, m	10.60	10.32
Equiaxed crystal density, /mm^2	0.83	0.89
Equiaxed crystal ratio, %	32.0	28.4

6. Conclusions

Water flux distributions of six spraying nozzles were measured using an apparatus. The nozzles were from two similar continuous casters. Two secondary cooling strategies were obtained, namely transverse uniform and non-uniform distribution secondary cooling strategies. A heat transfer model and a CAFE coupling model were established and assisted by the measured water flux distributions. The models were verified by comparing the surface temperature and the solidification structure to ensure their accuracy. The heat transfer and solidification structure of C80D steel billet were compared under the two cooling strategies. The following conclusions can be drawn:

1. For a single spraying nozzle, the transverse water flux distribution is non-uniform, and there is more water on the spraying center and less water on the spraying edge. However, a transverse uniform water flux distribution can be obtained in the effective range by adjusting the nozzle collocations.
2. The transverse non-uniform secondary cooling water is able to improve the cooling uniformity along the transverse direction. The corner temperature is effectively increased and the transverse temperature difference is reduced. Nevertheless, the reheating rate of the surface center is increased when the billet enters into the air-cooling zone.

3. Compared with the uniform cooling strategy, the cooling intensity of the non-uniform strategy on billet is enhanced. The shell thickness at the end of the secondary cooling zone is increased by 1.9 mm and the liquid pool length is decreased by 0.28 m.
4. The non-uniform cooling strategy promotes the growth of columnar crystal. The compactness of central equiaxed crystal is increased and the equiaxed crystal ratio is decreased. These effects are equivalent to increasing the secondary cooling intensity.
5. The non-uniform cooling strategy is beneficial to improving the surface quality, corner crack, and central segregation of billet, but may result in internal cracks. On the production of C80D steel, the non-uniform strategy is better than the uniform one.

Author Contributions: Conceptualization, Y.H. and Q.L.; data curation, Y.H. and X.W.; investigation, Y.H., X.W. and M.G.; project administration, Q.L.; validation, F.Z. and J.C.; writing—original draft, Y.H.; writing—review and editing, J.Z. and Q.L.

Funding: This research was funded by the Innovative and Entrepreneurial Talent Project in Jiangsu province, grant number 2016A426, and independent subject of State Key Laboratory of Advanced Metallurgy, University of Science and Technology Beijing, grant number 41617003.

Acknowledgments: The authors are grateful to the Innovative and Entrepreneurial Talent Project in Jiangsu province, and independent subject of State Key Laboratory of Advanced Metallurgy, University of Science and Technology Beijing, China, which enabled the research to be carried out successfully.

Conflicts of Interest: The authors declare no conflicts of interest.

References

1. Flesch, R.; Bleck, W. Crack susceptibility of medium and high alloyed tool steels under continuous casting conditions. *Steel Res.* **1998**, *69*, 292–299. [CrossRef]
2. Han, Z.; Cai, K.; Liu, B. Prediction and analysis on formation of internal cracks in continuously cast slabs by mathematical models. *ISIJ Int.* **2001**, *41*, 1473–1480. [CrossRef]
3. Lu, Y.J.; Wang, Q.; Li, Y.G.; He, S.P.; He, Y.M.; Pan, S.S.; Zhang, J.G.; Hu, B. Prevention of transverse corner cracks in continuously cast steel slabs using asymmetric secondary cooling nozzle. *Ironmak. Steelmak.* **2011**, *38*, 561–565. [CrossRef]
4. Raihle, C.; Fredriksson, H. On the formation of pipes and centerline segregates in continuously cast billets. *Metall. Mat. Trans. B* **1994**, *25*, 123–133. [CrossRef]
5. De Toledo, G.; Lainez, J.; Cirión, J. Model optimization of continuous casting steel secondary cooling. *Mater. Sci. Eng. A* **1993**, *173*, 287–291. [CrossRef]
6. Zeng, J.; Chen, W. Effect of secondary cooling conditions on solidification structure and central macrosegregation in continuously cast high-carbon rectangular billet. *High Temp. Mater. Processes* **2015**, *34*, 577–583. [CrossRef]
7. Dou, K.; Yang, Z.; Liu, Q.; Huang, Y.; Dong, H. Influence of secondary cooling mode on solidification structure and macro-segregation behavior for high-carbon continuous casting bloom. *High Temp. Mater. Process.* **2017**, *36*, 741–753. [CrossRef]
8. Camisani-Calzolari, F.; Craig, I.; Pistorius, P. Speed disturbance compenation in the secondary cooling zone in continuous casting. *ISIJ Int.* **2000**, *40*, 469–477. [CrossRef]
9. Zhang, J.; Chen, D.; Wang, S.; Long, M. Compensation control model of superheat and cooling water temperature for secondary cooling of continuous casting. *Steel Res. Int.* **2011**, *82*, 213–221. [CrossRef]
10. Long, M.; Chen, D.; Zhang, L.; Zhao, Y.; Liu, Q. A mathematical model for mitigating centerline macro segregation in continuous casting slab. *Met. Int.* **2011**, *16*, 19–33.
11. Shen, H.; Hardin, R.; MacKenzie, R.; Beckermann, C. Simulation using realistic spray cooling for the continuous casting of multi-component steel. *J. Mater. Sci. Technol.* **2002**, *18*, 311–314.
12. Ramstorfer, F.; Roland, J.; Chimani, C.; Mörwald, K. Investigation of spray cooling heat transfer for continuous slab casting. *Mater. Manuf. Process.* **2011**, *26*, 165–168. [CrossRef]
13. Ji, C.; Luo, S.; Zhu, M.; Sahai, Y. Uneven solidification during wide-thick slab continuous casting process and its influence on soft reduction zone. *ISIJ Int.* **2014**, *54*, 103–111. [CrossRef]

14. Wang, X.; Liu, Q.; Wang, B.; Wang, X.; Qing, J.-S.; Hu, Z.-G.; Sun, Y.H. Optimal control of secondary cooling for medium thickness slab continuous casting. *Ironmak. Steelmak.* **2011**, *38*, 552–560. [CrossRef]
15. Long, M.; Chen, D. Study on Mitigating center macro-segregation during steel continuous casting process. *Steel Res. Int.* **2011**, *82*, 847–856. [CrossRef]
16. Ma, J.; Xie, Z.; Jia, G. Applying of real-time heat transfer and solidification model on the dynamic control system of billet continuous casting. *ISIJ Int.* **2008**, *48*, 1722–1727. [CrossRef]
17. Hou, Z.; Jiang, F.; Cheng, G. Solidification structure and compactness degree of central equiaxed grain zone in continuous casting billet using cellular automaton-finite element method. *ISIJ Int.* **2012**, *52*, 1301–1309. [CrossRef]
18. Ma, J.; Wang, B.; Zhang, D.; Song, W. Optimization of secondary cooling water distribution for improving the billet quality for a small caster. *ISIJ Int.* **2018**, *58*, 915–920. [CrossRef]
19. Assuncao, C.; Tavares, R.; Oliveira, G. Comparison of uniform and non-uniform water flux density approaches applied on a mathematical model of heat transfer and solidification for a continuous casting of round billets. *Metall. Mat. Trans. B* **2015**, *46*, 366–377. [CrossRef]
20. Choudhary, S.; Mazumdar, D. Mathematical modelling of fluid flow, heat transfer and solidification phenomena in continuous casting of steel. *Steel Res. Int.* **1995**, *66*, 199–205. [CrossRef]
21. Cai, K.; Yang, J. Investigation of Heat Transfer in the Spray Cooling of Continuous Casting. *J. Univ. Sci. Technol. Beijing* **1989**, *11*, 510–515.
22. Nozaki, T.; Matsuno, J.; Murata, K.; Ooi, H.; Kodama, M. A secondary cooling pattern for preventing surface cracks of continuous casting slab. *Trans. Iron Steel Inst. Jpn.* **1978**, *18*, 330–338.
23. Thevoz, P.; Desbiolles, J.; Rappaz, M. Modeling of equiaxed microstructure formation in casting. *Metall. Trans. A* **1989**, *20*, 311–322. [CrossRef]
24. Kurz, W.; Giovanola, B.; Trivedi, R. Theory of microstructural development during rapid solidification. *Acta Metall.* **1986**, *34*, 823–830. [CrossRef]
25. Kurz, W.; Trivedi, R. Solidification microstructures: Recent developments and future directions. *Acta Metall.* **1990**, *38*, 1–17. [CrossRef]
26. Cornelissen, M. Mathematical-model for solidification of multicomponent alloys. *Ironmak. Steelmak.* **1986**, *13*, 204–212.
27. Jing, C.; Wang, X.; Jiang, M. Study on solidification structure of wheel steel round billet using FE-CA coupling model. *Steel Res. Int.* **2011**, *82*, 1173–1179. [CrossRef]
28. Yamazaki, M.; Natsume, Y.; Harada, H.; Ohsasa, K. Numerical simulation of solidification structure formation during continuous casting in Fe–0.7 mass% C alloy using cellular automaton method. *ISIJ Int.* **2006**, *46*, 903–908. [CrossRef]
29. Brimacombe, J.; Sorimachi, K. Crack formation in the continuous casting of steel. *Metall. Trans. B* **1977**, *8*, 489–505. [CrossRef]
30. Kulkarni, M.; Subash Babu, A. Optimization of continuous casting using simulation. *Mater. Manuf. Process.* **2005**, *20*, 595–606. [CrossRef]
31. M'Hamdi, M.; Combeau, H.; Lesoult, G. Modelling of heat transfer coupled with columnar dendritic growth in continuous casting of steel. *Int. J. Numer. Methods Heat Fluid Flow* **1999**, *9*, 296–317. [CrossRef]
32. Ludlow, V.; Normanton, A.; Anderson, A.; Thiele, M.; Ciriza, J.; Laraudogoitia, J.; Knoop, W. Strategy to minimise central segregation in high carbon steel grades during billet casting. *Ironmak. Steelmak.* **2005**, *32*, 68–74. [CrossRef]

© 2019 by the authors. Licensee MDPI, Basel, Switzerland. This article is an open access article distributed under the terms and conditions of the Creative Commons Attribution (CC BY) license (http://creativecommons.org/licenses/by/4.0/).

Article

Deformation Behavior of Internal Porosity in Continuous Casting Wide-Thick Slab during Heavy Reduction

Chenhui Wu [1,2], Cheng Ji [1,2,*] and Miaoyong Zhu [1,2]

1. Key Laboratory for Ecological Metallurgy of Multimetallic Ores (Ministry of Education), Shenyang 110819, China; wch_neu@126.com (C.W.); myzhu@mail.neu.edu.cn (M.Z.)
2. School of Metallurgy, Northeastern University, 3-11 Wenhua Road, Shenyang 110819, China
* Correspondence: jic@smm.neu.edu.cn; Tel.: +86-138-9881-2341

Received: 20 December 2018; Accepted: 23 January 2019; Published: 25 January 2019

Abstract: Heavy reduction (HR) is a novel technology that could effectively improve the internal porosities and other internal quality problems in continuously cast steel, during which a large reduction deformation is implemented at and after the strand solidification end. In the present paper, non-uniform solidification of the wide-thick slab was calculated with a two-dimensional (2D) heat transfer model. Based on the predicted temperature distribution at the solidification end of the casting strand, a three-dimensional (3D) thermal-mechanical coupled model was developed for investigating the deformation behavior of the internal porosities in wide-thick slab during HR. An Arrhenius-type constitutive model for the studied steel grade was derived based on the measured true stress-strain with single-pass thermosimulation compression experiments and applied to the 3D thermal-mechanical coupled model for improving the calculation accuracy. With the developed 3D thermal-mechanical coupled model, deformation behavior of the two artificial porosities located at the slab center of 1/2 width and 1/8 width during HR was investigated under different conditions of HR deformation, HR start position, and HR reduction mode. Based on the calculated porosity closure degree (η_s) and the corresponding equivalent strain (ε_{eq}) under different HR conditions, a prediction model that describes the quantitative relationship between η_s and ε_{eq} was derived for directly and accurately evaluating the process effect of HR on improving the internal porosities in wide-thick slab.

Keywords: continuous casting; wide-thick slab; non-uniform solidification; heavy reduction; porosity; deformation

1. Introduction

Due to solidification shrinkage and gas entrapment, internal porosity often occurs in casting steel. As one kind of the common internal defects, it seriously influences the mechanical properties of the final products, for example, decreasing the fatigue life and the yield strength, and should be eliminated in the subsequent rolling or forging process.

To provide theoretical guide for process design of rolling or forging, many investigations were carried out by previous researchers to clarify the closure mechanism of internal porosity in metal materials during the forging or rolling process. To quantitatively evaluate the porosity closure during forging process, Tanaka et al. [1] proposed the hydrostatic integration parameter, which was widely adopted as an indicator of the porosity closure degree by subsequent researchers [2–5]. In addition to the hydrostatic integration parameter, many other researchers [6–9] also adopted effective strain as the indicator of porosity closure degree during hot working process, and different threshold values mboxciteB7-metals-421132,B8-metals-421132,B9-metals-421132 of effective strain for eliminating the internal porosity were reported. Recently, a void aspect ratio evaluation index, defined as a function of stress deviator, effective strain, and effective stress, was proposed by Chen et al. [10], which could give an accurate description of the porosity evolution during forging process. After carrying out full-scale hot-rolling experiments, Ståhlberg et al. [11] concluded that temperature gradient between the lower temperature on workpiece surface and the higher temperature in its internal region, as obtained by water cooling, was advantageous for the elimination of internal porosity with a relatively small rolling reduction. The effect of temperature gradient was further discussed in the later studies of forging [4,12] or rolling [3,13,14] with numerical or experimental methods, and all of these investigations confirmed the promotion effect of temperature gradient on the porosity closure. During forging process, die shape is another critical process parameter that influences the process effect on eliminating the internal porosities in workpiece, and in order to design an optimum die geometry, some studies were conducted by previous researchers to investigate the effect of die shapes on porosity closure during open die forging [6,15,16], upset process [4], and hot radial forging [17], et al.

Internal porosities in casting steel usually could be eliminated by rolling or forging process. However, due to the increased solidification time for casting steel with a large section size, its internal porosities become more serious with the presence of coarser cast structure [4,18]. Under condition of large components that were produced by rolling or forging with a relatively low compression ratio [11,19–21], the serious internal porosities in large section size casting steel could not be easily eliminated, which seriously influences the mechanical properties of the final products. Meanwhile, as one of the main counter measures of the internal porosities, the traditional mechanical soft reduction (SR) [22,23] was proved to be insufficient on significantly improving the serious internal porosities in the casting steel with a large section size. In order to improve the serious internal defects in continuous casting steel significantly, some earlier researchers [24–27] proposed heavy reduction (HR) technology. By implementing a large reduction deformation around the strand solidification end, the internal quality of continuous casting steel could be significantly improved by HR, which effectively contributes to the complete elimination of internal porosity in the subsequent rolling or forging process.

As an effective counter measure of internal defects in continuous casting steel, HR has attracted more and more researchers' attention with the rapidly increased demand for large components in the large equipment manufacturing industry in recent years. Some theoretical and experimental investigations were carried out recently for studying the improving effect of HR on porosity and other internal defects in continuous casting bloom [28,29], billet [30], or slab [19–21], and some new HR technologies were then proposed and applied. By establishing a three-dimensional (3D) thermal-mechanical coupled model, the present authors [28] studied the deformation behavior of continuous casting bloom during HR and developed the two-stage sequential heavy reduction technology. Industrial trials indicated that the homogeneity and compactness of the continuous casting bloom could be obviously improved after the application of two-stage sequential heavy reduction technology. Based on numerical simulation results, two kinds of new HR technologies, named as START and HRPISP, were respectively proposed by Xu et al. [19] and Zhao et al. [20,21] for simultaneously improving the internal porosity and macro-segregation in continuous casting wide-thick slab, and the effectiveness of START was proved by experimental results in plant.

To provide theoretical basis for the development of the HR process and thus improve the internal porosities in wide-thick slab more effectively, the porosity deformation behavior during HR in the wide-thick slab continuous casting process was systematically investigated mainly by the numerical simulation method in the present work. A two-dimensional (2D) heat transfer model was established to calculate the non-uniform solidification process of the wide-thick slab. Based on the predicted heat transfer results by the 2D heat transfer model and the derived constitutive model for the studied steel grade during HR, a 3D thermal-mechanical coupled model, containing two artificial spheroidal porosities, respectively, located the slab center of 1/8 width ($P_{1/8}$) and 1/2 width ($P_{1/2}$), was established. With this 3D thermal-mechanical coupled model, the deformation behavior of $P_{1/8}$ and $P_{1/2}$ during HR was numerically investigated under different HR conditions, including the HR deformation, HR start position, and HR mode. Based on the predicted porosity deformation results under different HR conditions and the corresponding equivalent strain (ε_{eq}), a prediction model for the porosity closure behavior was derived to describe the quantitative relationship between the porosity closure degree (η_s) and ε_{eq}.

2. Finite Element Model

2.1. 2D Heat Transfer Model

During HR, the deformation behavior of the casting strand is closely related to its temperature distribution. In order to improve the calculation efficiency, a 2D heat-transfer model, as shown in Figure 1, was firstly developed with the commercial finite element software MSC.Marc (2013.0.0, MSC Software Corporation, Newport Beach, CA, USA) based on the practical casting conditions in Table 1 and some simplified conditions [31]. Heat transfer analysis was then carried out with this model, which determined the strand solidification end and the corresponding initial temperature field for the subsequent 3D thermal-mechanical coupled model in Section 2.2. Due to the symmetry of heat transfer behavior of the casting strand along its width direction, half of the wide-thick slab transverse section, as shown in Figure 1, was taken as the calculation domain. Four-nodes quadrilateral elements with a side length of 5 mm were applied to uniformly mesh the calculation domain, and the final 2D heat transfer model contains 11,200 elements and 11,457 nodes. Automatic time step with 0.1 s and 1 s taken as the minimum and maximum time step, respectively, was adopted during the calculation.

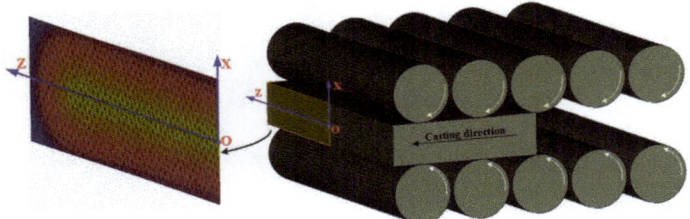

Figure 1. Schematic of the two-dimensional (2D) heat transfer model.

Table 1. The practical production conditions for the wide-thick slab continuous casting.

Item	Content
Type of the continuous casting machine	Straight-bow wide-thick slab continuous casting machine
Main chemical composition of the studied steel grade (wt %)	C: 0.17, Si: 0.15, Mn: 0.60, P: 0.015, S: 0.01
Slab transverse section size (mm × mm)	280 × 2000
Casting speed (m/min)	0.80
Casting temperature (°C)	1520–1550
Length of each cooling zone (m)	Effective mold height: 0.8, Secondary cooling zone: 19.7, Air cooling zone: 9.8
Specific water flow in the secondary cooling zone (L/kg)	0.85

Thermal material properties and the cooling boundary conditions are two critical factors that influence the calculation accuracy of the 2D heat transfer model. In order to improve the calculation accuracy, thermal material properties of the studied steel grade, such as the conductivity, density, and enthalpy, were calculated with weighted averaging of phase fraction method [32–34], and the final thermal material properties of the studied steel grade can be found in our previous work. [34] In addition to thermal material properties, cooling boundary conditions is another critical factor that directly determines the calculation accuracy of the 2D heat transfer model. When compared with the conventional continuous casting slab with a relatively small section size, solidification of the wide-thick slab is obviously non-uniform along its width direction due to the large section size and the non-uniform water flux distribution in the secondary cooling zone of the continuous casting machine, and the final solidification region of the wide-thick slab was located around 1/8 width of its transverse section [33,34]. In order to accurately determine the complicated cooling boundary conditions in the secondary cooling zone, the non-uniform cooling water flux distribution in this cooling zone was measured and applied during the calculation of cooling boundary conditions in the secondary cooling zone. More detailed information about the measured water flux distribution and the calculation method of the cooling boundary conditions in mold, secondary cooling zone, and air cooling zone for the studied wide-thick slab continuous casting machine could be found in our previous work [34].

2.2. 3D Thermal-Mechanical Coupled Model

In order to simulate the evolution of internal porosities in the wide-thick slab during HR, a 3D thermal-mechanical coupled model, as shown in Figure 2, was developed with MSC.Marc. Due to the symmetry of the strand deformation behavior during HR, a section of half the wide-thick slab along its width direction was taken as the calculation domain. The deformation of the wide-thick slab that is caused by HR is obviously much larger than the thermal deformation. Therefore, the much smaller thermal deformation of the wide-thick slab during HR was neglected in the 3D thermal-mechanical coupled model.

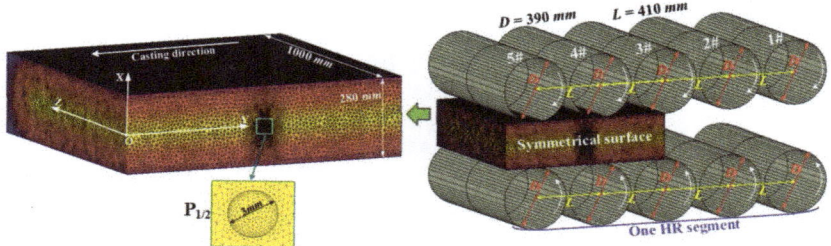

Figure 2. Schematic of the three-dimensional (3D) thermal-mechanical coupled model.

For the studied wide-thick slab continuous casting machine, HR can be implemented by one or several HR segments. Each HR segment, as shown in Figure 2, contains five pairs of rollers, and the roller diameter and roller pitch are 390 mm and 410 mm, respectively. During HR, roller gap linearly decreases from entrance (Roller 1#) to exit (Roller 5#) of the segment, and rollers of the HR segment are regarded as rigid bodies without considering their small deformation during HR. The friction factor between rollers and the wide-thick slab was set as 0.3 [13] and the adopted contact detection method between rollers and the casting strand was node to segment contact algorithm [28].

During the calculation of the 3D thermal-mechanical coupled model, the slab temperature field at each increment was firstly solved by the solver based on the corresponding cooling boundary conditions and the thermal material properties. Secondly, the mechanical properties in the 3D thermal-mechanical coupled model were then updated mainly based on the temperature field at the present increment and the derived constitutive equations (Equations (1) and (2)), and the deformation behaviour of the wide-thick slab was then solved based on the mechanical boundary conditions during HR.

The practical production results, as shown in Figure 3, indicate that serious porosities are centrally distributed around the slab centerline, and the porosity size is usually less than 5 mm. During the development of the 3D thermal-mechanical coupled model, porosity was simplified to be a spheroidal void with a diameter of 3 mm and located at the slab centerline.

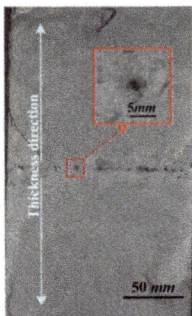

Figure 3. Macrographs of the wide-thick slab longitudinal section.

Due to the non-uniform solidification of the wide-thick slab, there still remains a small mushy region, as shown in Figure 4, around 1/8 width at the strand solidification end when the solid phase fraction (f_s) at the slab center of 1/2 width reaches 1.0. This means that the temperature distribution and variation around the region of 1/8 width differs from those at the other regions during HR at the strand solidification end, which will impact the porosity deformation behavior. Therefore, a total of two artificial spheroidal porosities (each with a diameter of 3 mm) were created at the slab centerline of 1/8 width ($P_{1/8}$) and 1/2 width ($P_{1/2}$) in the 3D thermal-mechanical coupled model for investigating the porosity deformation behavior in these two typical regions during HR. Figure 4 schematically shows the distribution of these two artificial porosities on the slab transverse section in the 3D thermal-mechanical coupled model, and the artificial porosity located at 1/2 width ($P_{1/2}$) in Figure 4 can be also seen in Figure 2 and it is located on the symmetrical surface of the 3D thermal-mechanical coupled model.

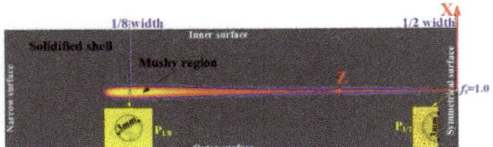

Figure 4. Solidification morphology of the wide-thick slab transverse section when solid phase fraction (f_s) at the slab center reaches 1.0, and the distribution of the two artificial porosities on slab the transverse section in the 3D thermal mechanical model.

In order to improve the calculation accuracy and efficiency, the calculation domain of the 3D thermal-mechanical coupled model was nonuniformly meshed with four-nodes tetrahedral elements. Fine elements with a side length of ~0.3 mm are distributed around the two artificial porosities, while coarser elements with a side length of ~10 mm are distributed around the slab surface. The final 3D thermal-mechanical coupled model contains 540,836 elements and 99,592 nodes. Automatic time step was adopted during the simulation, and the maximum and the minimum time step were 0.01 s and 1 s, respectively.

In order to accurately describe the metal flow behavior of the wide-thick slab during HR, the true stress-strain of the studied steel grade was measured at different temperatures and strain rates, and the measured results are presented in Figure 5.

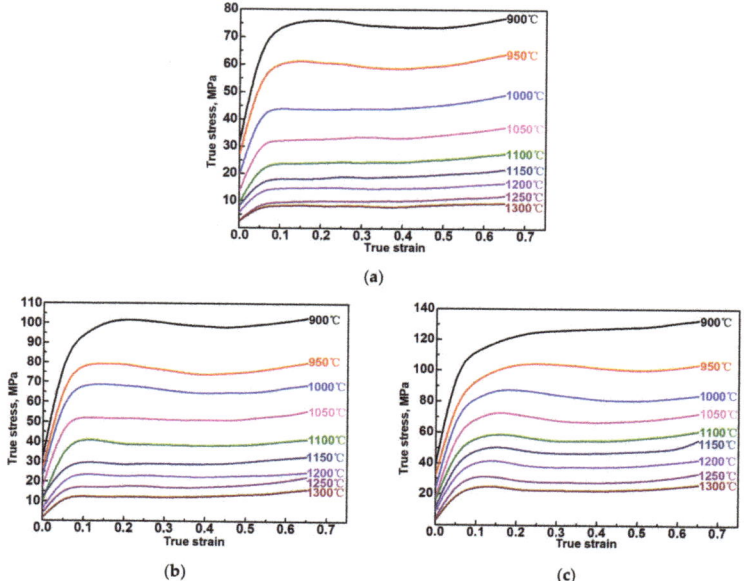

Figure 5. The measured true stress-strain at different temperature and strain rate of (a) 0.001 s^{-1}, (b) 0.01 s^{-1}, and (c) 0.1 s^{-1}.

Based on the measured results in Figure 5, an Arrhenius-type constitutive model was derived with the similar method adopted in our previous work for establishing an Arrhenius-type constitutive model of GCr15 steel [35]. The derived constitutive model of the studied steel was then applied to the 3D thermal-mechanical coupled model, and the strain and stress in the constitutive model were adopted as equivalent strain and equivalent stress in the 3D thermal-mechanical coupled model. The derived constitutive model can be expressed, as follows:

$$\begin{cases} \sigma = \frac{1}{\alpha} \left\{ \left(\frac{Z}{A}\right)^{1/n} + \left[\left(\frac{Z}{A}\right)^{2/n} + 1\right]^{1/2} \right\} \\ Z = \dot{\varepsilon} \cdot \exp\left(\frac{Q}{R(T+273)}\right) \end{cases}, \quad (1)$$

where σ is the stress, MPa; A and α are material constants; n is the material's stress index; Z is the Zenner-Hollomon parameter; $\dot{\varepsilon}$ is the strain rate, s^{-1}; Q is the activation energy of hot deformation, J mol^{-1}; R is the ideal gas constant, (8.314 J mol^{-1} K^{-1}); and, T is the temperature, °C. The strain-dependent parameters of α, A, n, and Q were derived based on the measured true stress-strain curves and could be calculated with Equation (2) and the corresponding parameters that are listed in Table 2.

$$\begin{cases} Q = B_0 + B_1\varepsilon^1 + B_2\varepsilon^2 + B_3\varepsilon^3 + B_4\varepsilon^4 + B_5\varepsilon^5 + B_6\varepsilon^6 \\ \ln A = C_0 + C_1\varepsilon^1 + C_2\varepsilon^2 + C_3\varepsilon^3 + C_4\varepsilon^4 + C_5\varepsilon^5 + C_6\varepsilon^6 \\ n = D_0 + D_1\varepsilon^1 + D_2\varepsilon^2 + D_3\varepsilon^3 + D_4\varepsilon^4 + D_5\varepsilon^5 + D_6\varepsilon^6 \\ \alpha = E_0 + E_1\varepsilon^1 + E_2\varepsilon^2 + E_3\varepsilon^3 + E_4\varepsilon^4 + E_5\varepsilon^5 + E_6\varepsilon^6 \end{cases}, \quad (2)$$

Table 2. Polynomial fitting coefficients of the material parameters (X represents B, C, D, and E in Equation (2)).

Items	Q (J mol^{-1})	$\ln A$	n	α
900 °C to 1300 °C				
X_0	730674.0	57.78	6.25	0.03965
X_1	−7080560.0	−607.39	−42.11	−0.35051
X_2	59878600.0	5157.77	295.89	3.09951
X_3	−237768000.0	−20523.01	−1053.31	−12.63650
X_4	488739000.0	42240.95	2024.39	26.65368
X_5	−502960000.0	−43514.69	−1978.96	−28.32745
X_6	204463000.0	17705.26	767.78	11.95993

For temperature of >1300 °C,

$$\sigma = \eta \sigma_{1300}, \tag{3}$$

where σ_{1300} is the flow stress at 1300 °C and a certain specified strain rate and strain and can be determined by Equation (1); η is the temperature-dependent attenuation coefficient and it can be expressed as:

$$\eta = \frac{1}{\sigma_{p1300}} c(T + 273)^d, \tag{4}$$

where σ_{p1300} is the peak stress of the measured true stress-strain curve at 1300 °C and 0.001 s^{-1} and equal to 8.431; and, T is the temperature of >1300 °C; c and d are equal to 5.898 × 1018 and −5.712, respectively, based on the variation trend of peak stress with temperature at the strain rate of 0.001 s^{-1}.

2.3. Model Validation

In order to show the accuracy of the derived Arrhenius-type constitutive model, flow stress under different temperature and strain rates was calculated with this constitutive model, and the calculated results are compared with the measured ones in Figure 6. Based on the measured and the calculated results in Figure 6, the standard statistical parameters of average absolute relative error (*AARE*) for the measured and the calculated values, which has been adopted in our previous work [35], was calculated with Equation (5). It is found that the value of *AARE* is about 4.7%, which proves the accuracy of the derived constitutive model.

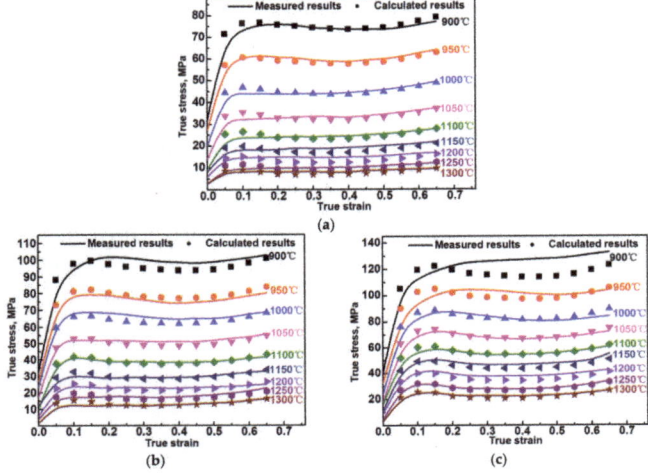

Figure 6. Comparison between the measured and the calculated results at different temperature and strain rate of (**a**) 0.001 s^{-1}, (**b**) 0.01 s^{-1}, and (**c**) 0.1 s^{-1}.

$$AARE(\%) = \frac{1}{N}\sum_{i=1}^{N}\left|\frac{E_i - P_i}{E_i}\right| \times 100\%, \quad (5)$$

where E_i is the measured value and P_i is the calculated value by the derived constitutive model; and, N is the total number of data sets in Figure 6.

In order to verify the accuracy of the 2D heat transfer model, the temperature of the slab inner surface at different strand positions were measured with a thermal infrared camera (A40, FLIR, FLIR Systems Inc., Goleta, CA, USA) when the wide-thick slab with a transverse section size of 2000 mm × 280 mm was cast at 0.8 m/min. The measured results are compared with the calculated ones in Figure 7, which indicates that the calculated temperature by the 2D heat transfer model agree well with the corresponding measured results and that the relative error between the calculated and the measured results is less than 2.2%.

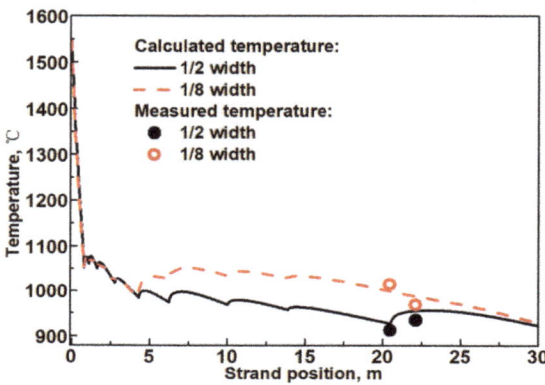

Figure 7. Comparison between the calculated and the measured temperature.

It should be noted that, due to the non-uniform cooling water flux distribution in the secondary cooling zone, the slab surface temperature at 1/8 width, as shown in Figure 7, is higher than that at 1/2 width. For this reason, the obvious mushy region, as has been presented in Figure 4, still could be observed around 1/8 width when f_s reaches 1.0 at the strand solidification end.

In order to verify the 3D thermal-mechanical coupled model, plant trial of HR was carried out. During the plant trial, the casting strand of the wide-thick slab moved through the HR segment from entrance (Roller 1#) to exit (Roller 5#) and f_s (solid phase fraction at the slab center) at the entrance of the HR segment is 1.0. The reduction force of the HR segment was measured in real time by the pressure sensors that were installed in the hydraulic cylinders of the HR segment. The calculated reduction force of the HR segment in the 3D thermal-mechanical coupled model can be determined by adding up the calculated reduction force of Roller 1#–5#. Figure 8 compares the actual measured reduction force of the HR segment during plant trial with the corresponding calculated results by the 3D thermal-mechanical coupled model. It can be seen that the calculated reduction force shows good agreement with the actual measured results. The relative error between the calculated and the actual measured results is less than 3.2%.

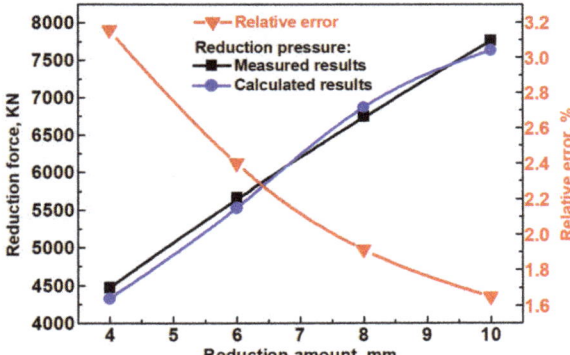

Figure 8. Comparison between the measured and the calculated reduction force.

3. Results and Discussion

Figure 9 schematically shows the porosity dimension along the slab thickness direction (corresponding to X axis), casting direction (corresponding to Y axis), and the slab width direction (corresponding to Z axis) for the two created artificial porosities in the 3D thermal-mechanical coupled. In order to quantitatively describe the porosity deformation behavior along three axis directions, the porosity deformation degree was defined:

$$\begin{cases} \Delta l_x = \ln\left(\frac{L'_x}{L_x}\right) \\ \Delta l_y = \ln\left(\frac{L'_y}{L_y}\right) \\ \Delta l_z = \ln\left(\frac{L'_z}{L_z}\right) \end{cases}, \tag{6}$$

where Δl_x, Δl_y, and Δl_z are the porosity deformation degree along the slab thickness direction, casting direction, and the slab width direction; L_x, L_y, and L_z are the porosity axis length along three axis directions before HR; L'_x, L'_y, and L'_z are porosity axis length along three axis directions after HR.

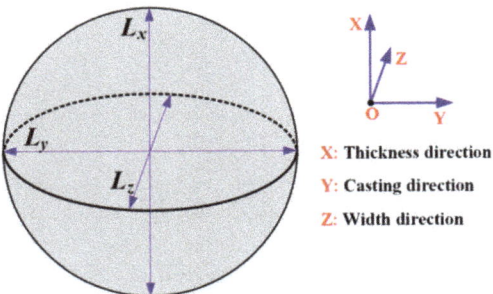

Figure 9. Schematic of the porosity dimension along three axis directions.

In order to quantitatively describe the overall deformation behavior of each artificial porosity in the 3D thermal-mechanical coupled model and evaluate the process effect of HR on improving the internal porosity, the porosity closure degree was defined based on the porosity axis length before and after HR:

$$\eta_s = \frac{2L_x}{(L_y + L_z)} - \frac{2L'_x}{(L'_y + L'_z)}, \tag{7}$$

where η_s is the porosity closure degree after HR and it ranges from 0 to 1. A larger value of η_s indicates a better process effect of HR on improving the internal porosity.

3.1. Porosity Deformation Behavior after Different HR Deformation

The porosity deformation behavior was firstly investigated under the condition that different HR deformation was uniformly implemented by the HR segment at the strand solidification end (f_s at the entrance of the HR segment is 1.0). The porosity deformation degree and the closure degree of $P_{1/2}$ and $P_{1/8}$ after different HR deformation are presented in Figure 10. The thickness reduction in each figure represents the magnitude of HR deformation implemented by the HR segment.

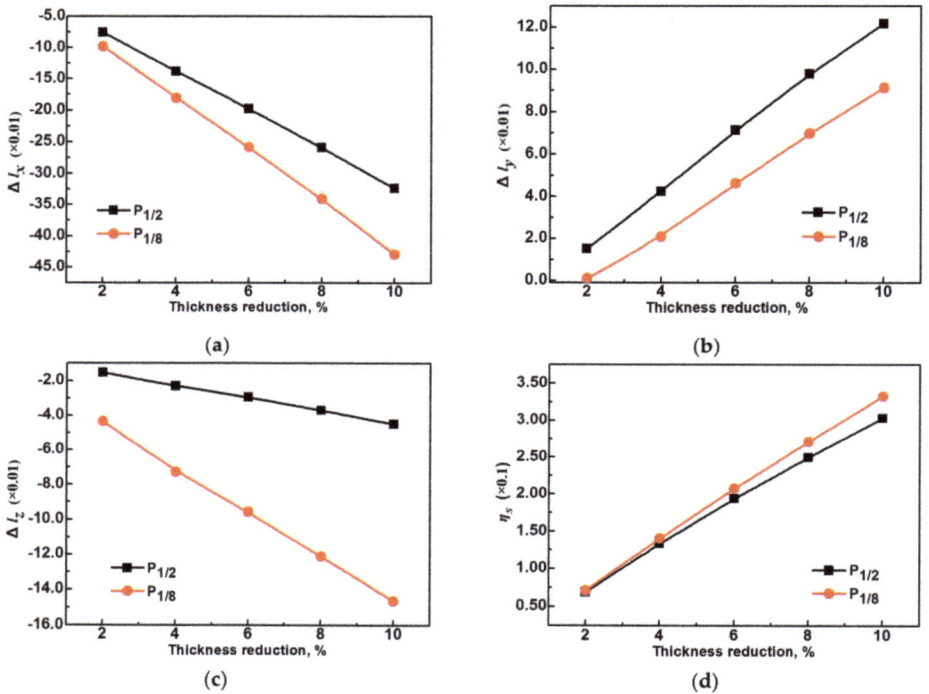

Figure 10. Porosity deformation degree along the (**a**) slab thickness direction, (**b**) casting direction, (**c**) width direction, and (**d**) the porosity closure degree after different heavy reduction (HR) deformation.

It can be seen from Figure 10a to c that the porosity deformation degree along the slab thickness direction (Δl_x), the casting direction (Δl_y), and the slab width direction (Δl_z) continuously increase with thickness reduction (represents the HR deformation implemented by the HR segment) increased. The values of Δl_x in Figure 10a and Δl_z in Figure 10c are negative, which is opposite to that of Δl_y in Figure 10b. This means that the porosity size decreases along the slab thickness direction and the slab width direction, and it meanwhile increases along the casting direction during HR. When compared with the porosity deformation degree along the casting direction (Δl_y) and the slab width direction (Δl_z), the magnitude of Δl_x is much larger, which indicates that the major deformation of the porosity is along the slab thickness direction. During HR, the internal porosity is continuously improved with the implemented HR deformation increase. As a result, an increasing trend for the porosity closure degree (η_s) can be observed in Figure 10d.

Figure 10 also indicates that there exists difference between the deformation behavior of $P_{1/2}$ and that of $P_{1/8}$, and the closure degree of $P_{1/8}$ is 9.7% larger than that of $P_{1/2}$ after 10% HR deformation. This indicates that the porosity at the slab center around 1/8 width can be improved more effectively during HR at the strand solidification end. Two possible factors may contribute to the difference of porosity deformation behavior at 1/8 width and 1/2 width: the different location of $P_{1/8}$ and $P_{1/2}$ and the different temperature distribution around 1/8 width and 1/2 width.

In order to investigate the influence of porosity location, the deformation behavior of $P_{1/8}$ and $P_{1/2}$ was calculated under the condition that the cooling water flux distribution in the secondary cooling zone of the wide-thick slab continuous casting machine and the corresponding solidification process of the wide-thick slab were assumed to be uniform along the slab width direction. As the porosity mainly deforms along the slab thickness direction, only the porosity deformation degree along the slab thickness direction (Δl_x) and the porosity closure degree at 1/8 width and 1/2 width are compared in Figure 11a and b, respectively. The comparison results in Figure 11 show that the difference between the porosity deformation behaviour at 1/8 width and 1/2 width during HR is very small. This indicates that the influence of porosity location on the porosity deformation behavior during HR is not obvious under uniform cooling condition and also simultaneously proves that the difference of porosity deformation behavior at 1/8 width and 1/2 width is mainly caused by the different temperature distribution around these two regions.

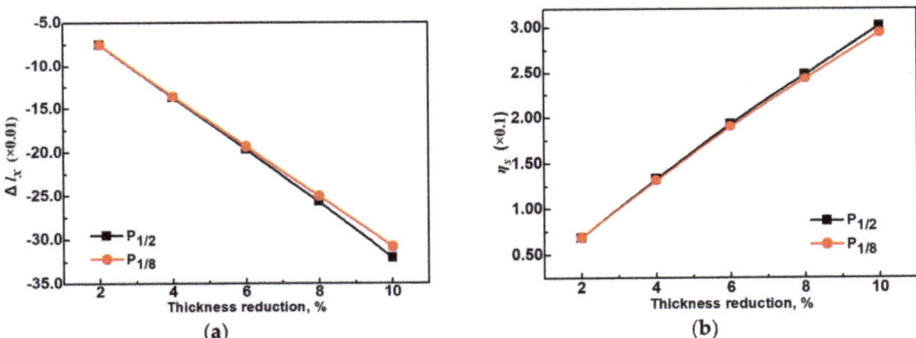

Figure 11. (a) Porosity deformation degree along the thickness direction and (b) porosity closure degree after different HR deformation implemented at the strand solidification end under condition of uniform solidification.

In order to further clarify the influence of temperature distribution on the porosity deformation behavior at 1/8 width and 1/2 width during HR, the variation of temperature, and the corresponding temperature difference at different typical locations ($L_s^{1/8}$, $L_c^{1/8}$, $L_s^{1/2}$, $L_c^{1/2}$) shown in Figure 12a during HR at the strand solidification end are compared in Figure 12b,c, respectively. $T_s^{1/8}$, $T_c^{1/8}$, $T_s^{1/2}$, $T_c^{1/2}$ in Figure 12b are the calculated temperature at $L_s^{1/8}$, $L_c^{1/8}$, $L_s^{1/2}$, $L_c^{1/2}$ respectively, and $\Delta T^{1/8}$ ($\Delta T^{1/8} = T_c^{1/8} - T_s^{1/8}$) and $\Delta T^{1/2}$ ($\Delta T^{1/2} = T_c^{1/2} - T_s^{1/2}$) in Figure 12c represent the temperature difference between the slab surface and center at 1/8 width and 1/2 width, respectively.

It can be seen from Figure 12b that temperature at different typical locations overall present a decreasing trend from entrance (Roller 1#) to exit (Roller 5#) of the HR segment during HR at the strand solidification end. However, when compared with the gradual temperature variation at the slab surface ($T_s^{1/8}$ and $T_s^{1/8}$), temperature variation at the slab center ($T_c^{1/8}$ and $T_c^{1/2}$) are much more remarkable. As a result, the variation trend of $\Delta T^{1/8}$ and $\Delta T^{1/2}$ in Figure 12c are similar with that of $T_c^{1/8}$ and $T_c^{1/2}$ in Figure 12b. As mentioned above, a small mushy region still remains around 1/8 width at the strand solidification end when f_s reaches 1.0. This means that the decrease of $T_c^{1/8}$ and $\Delta T^{1/8}$ during HR at the strand solidification end can be slowed down to some extent by the released latent heat from the remained small mushy region around 1/8 width. For this reason, the temperature difference at 1/8 width ($\Delta T^{1/8}$), as shown in Figure 12c, is larger than that at 1/2 width ($\Delta T^{1/2}$) during HR, except at the entrance (Roller 1#) of the HR segment. Combined with the larger closure degree of $P_{1/8}$ than that of $P_{1/2}$ in Figure 10d, it can be concluded that, due to the larger temperature difference at 1/8 width, porosity around this region can be improved more effectively by HR at the strand solidification end.

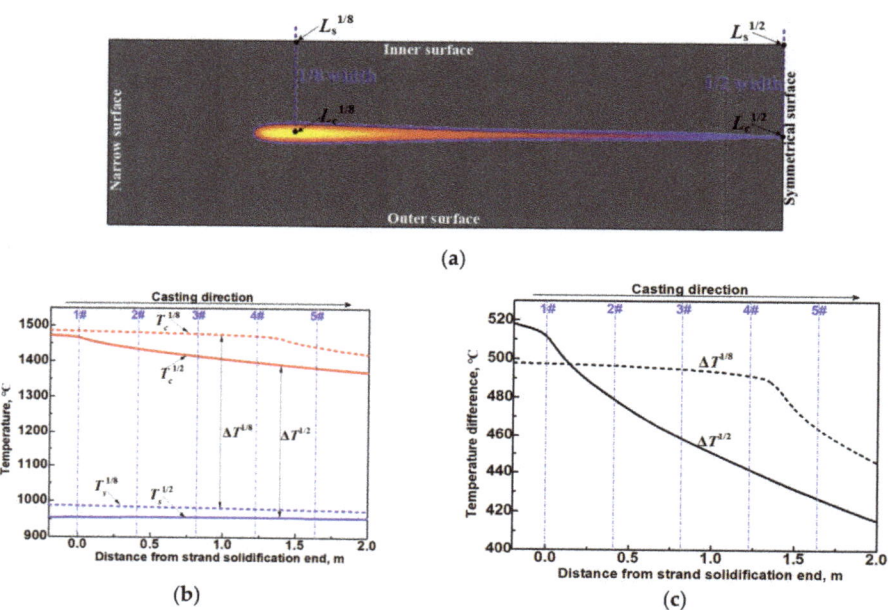

Figure 12. (a) Distribution of the typical locations on the slab transverse section and variation of the (b) temperature and (c) temperature difference at the typical locations during HR.

During the hot working process, the deformation degree at one position of the workpiece can be quantitatively evaluated by the corresponding equivalent strain, and for this reason, many previous researchers [6–9] adopted equivalent strain as an indicator of closure degree of internal porosity in workpiece.

Figure 13 compares the equivalent strain at $L_c^{1/8}$ (corresponding to the location of $P_{1/8}$) and $L_c^{1/2}$ (corresponding to the location of $P_{1/2}$) after different HR deformation implemented by the HR segment at the strand solidification end. It can be seen that, when the HR deformation increased, equivalent strain at $L_c^{1/8}$ and $L_c^{1/2}$ after HR continuously increase. As a result, the porosity at 1/8 width and 1/2 width can be continuously improved, and the corresponding porosity closure degree in Figure 10d shows a rising trend with HR deformation increased. However, it should be noted that the equivalent strain at $L_c^{1/8}$ is overall larger than that at $L_c^{1/2}$. This proves that, due to the larger temperature difference at 1/8 width, as mentioned above, HR deformation can transfer from the slab surface into its center more effectively for better improving the internal porosities around this region.

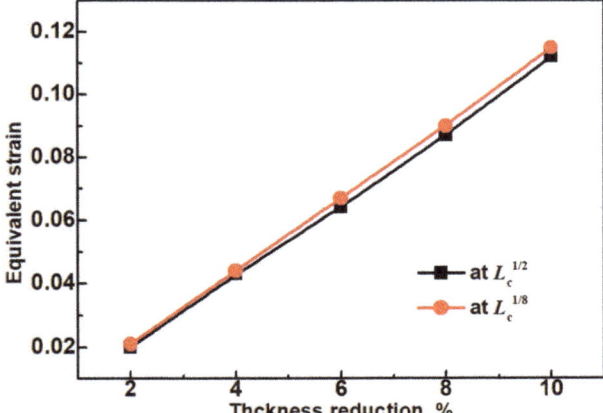

Figure 13. Equivalent strain at the slab center of 1/8 width ($L_c^{1/8}$) and 1/2 width ($L_c^{1/2}$) after different HR deformation implemented at the strand solidification end.

3.2. Influence of HR Position on the Porosity Deformation Behavior

For the studied wide-thick slab continuous casting machine, HR can be implemented by one or more HR segments, and the reduction position can be flexibly changed by adjusting the roller gap of the corresponding HR segment. In order to study the influence of HR position on the porosity deformation behavior, the porosity deformation behavior was calculated with 6% HR deformation implemented by one HR segment at different strand position.

Figure 14a–d show the calculated porosity deformation behavior, and the abscissa axis in each figure represents the HR start position (corresponding to Roll 1# of the HR segment) after the strand solidification end. With HR start position moving away after the strand solidification end, the porosity deformation degree along the slab thickness direction in Figure 14a and along the slab width direction in Figure 14c both decrease, while the porosity deformation degree along the casting direction in Figure 14b presents an increasing trend. This indicates that, with the HR starting position moving away after the strand solidification end, the porosity size after HR increases along three axis directions. Figure 14d shows the porosity closure degree after HR implemented at different strand positions. When compared with the porosity closure degree after HR implemented at the strand solidification end, the closure degree of $P_{1/8}$ and $P_{1/2}$ decrease by 9.3% and 6.3%, respectively, with the HR starting position moving away by 3 m after the strand solidification end, which indicates that the process effect of HR on improving the internal porosity becomes worse with the HR starting position moving away after the strand solidification end.

Figure 15a,b, respectively, presents the average temperature difference between the slab surface and center within the HR segment and the equivalent strain after 6% HR deformation implemented at different strand position. With the HR starting position moving away after the strand solidification end, the average temperature difference, which could promote the transfer of HR deformation from the slab surface into its center, significantly decreases. As a result, the equivalent strain, which represents the material deformation degree and is regarded as an indicator of porosity closure degree, after HR shown in Figure 15b continuously decreases, which explains the continuously decreasing trend of η_s in Figure 15d.

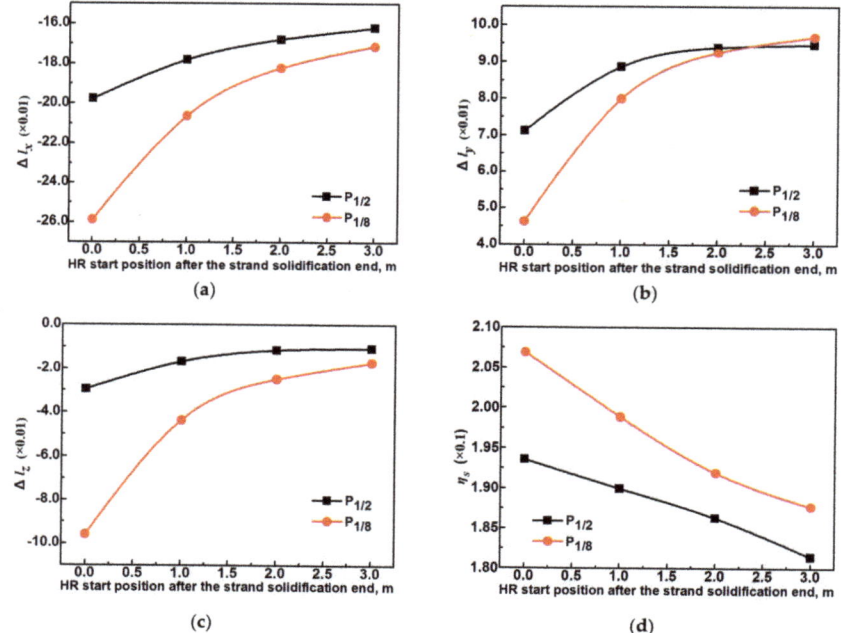

Figure 14. Porosity deformation degree along the (**a**) slab thickness direction, (**b**) casting direction, (**c**) slab width direction, and (**d**) the porosity closure degree after 6% HR deformation implemented at different strand position.

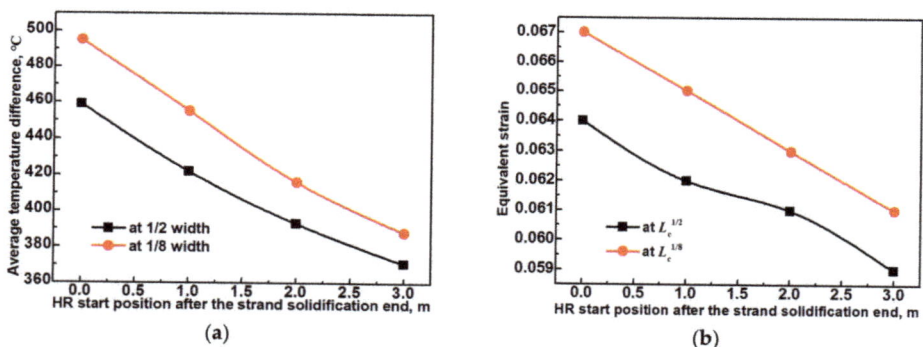

Figure 15. (**a**) The average temperature difference at 1/8 width and 1/2 width within the HR segment and (**b**) the equivalent strain at the slab center of 1/8 width ($L_c^{1/8}$) and 1/2 width ($L_c^{1/2}$) after 6% HR deformation implemented at different strand position.

With the HR starting position moving away after the strand solidification end, the temperature of the casting strand decreases, and its deformation-resistant ability during HR correspondingly increases. As a result, the required reduction force for the HR segment to implement the same HR deformation, as shown in Figure 16, significantly increases with the HR start position moving away after the strand solidification end. When compared with the required reduction force for the HR segment to implement 6% HR deformation at the strand solidification end, this value increases by ~20% with the HR starting position moving away by 3 m after the strand solidification end. This indicates that the reduction capacity of the HR segment will significantly decrease with the HR starting position moving away after the strand solidification end.

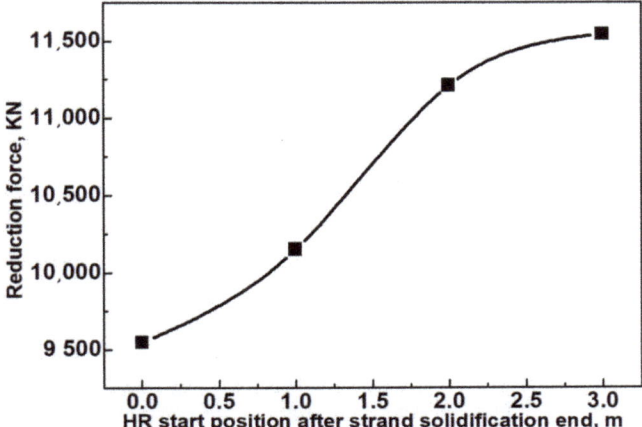

Figure 16. The required reduction force for the HR segment to implement 6% HR deformation at different strand position.

From the discussion above, it can be concluded that, due to the significant decrease of porosity closure degree and the reduction capacity of the HR segment, the HR efficiency on improving the internal porosities significantly decreases with the HR starting position moving away after the strand solidification end.

3.3. Influence of Reduction Mode on the Porosity Deformation Behavior

In order to implement HR more effectively and thus better improve the internal porosities, the influence of reduction mode on the porosity deformation behavior during HR was investigated. Table 3 compares the HR deformation distribution within the HR segment in five cases, and the variation of the corresponding slab thickness from entrance (Roller 1#) to exit (Roller 5#) of the HR segment are presented in Figure 17. The total HR deformation in each case is 6.0%. In Case 1, HR deformation is uniformly implemented with 1.2% HR deformation at each roller of the HR segment, which represents the traditional reduction mode and it is called UHR (Uniform Heavy Reduction) in the present work. In addition to UHR, a new reduction mode, called SPUHR (Single Point and Uniform Heavy Reduction), was proposed based on the mechanical structure of the HR segment. For SPUHR (corresponding to Case 2 to 5), a relatively larger HR deformation was implemented at Roller 1# by adjusting the hydraulic cylinders that were installed at the entrance of the HR segment. The residual HR deformation was then uniformly implemented from Roller 2# to 5# with a relatively smaller HR deformation at each roller than that at Roller 1#, and the HR deformation at each roller of 2# to 5# was equal due to the limitation of mechanical structure of the HR segment.

Table 3. HR deformation distribution within the HR segment in five cases.

Case	Reduction mode	Thickness reduction at each roller of the HR segment, %					Total thickness reduction, %
		1#	2#	3#	4#	5#	
1	UHR	1.20	1.20	1.20	1.20	1.20	6.0
2		1.80	1.05	1.05	1.05	1.05	6.0
3	SPUHR	2.40	0.90	0.90	0.90	0.90	6.0
4		3.00	0.75	0.75	0.75	0.75	6.0
5		3.60	0.60	0.60	0.60	0.60	6.0

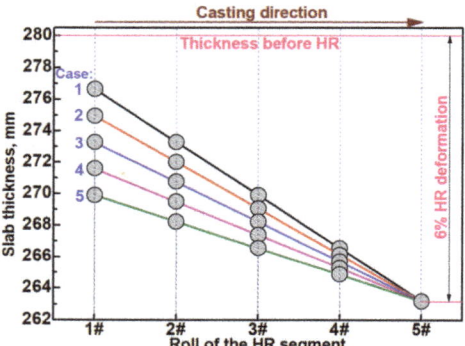

Figure 17. Variation of the slab thickness within the HR segment in Case 1 to 5.

After 6% HR deformation implemented at the strand solidification end by the HR segment with the reduction mode of UHR (Case 1) and SPUHR (Case 2 to 5), the porosity deformation behaviors are presented in Figure 18. When compared with Case 1, porosity deformation degree along the slab thickness direction in Figure 18a and that along the slab width direction in Figure 18c increase while the porosity deformation degree along the casting direction in Figure 18b decreases with HR deformation at Roller 1# increased in Case 2 to 5. This means that the porosity size after HR decreases along three axis directions with HR deformation at Roller 1# increased.

Figure 18d shows the porosity closure degree after HR in five cases. The porosity closure degree continuously increases with HR deformation at Roller 1# increased, and, when compared with closure degree of $P_{1/8}$ and $P_{1/2}$ in Case 1, these two values, respectively, increase by 6.2% and 8.2% with HR deformation at Roller 1# increased to 3.6% in Case 5. This indicates that the porosity can be improved more effectively by HR with the newly-proposed reduction mode of SPUHR and that the effect of SPUHR on improving the HR efficiency becomes more significant with the HR deformation at Roller 1# increased.

Figure 19 shows the equivalent stain at the slab center of 1/8 width and 1/2 width after HR in five cases. The continuously increasing trend of equivalent strain from Case 1 to 5 indicates that the HR deformation could transfer from the slab surface into its center more effectively with HR deformation as Roller 1# increased. As a result, the porosity could be improved more effectively by HR with HR deformation as Roller 1# increased, which explains the increasing trend of porosity closure degree in Figure 18d and it proves the effect of SPUHR on improving HR efficiency.

During HR at the strand solidification end, HR efficiency continuously decreases from the entrance (Roller 1#) to exit (Roller 5#) of the HR segment due to the decrease of temperature difference. Therefore, with more HR deformation being concentrated at Roller 1# for the newly-proposed reduction mode of SPUHR, the final HR efficiency will be improved. However, in addition to temperature difference, another potential factor that may influence the HR efficiency is the distribution of HR deformation within the HR segment. Although the total HR deformation in Case 1 to 5 is equal, the slab deformation behavior at each roller changes with the change of HR deformation distribution within the HR segment, which may influence the final deformation behavior of the slab and its internal porosities after HR, even ignoring the variation of the slab temperature field within the HR segment.

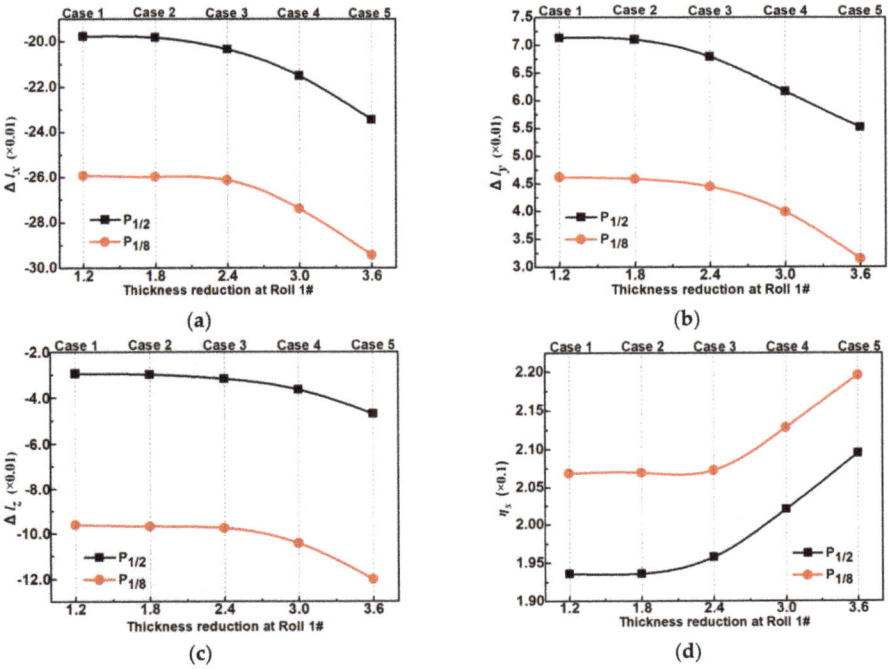

Figure 18. Porosity deformation degree along the (**a**) slab thickness direction, (**b**) casting direction, (**c**) slab width direction, and (**d**) the porosity closure degree after 6% HR deformation in Case 1 to 5.

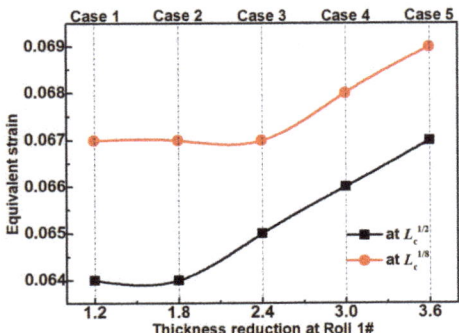

Figure 19. Equivalent strain at the slab center of 1/8 width ($L_c^{1/8}$) and 1/2 width ($L_c^{1/2}$) after 6% HR deformation in Case 1 to 5.

To evaluate the influence of HR deformation distribution within the HR segment on the HR efficiency. The porosity closure degree after 6% HR deformation in Case 1 to 5 was calculated with the assumption that the slab temperature field during HR within the HR segment was fixed, and the calculated results are shown in Figure 20. It can be seen that, although the variation of the slab temperature field was neglected during HR, the closure degree of $P_{1/8}$ and $P_{1/2}$ continuously increases by 5.9% and 5.2% from Case 1 to 5. This proves that HR deformation distribution within the HR segment is another factor that influences the HR efficiency and that the HR efficiency will be improved more significantly with more HR deformation concentrated at Roller 1#.

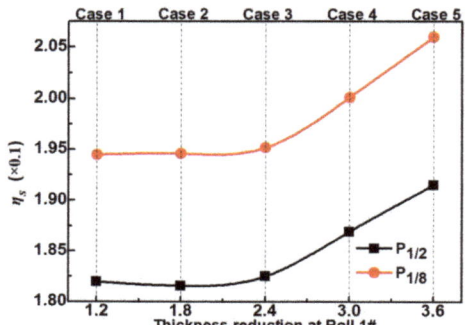

Figure 20. Porosity closure degree after HR in Case 1 to 5 ignoring the variation of the slab temperature field during HR.

4. Prediction Model for Porosity Closure Behavior

During HR, equivalent strain (ε_{eq}) distribution of the casting steel could be easily determined by conducting thermal-mechanical analysis with a corresponding thermal-mechanical coupled model. This means that the process effect of HR on improving the internal porosities that were distributed at different locations of the casting steel can be directly evaluated with ε_{eq} at the corresponding location if a quantitative relationship between the porosity deformation behavior and ε_{eq} during HR could be established.

In order to derive the relationship between the porosity deformation behavior and ε_{eq} for the wide-thick slab during HR, the calculated closure degree (η_s) of $P_{1/8}$ and $P_{1/2}$ and the corresponding ε_{eq} under different HR conditions in Section 3 are shown in Figure 21.

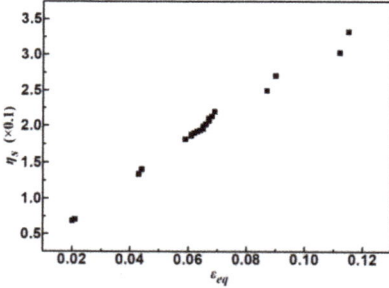

Figure 21. Relationship between the porosity closure degree (η_s) and the equivalent strain (ε_{eq}).

To quantitatively evaluate the correlation between η_s and ε_{eq}, the Pearson correlation coefficient [36] for the scattered data in Figure 21 was calculated with the following formula:

$$r = \frac{\sum (X_i - \overline{X})(Y_i - \overline{Y})}{\left[\sum (X_i - \overline{X})^2 \sum (Y_i - \overline{Y})^2\right]^{1/2}}, \quad (8)$$

where r is the Pearson correlation coefficient; X_i and Y_i denote ε_{eq} and the corresponding η_s for the scattered data in Figure 21; and, $\overline{X}$ and $\overline{Y}$ are, respectively, the mean value of ε_{eq} and η_s in Figure 21. The absolute value of r ranges from 0 to 1, and a larger absolute value of r indicates a closer relationship between X and Y.

The calculated result indicates that the r value for the scattered data in Figure 21 reaches as high as 0.9938, which proves that there exists a very close positive correlation between η_s and ε_{eq} for the wide-thick slab during HR. In order to quantitatively describe the relationship between η_s and ε_{eq}, polynomials with different orders were adopted to fit the scattered data in Figure 21 with the help of MATLAB. It was found that the quantitative relationship between η_s and ε_{eq} could be well described by a second order polynomial:

$$\eta_s = -5.23 \times \varepsilon_{eq}^2 + 3.39 \times \varepsilon_{eq} + 0.12 \times 10^{-2}, \tag{9}$$

Comparison results between the original data in Figure 21 and the fitting results with Equation (9) are presented in Figure 22. It can be seen that the fitting results agree well with the original data. The adjusted R square (R^2) reaches 0.9921, which proves the fitting accuracy.

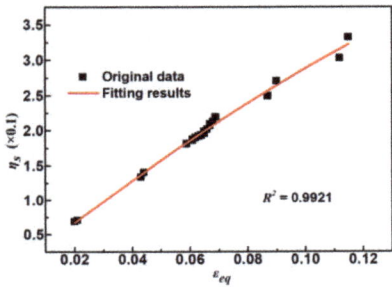

Figure 22. Comparison between the predicted and the fitting results for η_s and ε_{eq}.

5. Conclusions

The deformation behavior of the internal porosities in wide-thick slab during HR was numerically investigated. Some main conclusions are summarized, as follows:

(1) After different HR deformation, the internal porosity size decreases along the slab thickness direction and the slab width direction and meanwhile increases along the casting direction, and the porosity deformation degree along the slab thickness direction (Δl_x) is much larger than that along the casting direction (Δl_y) and the slab width direction (Δl_z). Due to the larger temperature difference at 1/8 width during HR, the closure degree (η_s) of $P_{1/8}$ is 9.7% larger than that of $P_{1/2}$ and it reaches 0.332 after 10% HR deformation.

(2) With HR start position moving away after the strand solidification end, Δl_x and Δl_z decrease, while Δl_y increases. After 6% HR deformation within the HR segment, η_s of $P_{1/8}$ and $P_{1/2}$ decrease by 9.3% and 6.3%, respectively, with the HR starting position moving away by 3 m after the strand solidification end. Meanwhile, the required reduction force for the HR segment increases by 20%. Therefore, HR efficiency on improving the internal porosities significantly decreases with the HR starting position moving away after the strand solidification end.

(3) When compared with the traditional reduction mode of UHR, the newly-proposed reduction mode of SPUHR could improve the HR efficiency. With more HR deformation being concentrated at entrance (Roller 1#) of the HR segment for SPUHR, Δl_x and Δl_z increase, while Δl_y decreases. η_s of $P_{1/8}$ and $P_{1/2}$ after total 6% HR deformation, respectively, increases by 6.2% and 8.2% with HR deformation at Roller 1# increased from 1.2% to 3.6%.

(4) A prediction model for the porosity closure behavior was derived based on the relationship between η_s and the corresponding equivalent strain (ε_{eq}), which can be expressed as:

$$\eta_s = -5.23 \times \varepsilon_{eq}^2 + 3.39 \times \varepsilon_{eq} + 0.12 \times 10^{-2}$$

Author Contributions: C.J. designed the paper and guided the development of the finite element models; M.Z. arranged the funding and revised the original manuscript; C.W. performed the numerical simulation and wrote the manuscript.

Funding: This research was funded by the National Natural Science Foundation of China No. 51474058 and U1708259, the Fundamental Research Funds for the Central Universities of China (N172504024).

Acknowledgments: Special thanks are due to the cooperating company for industrial trials and application.

Conflicts of Interest: The authors declare no conflict of interest.

References

1. Tanaka, M.; Ono, S.; Tsuneno, M.A. Numerical analysis on void crushing during side compression of round bar by flat dies. *J. Jpn. Soc. Technol. Plast.* **1987**, *28*, 238–244.
2. Nakasaki, M.; Takasu, I.; Utsunomiya, H. Application of hydrostatic integration parameter for free-forging and rolling. *J. Mater. Process. Technol.* **2006**, *177*, 521–524. [CrossRef]
3. Nalawade, R.S.; Patil, P.P.; Balachandran, G.; Balasubramanian, V. Void closure in a large cross section bars hot rolled from a low alloy steel ingot casting. *Trans. Indian Inst. Met.* **2016**, *69*, 1711–1721. [CrossRef]
4. Kakimoto, H.; Arikawa, T.; Takahashi, Y.; Tanaka, T.; Imaida, Y. Development of forging process design to close internal voids. *J. Mater. Process. Technol.* **2010**, *210*, 415–422. [CrossRef]
5. Wang, B.; Zhang, J.M.; Xiao, C.; Song, W.; Wang, S.X. Analysis of the evolution behavior of voids during the hot rolling process of medium plates. *J. Mater. Process. Technol.* **2015**, *221*, 121–127. [CrossRef]
6. Dudra, S.P.; Im, Y.T. Analysis of void closure in open-die forging. *Int. J. Mach. Tools Manuf.* **1990**, *30*, 65–75. [CrossRef]
7. Lee, Y.S.; Lee, S.U.; Van Tyne, C.J.; Joo, B.D. Internal void closure during the forging of large cast ingots using a simulation approach. *J. Mater. Process. Technol.* **2011**, *211*, 1136–1145. [CrossRef]
8. Park, J.J. Prediction of void closure in steel slabs by finite element analysis. *Met. Mater. Int.* **2013**, *19*, 259–265. [CrossRef]
9. Park, J.J. Finite-Element analysis of cylindrical-void closure by flat-die forging. *ISIJ Int.* **2013**, *53*, 1420–1426. [CrossRef]
10. Chen, M.S.; Lin, Y.C. Numerical simulation and experimental verification of void evolution inside large forgings during hot working. *Int. J. Plast.* **2013**, *49*, 53–70. [CrossRef]
11. Ståhlberg, U.; Keife, H. A study of hole closure in hot rolling as influenced by forced cooling. *J. Mater. Process. Technol.* **1992**, *30*, 131–135. [CrossRef]
12. Verstam, H.; Jarl, M. FEM-simulation of drawing out in open die forging. *Steel Res. Int.* **2004**, *75*, 812–817. [CrossRef]
13. Li, G.S.; Yu, W.; Cai, Q.W. Investigation of the evolution of central defects in ultra-heavy plate rolled using gradient temperature process. *Metall. Mater. Trans. B* **2015**, *46*, 831–840.
14. Li, G.S.; Yu, W.; Cai, Q.W. Investigation of reduction pretreatment process for continuous casting. *J. Mater. Process. Technol.* **2016**, *227*, 41–48. [CrossRef]
15. Chun, M.S.; Van Tyne, C.J.; Moon, Y.H. FEM analysis of void closure behaviour during open die forging of rectangular billets. *Steel Res. Int.* **2006**, *77*, 116–121. [CrossRef]
16. Banaszek, G.; Stefanik, A. Theoretical and laboratory modelling of the closure of metallurgical defects during forming of a forging. *J. Mater. Process. Technol.* **2006**, *177*, 238–242. [CrossRef]
17. Chen, J.; Chandrashekhara, K.; Mahimkar, C.; Lekakh, S.N.; Richards, V.L. Study of void closure in hot radial forging process using 3D nonlinear finite element analysis. *Int. J. Adv. Manuf. Technol.* **2012**, *62*, 1001–1011. [CrossRef]
18. Park, C.Y.; Yang, D.Y. A study of void crushing in large forgings II. estimation of bonding efficiency by finite-element analysis. *J. Mater. Process. Technol.* **1997**, *72*, 32–41. [CrossRef]
19. Xu, Z.G.; Wang, X.H.; Jiang, M. Investigation on improvement of center porosity with heavy reduction in continuously cast thick slabs. *Steel Res. Int.* **2017**, *88*, 231–242. [CrossRef]
20. Zhao, X.K.; Zhang, J.M.; Lei, S.W.; Wang, Y.N. The position study of heavy reduction process for improving centerline segregation or porosity with extra-thickness slabs. *Steel Res. Int.* **2014**, *85*, 645–658. [CrossRef]
21. Zhao, X.K.; Zhang, J.M.; Lei, S.W.; Wang, Y.N. Finite-Element analysis of porosity closure by heavy reduction process combined with ultra-heavy plates rolling. *Steel Res. Int.* **2014**, *85*, 1533–1543. [CrossRef]

22. Ji, C.; Luo, S.; Zhu, M.Y. Analysis and application of soft reduction amount for bloom continuous casting process. *ISIJ Int.* **2014**, *54*, 504–510. [CrossRef]
23. Luo, S.; Zhu, M.Y.; Ji, C. Theoretical model for determining optimum soft reduction zone of continuous casting steel. *Ironmak. Steelmak.* **2014**, *41*, 233–240. [CrossRef]
24. Nabeshima, S.; Nakato, H.; Fujii, T.; Kushida, K.; Mizota, H. Control of centerline segregation in continuously cast blooms by continuous forging process. *ISIJ Int.* **1995**, *35*, 673–679. [CrossRef]
25. Kojima, S.; Imai, T.; Mizota, H.; Fujimura, T.; Matsukawa, T. Improvement of centerline segregation in continuously cast strand by continuous forging process. *Tetsu Hagane* **1992**, *78*, 1794–1801. [CrossRef]
26. Hiraki, S.; Yamanaka, A.; Shirai, Y.; Satou, Y.; Kumakura, S. Development of new continuous casting technology (PCCS) for very thick plate. *Mater. Jpn.* **2009**, *48*, 20–22. [CrossRef]
27. Kawamoto, M. Recent development of steelmaking process in Sumitomo Metals. *J. Iron Steel Res. Int.* **2011**, *18*, 28–35.
28. Ji, C.; Wu, C.H.; Zhu, M.Y. Thermo-mechanical behavior of the continuous casting bloom in the heavy reduction process. *JOM* **2016**, *68*, 3107–3115. [CrossRef]
29. Zhao, J.P.; Liu, L.; Wang, W.W.; Lu, H. Effects of heavy reduction technology on internal quality of continuous casting bloom. *Ironmak. Steelmak.* **2017**. [CrossRef]
30. Dong, Q.P.; Zhang, J.M.; Wang, B.; Zhao, X.K. Shrinkage porosity and its alleviation by heavy reduction in continuously cast strand. *J. Mater. Process. Technol.* **2016**, *238*, 81–88. [CrossRef]
31. Wang, H.M.; Li, G.R.; Lei, Y.C.; Zhao, Y.T.; Dai, Q.X.; Wang, J.J. Mathematical heat transfer model research for the improvement of continuous casting slab temperature. *ISIJ Int.* **2005**, *45*, 1291–1296. [CrossRef]
32. Li, C.; Thomas, B.G. Thermomechanical finite-element model of shell behavior in continuous casting of steel. *Metall. Mater. Trans. B* **2004**, *35B*, 1151–1172. [CrossRef]
33. Ji, C.; Luo, S.; Zhu, M.Y.; Sahai, Y. Uneven solidification during wide-thick slab continuous casting process and its influence on soft reduction zone. *ISIJ Int.* **2014**, *54*, 103–111. [CrossRef]
34. Wu, C.H.; Ji, C.; Zhu, M.Y. Analysis of the thermal contraction of wide-thick continuously cast slab and the weighted average method to design a roll gap. *Steel Res. Int.* **2017**, *88*, 1600514. [CrossRef]
35. Ji, C.; Wang, Z.L.; Wu, C.H.; Zhu, M.Y. Constitutive modeling of the flow stress of GCr15 continuous casting bloom in the heavy reduction process. *Metall. Mater. Trans. B* **2018**, *49*, 767–782. [CrossRef]
36. Rodgers, J.L.; Nicewander, W.A. Thirteen ways to look at the correlation coefficient. *Am. Stat.* **1988**, *42*, 59–66. [CrossRef]

© 2019 by the authors. Licensee MDPI, Basel, Switzerland. This article is an open access article distributed under the terms and conditions of the Creative Commons Attribution (CC BY) license (http://creativecommons.org/licenses/by/4.0/).

MDPI
St. Alban-Anlage 66
4052 Basel
Switzerland
Tel. +41 61 683 77 34
Fax +41 61 302 89 18
www.mdpi.com

Metals Editorial Office
E-mail: metals@mdpi.com
www.mdpi.com/journal/metals

www.ingramcontent.com/pod-product-compliance
Lightning Source LLC
LaVergne TN
LVHW070747100526
838202LV00013B/1321